Second Edition

Human Performance, Workload, and Situational Awareness Measures Handbook

Second Edition

HUMAN PERFORMANCE, WORKLOAD, and SITUATIONAL AWARENESS MEASURES HANDBOOK

VALERIE J. GAWRON

CRC Press is an imprint of the
Taylor & Francis Group, an informa business

CRC Press
Taylor & Francis Group
6000 Broken Sound Parkway NW, Suite 300
Boca Raton, FL 33487-2742

Printed in the United States of America on acid-free paper
10 9 8 7 6 5 4 3 2 1

International Standard Book Number-13: 978-1-4200-6449-0 (Hardcover)

Library of Congress Cataloging-in-Publication Data

Gawron, Valerie J.
Human performance, workload, and situational awareness measures handbook / Valerie Jane Gawron. -- 2nd ed.
p. cm.
Previously published under title: Human performance measures handbook. Mahwah, NJ : Lawrence Erlbaum Associates, 2000.
Includes bibliographical references and index.
ISBN 978-1-4200-6449-0 (alk. paper)
1. Human engineering--Handbooks, manuals, etc. 2. Human-machine systems--Handbooks, manuals, etc. I. Gawron, Valerie J. Human performance measures handbook. II. Title.

T59.7.G38 2008
620.8'2--dc22 2008000781

Visit the Taylor & Francis Web site at
http://www.taylorandfrancis.com

and the CRC Press Web site at
http://www.crcpress.com

Dedication

To my parents:
Jane Elizabeth Gawron 12 June 1926 to 17 March 2002
and
Stanley Carl Gawron 17 March 1921 to 9 February 2000

Contents

List of Illustrations

List of Tables

Preface

This human performance measures handbook was developed to help researchers and practitioners select measures to be used in the evaluation of human/machine systems. It can also be used to supplement classes at both the undergraduate and graduate levels in ergonomics, experimental psychology, human factors, human performance, measurement, and system test and evaluation. The handbook begins with an overview of the steps involved in developing a test to measure human performance, workload, and/or situational awareness. This is followed by a definition of human performance and a review of human performance measures. Workload and situational awareness are similarly treated in subsequent chapters.

Acknowledgments

This book began while I was supporting numerous test and evaluation projects of military and commercial transportation systems. Working with engineers, operators, managers, programmers, and scientists showed a need for both educating them on human performance measurement and providing guidance for selecting the best measures for the test. I thank Dr. Dave Meister who provided great encouragement to me to write this book based on his reading of my "Measure of the Month" article in the *Test and Evaluation Technical Group Newsletter.* He, Dr. Tom Enderwick, and Dr. Dick Pew also provided a thorough review of the first draft of the first edition of this book. For these reviews I am truly grateful. I miss you, Dave.

1 Introduction

Human factors specialists, including industrial engineers, engineering psychologists, human factors engineers, and many others, constantly seek better (more efficient and effective) ways to characterize and measure the human element of a physical system, so that we can build trains, planes, and automobiles with superior human–system interfaces. Yet, the human factors specialist is often frustrated by the lack of readily accessible information on human performance, workload, and situational awareness (SA) measures. This book guides the reader through the critical process of selecting the appropriate measures of human performance, workload, and SA, and later provides specific examples of such measures.

There are two types of evaluations of human performance. The first type is subjective measures. These are characterized by humans providing opinions through interviews and questionnaires or by observing others' behavior. There are several excellent references on these techniques (e.g., Meister, 1986). The second type of evaluation of human performance is the experimental method. Again, there are several excellent references (e.g., Keppel, 1991; Kirk, 1995). The experimental method is the focus of this book. Chapter 1 is a short tutorial on experimental design; chapter 2 describes measures of human performance; chapter 3, measures of workload; and chapter 4, measures of SA.

For the tutorial, the task of selecting among aircraft cockpit displays is used as an example. For readers familiar with the general principles of experimentation, this would be an interesting application of academic theory. For readers who may not be so familiar, it underscores the importance of selecting the right measures when preparing to carry out an experiment.

The need for efficient and effective selection of the appropriate human performance, workload, and SA measures has never been greater. However, little guidance is available to support this selection process. This book was written to meet this need. The book begins with an example in which an experimenter must select measures of performance and workload to evaluate a cockpit display. Next, human performance is defined and measures presented. Each measure is described, along with its strengths and limitations, data requirements, threshold values, and sources of further information. After all the performance measures are described, a procedure for selecting among them is presented. In the next section, workload is defined, and workload measures are described in the same format as performance measures. In the last section, similar information is provided for SA measures. To make this desk reference easier to use, extensive author and subjective indices are provided.

1.1 THE EXAMPLE

An experiment is a comparison of two or more ways of performing tasks. The "tasks" being performed are called *independent variables*. The "ways" of performing tasks are called *experimental conditions*. The measures used for comparison are called *dependent variables*. Designing an experiment requires the following: defining the independent variables, developing the experimental conditions, and selecting the dependent variables. Ways of meeting these requirements are described in the following steps.

Step 1: Define the Question

Clearly define the question to be answered by the results of the experiment. Let us work through an example. Suppose a moving map display is being designed and the lead engineer wants to know if the map should be designed as track up, north up, or something else. He comes to you for an answer. You have an opinion but no hard evidence. You decide to run an experiment. Start by working with the lead engineer to define the question. First, what are the ways of displaying navigation information, that is, what are the experimental conditions to be compared? The lead engineer responds, "Track up, north up, and maybe something else." If he cannot define something else, you cannot test it. So, now you have two experimental conditions: track up versus north up. These conditions form the two levels of your first independent variable, direction of map movement.

Step 2: Check for Qualifiers

Qualifiers are independent variables that qualify or restrict the generalizability of your results. In our example, an important qualifier is the type of user of the moving map display. Will the user be a pilot (who is used to track up) or a navigator (who has been trained with north-up displays)? If you run the experiment with pilots, the most you can say from your results is that one type of display is best *for pilots*. There is your qualifier. If your lead engineer is designing moving map displays for both pilots and navigators, you have only given him half an answer; or worse, if you did not think about the qualifier of type of user, you may have given him an incorrect answer. So check for qualifiers and use the ones that will impact decision making as independent variables.

In our example, the type of user will have an effect on decision making, so it should be the second independent variable in the experiment. Also, in our example, the size of the display will not impact decision making, because the lead engineer only has room for an 8 in. (inch) display in the instrument panel. Therefore, size of the display should *not* be included as an independent variable.

Step 3: Specify Conditions

Specify the exact conditions to be compared. In our example, the lead engineer is interested in track up versus north up. So, the movement of the map will vary between the two conditions, but everything else about the displays (e.g., scale factor, display resolution, color quality, size of the display, and so forth) should be exactly the same. This way, if the subject's performance using the two types of displays is different, that difference can be attributed only to the type of display and not to some other difference between the displays.

Step 4: Match Subjects

Match the subjects to the end users. If you want to generalize the results of your experiment to what will happen in the real world, try to match the subjects to the users of the system in the real world. This is extremely important because subjects' past experiences may greatly affect their performance in an experiment. In our example, we added a second independent variable to our experiment specifically because of subjects' previous experiences (that is, pilots are used to track up, and navigators are trained with north up). If the end users of the display are pilots, we should use pilots as our subjects. If the end users are navigators, we should use navigators as our subjects. Other subject variables may also be important; in our example, age and training are both very important. Therefore, you should identify what training the user of the map display must have and provide that same training to the subjects before the start of data collection.

Age is important because pilots in their forties may have problems focusing on near objects such as map displays. Previous training is also important: F-16 pilots have already used moving map displays, whereas C-130 pilots have not. If the end users are pilots in their twenties with F-16 experience and your subjects are pilots in their forties with C-130 experience, you may be giving the lead engineer the wrong answer to his question as to which is the better display format.

Step 5: Select Performance Measures

Your results are influenced to a large degree by the performance measures you select. Performance measures should be relevant, reliable, valid, quantitative, and comprehensive. Let us use these criteria to select performance measures for our example problem.

Criterion 1: Relevance. Relevance to the question being asked is the prime criterion to be used when selecting performance measures. In our example, the lead engineer's question is, "What type of display format is better?" *Better* can refer to staying consistently on course (accuracy), but it can also refer to getting to the waypoints faster (time). Subjects' ratings of which display format they prefer do not answer the question of which display is better from a performance standpoint, because preference ratings can be affected by factors other than performance.

Criterion 2: Reliability. This refers to the repeatability of the measurements. For recording equipment, reliability is dependent on careful calibration of equipment to ensure that measurements are repeatable and accurate; (i.e., an actual course deviation of 50.31 ft (foot) should always be recorded as 50.31 ft). For rating scales, reliability is dependent on the clarity of the wording. Rating scales with ambiguous wording will not give reliable measures of performance. For example, if the question on the rating scale is, "Was your performance okay?," the subject may respond "no" after his first simulated flight but "yes" after his second, simply because he is more comfortable with the task. If you now let him repeat his first flight, he may respond "yes." In this case, you are getting a different answer to the same question although the conditions are identical. Subjects will give more reliable responses to less ambiguous questions such as "Did you deviate more than 100 ft from course in this trial?" Even so, you may still get a first "no" and a second "yes" to the more precise question, indicating that some learning had improved his performance the second time.

Subjects also need to be calibrated. For example, if you are asking which of eight flight control systems is best and your metric is an absolute rating (e.g., Cooper –Harper Handling Qualities Rating), your subject needs to be calibrated with both a "good" aircraft and a "bad" aircraft at the beginning of the experiment. He or she may also need to be recalibrated during the course of the experiment. The symptoms that suggest the need to recalibrate your subject are the same as those that indicate that you should recalibrate your measuring equipment: (a) all the ratings are falling in a narrower band than you expect, (b) all the ratings are higher or lower than you expect, and (c) the ratings are generally increasing (or decreasing) across the experiment, independent of experimental conditions. In these cases, give the subject a flight control system that he or she has already rated. If this second rating is substantially different from the one previously given to you for the same flight control system, you need to recalibrate your subjects with an aircraft that pulls their ratings away from the average: bad aircraft if all the ratings are near the top, and good aircraft if all the ratings are near the bottom.

Criterion 3: Validity. Validity refers to measuring what you really think you are measuring. Validity is closely tied to reliability. If a measure is not reliable, it can never be valid. The converse is not necessarily true. For example, if you ask a subject to rate his or her workload from 1 to 10 but do not define what you mean by workload, he or she may rate the perceived difficulty of the task rather than the amount of effort expended in performing the task.

Criterion 4: Quantitativeness. Quantitative measures are easier to analyze than qualitative measures. They also provide an estimate of the size of the difference between experimental conditions. This is often very useful in performing trade-off analyses of performance versus cost of system designs. This criterion does not preclude the use of qualitative measures, however, because qualitative measures often improve the understanding of experiment results. For qualitative measures, an additional issue must be considered—the type of rating scale. Nominal scales assign an adjective to the system being evaluated, (e.g., easy to use). "A nominal scale is categorical in nature, simply identifying differences among things on some characteristic. There is no notion of order, magnitude or size" (Morrow, Jackson, Disch, and Mood, 1995, p. 28). Ordinal scales rank systems being evaluated on a single or a set of dimensions (e.g., the north-up is easier than the track-up display). "Things are ranked in order, but the difference between ranked positions are not comparable" (Morrow, Jackson, Disch, and Mood, 1995, p. 28).

Interval scales have equal distances between the values being used to rate the system under evaluation. For example, a bipolar rating scale is used in which the two poles are *extremely easy to use* and *extremely difficult to use*. In between these extremes are the words *moderately easy, equally easy,* and *moderately difficult*. The judgment is that there is an equal distance between any two points on the scale. The perceived difficulty difference between *extremely* and *moderately* is the same as between *moderately* and *no difference*. However, "the zero point is arbitrarily chosen" (Morrow, Jackson, Disch, and Mood, 1995, p. 28). The fourth type of scale is a ratio scale, which possesses a true zero (Morrow, Jackson, Disch, and Mood, 1995, p. 29). More detailed descriptions of scales are presented in Baird and Noma (1978), Torgerson (1958), and Young (1984).

Criterion 5: Comprehensiveness. This denotes the ability to measure all aspects of performance. Recording multiple measures of performance during an experiment is cheaper than setting up a second experiment to measure something that you missed in the first experiment. So, measure all aspects of performance that may be influenced by the independent variables. In our example, subjects can trade off accuracy for time (e.g., cut a leg to reach a waypoint on time) and vice versa (e.g., go slower to stay on course better), so we should record both accuracy and time measures.

Step 6: Use Enough Subjects

Use enough subjects to statistically determine if the values of the dependent variables differ depending on the experimental conditions. In our example, is the performance of subjects using the track-up display statistically different from those using the north-up display? Calculating the number of subjects you need is very simple. First, predict how well subjects will perform in each condition. You can do this using your own judgment, previous data from similar experiments, or from pretest data using your experimental setup. In our example, how much error will there be in waypoint arrival times using the track-up display and the north-up display, respectively? From previous studies, you may think that the average error for pilots using the track-up display will be 1.5 s (seconds) and using the north-up display, 2 s. Similarly, the navigators will have about 2 s error using the track-up display and 1.5 s error with the north-up display. For both sets of subjects and both types of displays, you think the standard deviation will be about 0.5 s.

Now we can calculate the effect size, that is, the difference between performances in each condition:

$$\text{Effect size} = \frac{\text{performance in track up} - \text{performance in north up}}{\text{Standard deviation}}$$

$$\text{Effect size for pilots} = \frac{1.5 - 2}{0.5} = 1$$

$$\text{Effect size for navigators} = \frac{2 - 1.5}{0.5} = 1$$

In figure 1 we can now read the number of subjects needed to discriminate between the two conditions. For an effect size of 1, the number of subjects needed is 18. Therefore, we need 18 pilots and 18 navigators in our experiment. Note that although the function presented in figure 1 is not etched in stone, it is based on over 100 years of experimentation and statistics.

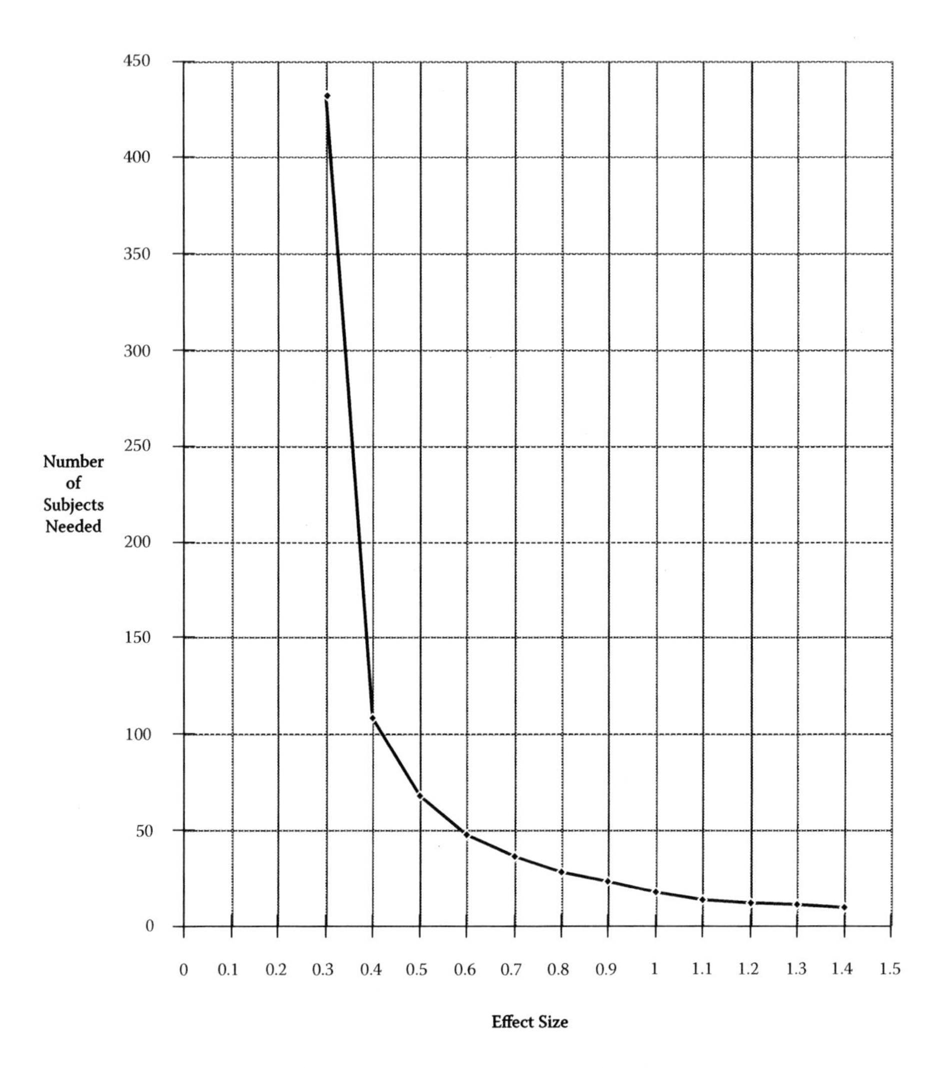

FIGURE 1 Number of subjects needed as a function of effect size.

Note that you should estimate your effect size in the same units that you will use in the experiment. Also note that because effect size is calculated as a ratio, you will get the same effect size (and hence the same number of subjects) for equivalent measures. Finally, if you have no idea of the effect size, try the experiment yourself and use your own data to estimate the effect size.

Step 7: Select Data Collection Equipment

Now that you know the size of the effect of the difference between conditions, check that the data collection equipment you have selected can reliably measure performance at least one order of magnitude smaller than the smallest discriminating decimal place in the size of the expected difference between conditions. In our example, the expected size in one condition was 1.5 s. The smallest discriminating decimal place (1.5 versus 2.0) is 1 ms (millisecond). One order of magnitude smaller is 1/100th of a second. Therefore, the recording equipment should be accurate to 1/100th of a second. In our example, can the data collection equipment measure time in 1/100th of a second?

Step 8: Match Trials

Match the experimental trials to the end usage. As in step 4, if you want to generalize the results of your experiment to what will happen in the real world, you must match the experimental trials to the real world. (Note, a single trial is defined as continuous data collection under the same experimental conditions. For example, three successive instrument approaches with the same flight-control configuration constitute one trial). The following are the important characteristics to match:

Characteristic 1: Length of the Trial. Over the length of a trial, performance improves due to warm-up effects and learning and then degrades as fatigue sets in. If you measure performance in the experiment for 10 min (minute) but in the real world pilots and navigators perform the task for 2 h (hour), your results may not reflect the peak warm-up or the peak fatigue. Consequently, you may give the lead engineer the wrong answer. So, always try to match the length of each experimental trial to the length of the task in the real world.

Characteristic 2: Level of Difficulty. If you make the experimental task too easy, all the subjects will get the same performance score: 0 errors. If all the performance scores are the same, you will not be able to distinguish between experimental conditions. To avoid this problem, make the task realistically difficult. In general, the more difficult the task in the experiment, the more likely you are to find a statistical difference between experimental conditions. This is because difficulty enhances discriminability between experimental conditions. However, there are two exceptions that should be avoided in any experiment. First, if the experimental task is too difficult, the performance of all the subjects will be exactly the same: 100% errors. You will have no way of knowing which experimental condition is better, and the purpose of the experiment will be defeated. Second, if you increase the difficulty of the task beyond that which can ever be expected in the real world, you may have biased your results. In our example, you may have found that track-up displays are better than north-up displays in mountainous terrain, flying under 100 ft above ground level (AGL) at speeds exceeding 500 kn (knot) with wind gusts over 60 kn. But how are they in hilly terrain, flying at 1000 ft AGL at 200 kn with wind gusts between 10 and 20 kn, that is, in the conditions in which they will be used nearly 70% of the time? You cannot answer this question from the results of your experiment—or if you give an answer, it may be incorrect. Therefore, the conditions of the experiment should be real-world conditions.

Characteristic 3: Environmental Conditions. Just as in step 4, where you tried to match the subjects to the end users, you should also try to match the environmental conditions of the laboratory (even if that laboratory is an operational aircraft or an in-flight simulator) to the environmental conditions of the real world. This is extremely important because environmental conditions can have a greater effect on performance than the independent variables in your experiment. Important environmental conditions that should be matched include lighting, temperature, noise, and task load. Lighting conditions should be matched in luminance level (possible acuity differences), position of the light source (possible glare), and type of light source (incandescent lights have "hot spots" that can create point sources of glare; fluorescent lights provide even, moderate light levels; sunlight can mask some colors and create large glare spots). Temperatures above 80°F (Fahrenheit) decrease the amount of effort subjects expend; temperatures below 30°F make fine motor movements (e.g., setting radio frequencies) difficult. Noise can either enhance or degrade performance: enhancements are due to increased attention, whereas degradations are due to distractions. Meaningful noise (e.g., a conversation) is especially distracting. Task load refers to both the number and types of tasks that are being performed at the same time as your experimental task. In general, the greater the number of tasks that are being performed simultaneously and the greater the similarity of the tasks that are being performed simultaneously, the worse the performance on the experimental task. The classic example is monitoring three radio channels simultaneously. If the volume or quality of the communications is not varied (thus making the tasks less similar), this task is extremely difficult.

Step 9: Select Data-Recording Equipment

In general, the data-recording equipment should be able to record data for 1.5 times the length of the experimental trial. This allows for false starts without changing the data tape, disk, or other storage medium. The equipment should have separate channels for each continuous dependent variable (e.g., altitude and airspeed) and as many channels as necessary to record the discrete variables (e.g., reaction time [RT] to a simulated fire) without any possibility of recording the discrete variables simultaneously on the same channel (thus losing valuable data).

Step 10: Decide Subject Participation

Decide if each subject should participate on all levels. There are many advantages of having a single subject participate in more than one experimental condition: (a) reduced recruitment costs, (b) decreased total training time, and (c) better matching of subjects across experimental conditions. However, there are some conditions that preclude using the same subject in more than one experimental condition. The first is previous training. In our example, pilots and navigators have had very different training. The differences in their training may affect their performance; therefore, they cannot participate in both roles: pilot and navigator. Second, some experimental conditions can make the subjects' performance worse than even untrained subjects in another experimental condition. This effect is called *negative transfer*. Negative transfer is especially strong when two experimental conditions require a subject to give a different response to the same stimulus. For example, suppose that the response to a fire alarm in experimental condition 1 is pull the T-handles, then feather the engine, and in experimental condition 2, the response is feather the engine and then pull the T-handle. Subjects who have not participated in any experimental condition are going to have faster reaction times (RTs) and fewer errors than subjects who have already participated in either experimental condition

1 or 2. Whenever there is negative transfer (easy to find by comparing performance of new subjects to subjects who have already participated in another condition), use separate subjects.

Learning is another important condition governing whether or not to use the same subjects. Subjects who participate in more than one experimental condition are constantly learning about the task that they are performing. If you plot the subject's performance (where high scores mean good performance) on the ordinate and the number of trials he or she has completed along the abscissa, you will find a resulting J-curve, where a lot of improvement in performance occurs in the first few trials and very little improvement occurs in the final trials. The point at which there is very little improvement is called *asymptotic learning*. Unless subjects are all trained to asymptote before the first trial, their performance will improve over the entire experiment regardless of the differences in the experimental conditions. Therefore, the "improvement" you see in later experimental conditions may have nothing to do with the experimental condition but rather with how long the subject has been performing the task in the entire experiment.

A similar effect occurs in simple, repetitive, mental tasks and all physically demanding tasks. This effect is called *warm-up*. If the subject's performance improves over trials regardless of the experimental conditions, you may have a warm-up effect. This effect can be eliminated by having subjects perform preliminary trials until their performance on the task matches their asymptotic learning.

The final condition is fatigue. If the same subject is performing more than one trial, fatigue effects may begin to mask the differences in the experimental conditions. You can check for fatigue effects in four ways: by performing a number of trials yourself (how are you feeling?); by observing your subjects (are they showing signs of fatigue?); by comparing performance in the same trial number but different conditions across subjects (is everyone doing poorly after three trials?); and by asking the subjects how they are feeling.

Step 11: Order the Trials

In step 10, we described order or carryover effects. Even if these do not occur to a great degree or if they do not seem to occur at all, it is still important to order your data collection trials so as to minimize order and carryover effects. Another important carryover effect is the experimenter's experience—during your first trial, experimental procedures may not yet be smoothed out. By the 10th trial, everything should be running efficiently, and you may even be anticipating subjects' questions before they ask them. The best way to minimize order and carryover effects is to use a Latin-square design. This design ensures that every experimental condition precedes and succeeds every other experimental condition an equal number of times.

Once the Latin square is generated, check the order for any safety constraints (e.g., landing a level 3 aircraft in maximum turbulence or severe crosswinds). Adjust this order as necessary to maintain safety. The resulting numbers indicate the order in which you should collect your data. For example, subject 1 gets north up followed by track up. Subject 2 gets the opposite sequence. Once you have completed data collection for the pilots, you can collect data on the navigators. It does not matter in what order you collect the pilot and navigator data because the pilot data will never be compared to the navigator data; that is, you are not looking for an interaction between the two independent variables. If the second independent variable in the experiment had been size (e.g., the lead engineer gives you the option for an 8 in. or 12 in. display), the interaction would have been of interest. For example, are 12 in., track-up displays better than 8 in., north-up displays? If we had been interested in this interaction, a Latin square for four conditions: condition 1, 8 in., north up; condition 2, 8 in., track up; condition 3, 12 in., north up; and condition 4, 12 in., track up would have been used.

Step 12: Check for Range Effects

Range effects occur when your results differ, depending on the range of experimental conditions that you use. For example, in experiment 1 you compare track-up and north-up displays, and find that for pilots track-up displays are better. In experiment 2, you compare track-up, north-up, and horizontal situation indicator (HSI) displays. This time you find no difference between track-up and north-up displays, but both are better than a conventional HSI. This is an example of a range effect: when you compare across one range of conditions, you get one answer; when you compare across a second range of conditions, you get another answer. Range effects are especially prevalent when you vary environmental conditions such as noise level and temperature. Range effects cannot be eliminated. This makes selecting a range of conditions for your experiment especially important.

To select a range of conditions, first return to your original question. If the lead engineer is asking which of two displays to use, experiment 1 is the right experiment. If he is asking whether track-up or north-up displays are better than an HSI, experiment 2 is correct. Second, you have to consider how many experimental conditions your subjects are experiencing. If it is more than seven, your subject is going to have a hard time remembering what each condition was, but his performance will still show the effect. To check for a "number of trials" effect, plot the average performance in each trial against the number of the trials the subject has completed. If you find a general decrease in performance, it is time to either reduce the number of experimental conditions that the subject experiences or provide long rest periods.

1.2 SUMMARY

The quality and validity of the data are improved by incorporating the following steps in the experimental design:

Step 1: Clearly define the question to be answered.
Step 2: Check for qualifiers.
Step 3: Specify the exact conditions to be compared.
Step 4: Match the subjects to the end users.
Step 5: Select performance measures.
Step 6: Use enough subjects.
Step 7: Select data collection equipment.

Step 8: Match the experimental trials to the end usage.
Step 9: Select data-recording equipment.
Step 10: Decide if each subject should participate in all levels.
Step 11: Order the trials.
Step 12: Check for range effects.

Step 5 is the focus of the remainder of this book.

REFERENCES

Baird, J.C. and Noma, E. *Fundamentals of scaling and psychophysics.* New York: Wiley, 1978.
Keppel, G. *Design and analysis: A researcher's handbook.* Englewood Cliffs, NJ: Prentice Hall, 1991.
Kirk, R.R. *Experimental design: Procedures for the behavioral sciences.* Pacific Grove, CA: Brooks/Cole Publishing Company, 1995.
Meister, D. *Human factors testing and evaluation.* New York: Elsevier, 1986.
Morrow, J.R., Jackson, A.W., Disch, J.G., and Mood, D.P. *Measurement and evaluation in human performance.* Champaign, IL: Human Kinematics, 1995.
Torgerson, W.S. *Theory and methods of scaling.* New York: Wiley, 1958.
Young, F.W. Scaling. *Annual Review of Psychology.* 35, 55–81, 1984.

2 Human Performance

Human performance is the accomplishment of a task by a human operator or by a team of human operators. Tasks can vary from the simple (card sorting) to the complex (landing an aircraft). Humans can perform the task manually or monitor an automated system. In every case, human performance can be measured, and this handbook can help with the measurement process.

Performance measures can be classified into six categories. The first is accuracy, in which the measure assesses the degree of correctness. Such measures assume that there is a correct answer. Human performance measures in this category are presented in section 2.1. The second category of human performance measures is time. Measures in this category assume that tasks have a well-defined beginning and end so that the duration of task performance can be measured. Measures in this category are listed in section 2.2. The third category is task battery. Task batteries are collections of two or more tasks performed in series or in parallel to measure a range of abilities or effects. These batteries assume that human abilities vary across types of tasks or are differentially affected by independent variables. Examples are given in section 2.3.

The fourth category of human performance measures is domain-specific measures, which assess abilities to perform a family of related tasks. These measures assume that performance varies across segments of a mission or on the use of different controllers. Examples in this category are presented in section 2.4. The fifth category is critical incident measures, which are typically used to assess worst-case performance (see section 2.5). The final category is team performance measures. These assess the abilities of two or more persons working in unison to accomplish a task or tasks. These measures assume that human performance varies when humans work as a team. Team performance measures are described in section 2.6. There are also measures that fall into more than one category. This is true of all the team measures (section 2.6).

For uniformity and ease of use, each discussion of a measure of human performance has the same sections:

1. general description of the measure;
2. strengths and limitations or restrictions of the measure, including any known proprietary rights or restrictions as well as validity and reliability data;
3. data collection, reduction, and analysis requirements;
4. thresholds, the critical levels of performance above or below which the researcher should pay particular attention; and
5. sources of further information and references.

2.1 ACCURACY

The first category of human performance measures is accuracy, in which the measure assesses the degree of correctness. Such measures assume that there is a correct answer.

General description. Accuracy is a measure of the quality of a behavior. Measures of accuracy include correctness score (section 2.1.3), number correct (section 2.1.7), percent correct (section 2.1.9), percent correct detections (section 2.1.10), and probability of correct detections (section 2.1.12). Error can also be used to measure accuracy—or the lack thereof. Error measures include absolute error (section 2.1.1), average range scores (section 2.1.2), deviations (section 2.1.4), error rate (section 2.1.5), false alarm rate (section 2.16), number of errors (section 2.1.8), percent errors (section 2.1.11), and root-mean-square error (section 2.1.13). Errors can be of omission (i.e., leaving a task out) or commission (i.e., doing the task but not correctly). Sometimes, accuracy and error measures are combined to provide ratios (section 2.1.13).

Strengths and limitations. Accuracy can be measured on a ratio scale and is thus mathematically robust. However, distributions of the number of errors or the number correct may be skewed and thus may require mathematical transformation into a normal distribution. In addition, some errors rarely occur and are therefore difficult to investigate (Meister, 1986). There is also a speed–accuracy trade-off that must be considered (Drinkwater, 1968). Data collection requirements as well as thresholds are discussed for each accuracy measure. Source information is presented at the end of section 2.1.

2.1.1 Absolute Error

Mertens and Collins (1986) used absolute and root-mean-square error on a two-dimensional compensatory tracking task to evaluate the effects of age (30 to 39 versus 60 to 69 years old), sleep (permitted versus deprived), and altitude (ground versus 3810 m [meters]). Performance was not significantly affected by age but was significantly degraded by sleep deprivation and altitude. Similar results occurred for a problem-solving task.

Elvers, Adapathya, Klauer, Kancler, and Dolan (1993) reported that as the probability of a task requiring the subject to determine a volume rather than a distance increased, absolute error of the distance judgment increased.

2.1.2 Average Range Score

Rosenberg and Martin (1988) used average range scores ("largest coordinate value minus smallest coordinate value," p. 233) to evaluate a digitizer puck (i.e., a cursor-positioning device for digitizing images). Type of optical sight had no effect; however, magnification improved performance.

2.1.3 Correctness Score

Correctness scores were developed to evaluate human problem solving. A score using the following five-point rating scale was awarded based on a subject's action:

0—Subject made an incorrect or illogical search.
1—Subject asked for information with no apparent connection to the correct response.
2—Subject asked for incorrect information based on a logical search pattern.
4—Subject was on the right track.
5—Subject asked for the key element (Giffen and Rockwell, 1984).

Giffen and Rockwell (1984) used correctness scores to measure a subject's problem-solving performance. These authors used stepwise regression to predict correctness scores for four scenarios: "(1) an oil-pressure gauge line break, (2) a vacuum-pump failure, (3) a broken magneto drive gear, and (4) a blocked static port" (p. 575). Demographic data, experience, knowledge scores, and information-seeking behavior were only moderately related to correctness scores. The subjects were 42 pilots with 50 to 15,000 flight hours of experience. This measure requires well-defined search patterns. The thresholds are 0 (poor performance) to 5 (excellent performance).

2.1.4 Deviations

Ash and Holding (1990) used timing accuracy (i.e., difference between the mean spacing between notes played and the metronome interval) to evaluate keyboard-training methods. There was a significant training-method effect but no significant trial or order effects on this measure.

Yeh and Silverstein (1992) asked subjects to make spatial judgments of simplified aircraft-landing scenes. They reported that subjects were less accurate in making altitude judgments relative to depth judgments. However, altitude judgments (mean percent correct) were more accurate as altitude increased and with the addition of binocular disparity.

2.1.5 Error Rate

Error rate has been defined as the number of incorrectly answered and unanswered problems divided by the total number of problems presented. Wierwille, Rahimi, and Casali (1985) reported that error rate was significantly related to the difficulty of a mathematical problem-solving task. Error rates were higher for rapid communication than for conventional visual displays (Payne and Lang, 1991).

Error rates did not discriminate between observers exposed to strobe lights and those who were not exposed (Zeiner and Brecher, 1975). Nor were error rates significantly different between five feedback conditions (normal, auditory, color, tactile, and combined) for computer mouse use (Akamatsu, MacKenzie, and Hasbroucq, 1995). Akamatsu, MacKenzie, and Hasbroucq (1995) reported no significant differences in error rate between tactile, auditory, and visual feedback pointing systems.

In another measure of error rate, Korpardekar and Mital (1994) reported that normalized errors (number of errors divided by time) were significantly different for work break schedules. The fewest errors for directory assistance operators occurred in the 120 min of continuous work with no breaks than the 30 min with a 5 min break or the 60 min work with a 10 min break.

2.1.6 False Alarm Rate

Mullin and Corcoran (1977) reported that false alarm rate decreased over time in an auditory vigilance task. Amplitude of the auditory signal or time of day (08:30 versus 20:30) had no significant effect.

Lanzetta, Dember, Warm, and Berch (1987) reported that the false alarm rate significantly decreased as the presentation rate increased (9.5%, 6/min; 8.2%, 12/min; 3.2%, 24/min; 1.5%, 48/min).

Loeb, Noonan, Ash, and Holding (1987) found the number of false alarms was significantly lower in a simple than a complex task.

Galinsky, Warm, Dember, Weiler, and Scerbo (1990) used false alarm rate to evaluate periods of watch (i.e., five 10-min intervals). False alarm rate significantly increased as event rate decreased (5 versus 40 events per minute).

Colquhoun (1961) reported that 15 of 21 subjects in a visual detection task had no false alarms, whereas the remaining 6 varied considerably in the false alarm rate. There was a trend for false alarm rate to decrease over time. There was a significant positive correlation between number of false alarms and number of detected signals.

2.1.7 Number Correct

Craig, Davies, and Matthews (1987) reported a significant decrease in the number of correct detections as event rate increased, signal frequency decreased, time on task increased, and the stimulus degraded.

Tzelgov, Henik, Dinstein, and Rabany (1990) used the number correct to evaluate two types of stereo picture compression. This measure was significantly different between tasks (higher in object decision task than in-depth decision task), depth differences (greater at larger depth differences), and presentation condition. There were also significant interactions.

Loeb, Noonan, Ash, and Holding (1987) found that cueing, constant or changing target, or brief or persistent target had no significant effects on number correct. Smith and Miles (1986) reported a decrease in the number of targets detected between those who ate lunch prior to the task as compared to subjects who did not. Fleury and Bard (1987) reported that the number correct in a visual detection task was decreased after aerobic effort as compared to pretest inactivity. Raaijmaker and Verduyn (1996) used the number of problems correctly solved on a fault diagnosis task to evaluate a decision aid. In an unusual application, Millar and Watkinson (1983) reported the number of correctly recognized words from those presented during surgery beginning after the first incision.

Van Orden, Benoit, and Osga (1996) used the average number correct to evaluate the effect of cold stress on a command-and-control task. There were no significant differences.

For tasks with low error rates, the number completed has been used instead of the number correct. One example is the serial search task, in which the subject looks for pairs of digits occurring among a page full of digits (Harris and Johnson, 1978).

2.1.8 Number of Errors

The number of errors has been sensitive to differences in the visual environment. In an early study, Baker, Morris, and Steedman (1960) reported increases in the number of errors in a visual target detection task as the number of irrelevant items increased and the difference between the resolution of the reference and the target increased. Downing and Sanders (1987) reported that significantly more errors were made in a simulated control room emergency with a mirror than with a non-mirror image control panel layout. Moseley and Griffin (1987) reported a larger number of reading errors for characters with high spatial complexity. Kuller and Laike (1998) reported an increase in the number of proofreading errors for individuals with high critical flicker fusion frequency when reading under conventional fluorescent lighting ballasts rather than high-frequency fluorescent ballasts. Chapanis (1990) used the number of errors to evaluate short-term memory for numbers. He reported large individual differences (71 to 2231 errors out of 8000 numbers). He also reported a significant serial position effect (70% of the subjects made the greatest number of errors at the seventh position). Women made significantly more errors than men.

There have also been significant differences in the number of errors related to modality. Ruffell-Smith (1979) asked subjects to solve visually or aurally presented mathematical equations without using paper, calculator, or computer. The number of errors was then compared to the performance of 20 three-person airline crews on a heavy versus a light two-leg flight. There were more errors in computation during the heavy-workload condition. Vermeulen (1987) used the number of errors to evaluate presentation modes for a system-state identification task and a process-control task. There was no mode effect on errors in the first task; however, inexperienced personnel made significantly more errors than experienced personnel. For the second task, the functional presentation mode was associated with fewer errors than the topographical presentation. Again, inexperienced personnel made more errors than experienced personnel.

Also significant were effects related to alcohol and cold. Billings, Demosthenes, White, and O'Hara (1991) reported significant increases in the number of errors pilots made during simulated flight after alcohol dosing (blood alcohol levels of 0.025, 0.05, and 0.075). Enander (1987) reported increase in the number of errors in moderate (+5°C) cold. However, Van Orden, Benoit, and Osga (1996) used the average number of incorrect responses to evaluate the effect of cold stress on a command-and-control task. There were no significant differences.

There have been inconsistent results with other stressors as well. McCann (1969) found that noise had no effect on the number of errors, but she did find a significant increase in the number of omissions when intermittent noise was introduced into an audio-visual checking task. Mertens and Collins (1986) reported that performance was not related to age (30 to 39 versus 60 to 69 years old). However, performance degraded as a function of sleep deprivation and altitude (0 versus 3810 m).

For speech recognition systems and feedback types, the number of errors did not always discriminate between the levels of the independent variables. Casali, Williges, and Dryden (1990) used the number of uncorrected errors to evaluate a speech recognition system. There was significant recognition accuracy and available vocabulary effects but a significant age effect was lacking. Frankish and Noyes (1990) used the number of data entry errors to evaluate four types of feedback: (1) concurrent visual feedback, (2) concurrent spoken feedback, (3) terminal visual feedback, and (4) terminal spoken feedback. There were no significant differences.

2.1.9 Percent Correct

Percent correct has been used to evaluate the effects of environmental stressors, visual display characteristics, decision aids, vigilance, task, and training.

2.1.9.1 Effects of Environmental Stressors on Percent Correct

Harris and Shoenberger (1980) used percent correct on a counting task to assess the effects of noise (65 or 100 dBA [decibel (A scale)]) and vibration (0 or 0.36 Hz [Hertz]). There were significant main and interactive effects of these two environmental stressors. For the interaction, performance was worse in 65 dBA than 100 dBA in vibration. Performance was worse in 100 dBA when vibration was not present.

Lee and Fisk (1993) reported extremely small changes (1 to 290) in percent correct as a function of the consistency in targets in a visual search task. Harris and Johnson (1978) asked subjects to count the frequency of visual flashes from three light sources while being exposed to 65, 110, or 125 dB sound for 15 or 30 min. Sound had no effect on percent correct.

2.1.9.2 Effects of Visual Display Characteristics on Percent Correct

Imbeau, Wierwille, Wolf, and Chun (1989) used the percentage of correct answers to evaluate instrument panel lighting in automobiles. They reported that accuracy decreased as character size decreased.

Coury, Boulette, and Smith (1989) reported that percent correct was significantly greater for bar graph displays than for digital or configurable displays after extended practice (eight trial blocks).

Matthews, Lovasik, and Mertins (1989) reported a significantly lower percentage of correct responses for red on green displays (79.1%) and red on blue displays (75.5%) than for monochromatic (green, red, or blue on black), achromatic (white on black), or blue on green displays (85.5%). In addition, performance was significantly worse on the first (79.8%) and last (81.8%) half hours than for the middle three hours (83.9%).

Brand and Judd (1993) reported a significantly lower percentage of correct responses for keyboard entry as the angle of the hardcopy which they were to enter by keyboard increased (89.8% for 90°, 91.0% for 30°, and 92.4% for 12°). Experienced users had significantly higher percentages of correct responses (94%) than naive users (88%).

Tullis (1980) used percent correct to evaluate four display formats (narrative text, structured text, black and white graphics, and color graphics). There was no significant difference in percent correct. There was, however, a significant difference in mean response time. Chen and Tsoi (1988) used the percentage of correct responses to comprehension questions to evaluate readability of computer displays. Performance was better in the slow (100 words per minute) than in the fast (200 words per minute) condition. It was also better when there were jumps of one rather than five or nine character spaces. However, there was no significant difference between 20- or 40-character windows. Chen, Chan, and Tsoi (1988), however, reported no significant effect of jump length but a significant effect of window size. Specifically, there was a higher comprehension score for the 20-character than for the 40-character window. The significant interaction indicated that this advantage only occurred in the one-jump condition.

2.1.9.3 Effects of Decision Aids on Percent Correct

Adelman, Cohen, Bresnick, Chinnis, and Laskey (1993) used the mean percentage of correct responses to evaluate operator performance with varying types of expert systems. The task was simulated in-flight communication. The expert systems were of three types: (1) with rule-generation capability, (2) without rule-generation capability, and (3) totally automated. The operator could screen, override, or provide a manual response. Rule-generation resulted in significantly better performance than no-rule-generating capability.

2.1.9.4 Effects of Vigilance on Percent Correct

Lanzetta, Dember, Warm, and Berch (1987) reported a significantly higher percentage of correct detections in a simultaneous (78%) than in a sequential (69%) vigilance task. In addition, the percent correct generally decreased as the presentation rate increased (79%, 6/min; 82%, 12/min; 73%, 23/min; 61%, 48/min). This difference was significant. Percent correct also significantly decreased as a function of time on watch (87%, 10 min; 75%, 20 min; 68%, 30 min; 65%, 40 min).

Galinsky, Warm, Dember, Weiler, and Scerbo (1990) used percentage of correct detections to evaluate periods of watch (i.e., five 10-min intervals). This percentage decreased as event rate increased from 5 to 40 events per minute in two (continuous auditory and continuous visual) conditions but not in a third (sensory alternation) condition. In an unusual application, Doll and Hanna (1989) forced subjects to maintain a constant percentage of correct responses during a detection task.

2.1.9.5 Effects of Task on Percent Correct

Kennedy, Odenheimer, Baltzley, Dunlap, and Wood (1990) used percent correct on four diverse tasks (pattern comparison, grammatical reasoning, manikin, and short-term memory) to assess the effects of scopolamine and amphetamine. They found significant drug effects on all but the grammatical reasoning task, although the effect on it approached statistical significance ($p < 0.0683$).

Kimchi, Gopher, Rubin, and Raij (1993) reported a significantly higher percentage of correct responses on a locally directed than a globally directed task in a divided attention condition. There were no significant attention effects on percent correct on the globally directed task.

Arnaut and Greenstein (1990) reported no significant difference in the percentage of control responses resulting in errors as a function of level of control input.

2.1.9.6 Effects of Training on Percent Correct

Fisk and Hodge (1992) reported a significant decrease in percent correct in a visual search task after 30 d (day) in only one of five groups (the same category and exemplars were used in training). There were no differences for new, highly related, moderately related, or unrelated exemplars.

Briggs and Goldberg (1995) used percentage of correct recognition of armored tanks to evaluate training. There were significant differences in presentation time (longer times were associated with higher accuracies), view (flank views were more accurate than frontal views), model (M1 had the highest accuracy, British Challenger had the worst accuracy), and subjects. Component shown or friend versus foe had no significant effects.

2.1.10 Percent Correct Detections

In an early study, O'Hanlon, Schmidt, and Baker (1965) reported no decrease in detecting auditory tones over a 90-min trial. Christensen, Gliner, Horvath, and Wagner (1977) used a visual reaction time (RT) task to evaluate the effects of hypoxia. In their study ten subjects performed the task while exposed to filtered air, carbon monoxide, low oxygen, and carbon monoxide combined with low oxygen. Subjects had significantly more correct detections breathing filtered air than while breathing the combination of carbon monoxide and low oxygen. The lowest percentage of correct detections occurred for the low oxygen condition. In addition there was a significant decrease in the percent correct detections over the first 15 min interval of a 2 h performance period.

Chan and Courtney (1998) reported a significant improvement in the percentage of peripheral visual targets presented when the targets were magnified as compared with when they were not magnified. The difference was especially apparent at high levels of eccentricity (i.e., 3.5 and 5.0) than at low levels of eccentricity (1.0, 1.6).

Donderi (1994) used percentage of targets detected to evaluate types of search for life rafts at sea. The daytime percentage of correct detections was positively correlated with low-contrast visual acuity and negatively correlated with vision test scores. Chong and Triggs (1989) used the percentage of correct responses to evaluate the effects of type of windscreen post on target detections. There were significantly smaller percentages of correct detections for solid or no posts than for open posts.

2.1.11 Percent Errors

Hancock and Caird (1993) reported significant increases in the percentage of errors as the shrink rate of a target decreased. The greatest percentage of errors occurred for paths with 4 steps rather than 2, 8, or 16 steps. Salame and Baddeley (1987) reported increases in percent errors as the serial position of visually presented digit sequences increased through 7 and then declined through 9.

Maddox and Turpin (1986) used errors and error category to evaluate performance using mark-sensed forms. Error categories were the following: (1) multiple entries in the same row or column, (2) substituting an incorrect number, (3) transposing two or more numbers, and (4) omitting one or more numbers. None of these measures were affected by number orientation (horizontal or vertical), number ordering (bottom-to-top or top-to-bottom), and handedness of users (Maddox and Turpin, 1986). The range of percent errors was 2.2 to 6.3% with an average of 4% per subject (Maddox and Turpin, 1986). Percentage of total errors by error category were: (1) 69.3%, (2) 17.3%, (3) 12%, and (4) 1.4% (Maddox and Turpin, 1986).

2.1.12 Probability of Correct Detections

The probability of correct detections of target words was significantly higher for natural speech than for synthetic speech (Ralston, Pisoni, Lively, Greene, and Mullennix, 1991). Using the same measure, Zaitzeff (1969) reported greater cumulative target acquisition probability for two- than for one-person crews.

2.1.13 Ratio of Number Correct/Number Errors

Ash and Holding (1990) used number of errors divided by number of correct responses to evaluate keyboard-training methods. They reported significant learning between the first and third trial blocks, significant differences between training methods, and a significant order effect.

2.1.14 Root-Mean-Square Error

Root-mean-square error (rmse) has been used extensively to evaluate differences between types of tracking tasks. In an early study, Kvalseth (1978) reported no significant difference in rmse between pursuit and compensatory tracking tasks. Pitrella and Kruger (1983) developed a tracking performance task to match subjects in tracking experiments.

Eberts (1987) used rmse on a second-order, compensatory tracking task to evaluate the effects of cueing. Cueing significantly decreased error. Vidulich (1991) reported test-retest reliability of rmse on a continuous tracking task of +0.945. The reliability of percent correct was +0.218.

Rmse has also been used to assess environmental effects. For example, Frazier, Repperger, Toth, and Skowronski (1982) compared rmse for roll-axis tracking performance while subjects were exposed to gravitational (G) forces. Rmse increased 19% from the 1 G when subjects were exposed to +5 G_z for 95 s, 45% when exposed to combined +5G_z/±1 G_y, and 70% for +5 G_z/± G_y. In another G study, Albery and Chelette (1998) used rmse scores on tracking a visual target while exposed to +5 to +9 G_z wearing one of six g-suits. There were no significant differences between the six suits on rmse scores. Pepler (1958) used time and distance of target to assess the effect of temperature and reported lagging, overshooting, leading, and approaching effects on tracking over temperatures of 76°F. He tested up to 91°. Bohnen and Gaillard (1994) reported significant effects of sleep loss on tracking performance.

Cohen, Otakeno, Previc, and Ercoline (2001) reported no significant differences in rmse on a compensatory tracking task using three different types of displays: (1) compressed pitch ladder, (2) aircraft viewed from the rear, and (3) aircraft viewed from the side.

Although rmse has been used extensively, the measure is not without its critics. For example, Hubbard (1987) argued that rmse is not an adequate measure of pilot performance, because deviations in one direction (e.g., down in altitude) do not have the same consequence as in the opposite direction (e.g., up in altitude).

Other measures of tracking errors have been used, such as an integral of vertical and lateral error over time (Torle, 1965). This measure was used to investigate the effects of backlash, friction, and presence of an armrest in an aircraft simulator. It was sensitive to all three independent variables. Beshir, El-Sabagh, and El-Nawawi (1981) developed a tracking error score that was significantly degraded by increases in ambient temperature. Beshir (1986) used time on target (in seconds) per minute and reported decreased performance after 15 min on task and at 18.3°C than at 21.1°C.

McLeod and Griffin (1993) partitioned the total rmse into a component linearly correlated with movement of the target (input correlated error) and a component not linearly correlated with target movement (remnant). There was a significant session (more in session 1 than in sessions 2) as well as a duration effect (increase over 18 min) but no effect due to vibration of 1 octave centered on 4 Hz.

Gerard and Martin (1999) used the number of contacts of a ring suspended on a rod with a wire as a measure of tracking performance. They reported that previous vibration exposure resulted in an increased number of tracking errors.

Data requirements. All correct answers must be identified prior to the start of the experiment. Errors should be reviewed to ensure that they are indeed errors and not alternative versions of the correct answers.

Thresholds. During data reduction, negative numbers of errors or of correct answers should be tested for accuracy. Percent correct or percent errors greater than 100% should also be tested for accuracy.

SOURCES

Adelman, L., Cohen, M.S., Bresnick, T.A., Chinnis, J.O., and Laskey, K.B. Real-time expert system interfaces, cognitive processes, and task performance: An empirical assessment. *Human Factors*. 35(2), 243–261, 1993.

Akamatsu, M., MacKenzie, I.S., and Hasbroucq, T. A comparison of tactile, auditory, and visual feedback in a pointing task using a mouse-type device. *Ergonomics*. 38(4), 816–827, 1995.

Albery, W.B. and Chelette, T.L. Effect of G suit on cognitive performance. *Aviation, Space, and Environmental Medicine*. 69(5), 474–479, 1998.

Arnaut, L.Y. and Greenstein, T.S. Is display/control gain a useful metric for optimizing an interface? *Human Factors*. 32(6), 651–663, 1990.

Ash, D.W. and Holding, D.H. Backward versus forward chaining in the acquisition of a keyboard skill. *Human Factors*. 32(2), 139–146, 1990.

Baker, C.A., Morris, D.F., and Steedman, W.C. Target recognition on complex displays. *Human Factors*. 2(2), 51–61, 1960.

Beshir, M.Y. Time-on-task period for unimpaired tracking performance. *Ergonomics*. 29(3), 423–431, 1986.

Beshir, M.Y., El-Sabagh, A.S., and El-Nawawi, M.A. Time on task effect on tracking performance under heat stress. *Ergonomics*. 24(2), 95–102, 1981.

Billings, C.E., Demosthenes, T., White, T.R., and O'Hara, D.B. Effects of alcohol on pilot performance in simulated flight. *Aviation, Space, and Environmental Medicine*. 62(3), 233–235, 1991.

Bohnen, H.G.M. and Gaillard, A.W.K. The effects of sleep loss in a combined tracking and time estimation task. *Ergonomics*. 37(6), 1021–1030, 1994.

Brand, J.L. and Judd, K.W. Angle of hard copy and text-editing performance. *Human Factors*. 35(1), 57–70, 1993.

Briggs, R.W. and Goldberg, J.H. Battlefield recognition of armored vehicles. *Human Factors*. 37(3), 596–610, 1995.

Casali, S.P., Williges, B.H., and Dryden, R.D. Effects of recognition accuracy and vocabulary size of a speech recognition system on task performance and user acceptance. *Human Factors*. 32(2), 183–196, 1990.

Chan, H.S., and Courtney, A.J. Stimulus size scaling and foveal load as determinants of peripheral target detection. *Ergonomics*. 41(10), 1433–1452, 1998.

Chapanis, A. Short-term memory for numbers. *Human Factors*. 32(2), 123–137, 1990.

Chen, H., Chan, K., and Tsoi, K. Reading self-paced moving text on a computer display. *Human Factors*. 30(3), 285–291, 1988.

Chen, H. and Tsoi, K. Factors affecting the readability of moving text on a computer display. *Human Factors*. 30(1), 25–33, 1988.

Chong, J. and Triggs, T.J. Visual accommodation and target detection in the vicinity of a window post. *Human Factors*. 31(1), 63–75, 1989.

Christensen, C.L., Gliner, J.A., Horvath, S.M., and Wagner, J.A. effects of three kinds of hypoxias on vigilance performance. *Aviation, Space, and Environmental Medicine*. 48(6), 491–496, 1977.

Cohen, D., Otakeno, S., Previc, F.H., and Ercoline, W.R. Effect of "inside-out" and "outside-in" attitude displays on off-axis tracking in pilots and nonpilots. *Aviation, Space, and Environmental Medicine*. 72(3), 170–176, 2001.

Colquhoun, W.P. The effect of 'unwanted' signals in performance in a vigilance task. *Ergonomics*. 4(1), 41–52, 1961.

Coury, B.G., Boulette, M.D., and Smith, R.A. Effect of uncertainty and diagnosticity on classification of multidimensional data with integral and separable displays of system status. *Human Factors*. 31(5), 551–569, 1989.

Craig, A., Davies, D.R., and Matthews, G. Diurnal variation, task characteristics, and vigilance performance. *Human Factors*. 29(6), 675–684, 1987.

Doll, T.J. and Hanna, T.E. Enhanced detection with bimodal sonar displays. *Human Factors*. 31(5), 539–550, 1989.

Donderi, D.C. Visual acuity, color vision, and visual search performance. *Human Factors*. 36(1), 129–144, 1994.

Downing, J.V. and Saunders, M.S. The effects of panel arrangement and focus of attention on performance. *Human Factors*. 29(5), 551–562, 1987.

Drinkwater, B.L. Speed and accuracy in decision responses of men and women pilots. *Ergonomics*. 11(1), 61–67, 1968.

Eberts, R. Internal models, tracking strategies, and dual-task performance. *Human Factors*. 29(4), 407–420, 1987.

Elvers, G.C., Adapathya, R.S., Klauer, K.M., Kancler, D.E., and Dolan, N.J. Effects of task probability on integral and separable task performance. *Human Factors*. 35(4), 629–637, 1993.

Enander, A. Effects of moderate cold on performance of psychomotor and cognitive tasks. *Ergonomics.* 30(10), 1431–1445, 1987.

Fisk, A.D. and Hodge, K.A. Retention of trained performance in consistent mapping search after extended delay. *Human Factors.* 34(2), 147–164, 1992.

Fleury, M. and Bard, C. Effects of types of physical activity on the performance of perceptual tasks in peripheral and central vision and coincident timing. *Ergonomics.* 30(6), 945–958, 1987.

Frankish, C. and Noyes, J. Sources of human error in data entry tasks using speech input. *Human Factors.* 32(6), 697–716, 1990.

Frazier, J.W., Repperger, D.N., Toth, D.N., and Skowronski, V.D. Human tracking performance changes during combined + G_z and ± G_y stress. *Aviation, Space, and Environmental Medicine.* 53(5), 435–439, 1982.

Galinsky, T.L., Warm, J.S., Dember, W.N., Weiler, E.M., and Scerbo, M.W. Sensory alternation and vigilance performance: The role of pathway inhibition. *Human Factors.* 32(6), 717–728, 1990.

Gerard, M.J. and Martin, B.J. Post-effects of long-term hand vibration on visuo-manual performance in a tracking task. *Ergonomics.* 42(2), 314–326, 1999.

Griffin, W.C. and Rockwell, T.H., Computer-aided testing of pilot response to critical in-flight events. *Human Factors.* 26(5), 573–581, 1984.

Hancock, P.A. and Caird, J.K. Experimental evaluation of a model of mental workload. *Human Factors.* 35(3), 413–319, 1993.

Harris, C.S. and Johnson, D.L. Effects of infrasound on cognitive performance. *Aviation, Space, and Environmental Medicine.* 49(4), 582–586, 1978.

Harris, C.S. and Shoenberger, R.W. Combined effects of broadband noise and complex waveform vibration on cognitive performance. *Aviation, Space, and Environmental Medicine.* 51(1), 1–5, 1980.

Hubbard, D.C. Inadequacy of root mean square as a performance measure. *Proceedings of the 4th International Symposium on Aviation Psychology.* 698–704, 1987.

Imbeau, D., Wierwille, W.W., Wolf, L.D., and Chun, G.A. Effects of instrument panel luminance and chromaticity on reading performance and preference in simulated driving. *Human Factors.* 31(2), 147–160, 1989.

Kennedy, R.S., Odenheimer, R.C., Baltzley, D.R., Dunlap, W.P., and Wood, C.D. Differential effects of scopolamine and amphetamine on microcomputer-based performance tests. *Aviation, Space, and Environmental Medicine.* 61(7), 615–621, 1990.

Kimchi, R., Gopher, D., Rubin, Y., and Raij, D. Performance under dichoptic versus binocular viewing conditions: Effects of attention and task requirements. *Human Factors.* 35(1), 35–56, 1993.

Korardekar, P. and Mital, A. The effect of different work-rest schedules on fatigue and performance of a simulated directory assistance operator's task. *Ergonomics.* 37(10), 1697–1707, 1994.

Kuller, R. and Laike, T. The impact of flicker from fluorescent lighting on well-being, performance and physiological arousal. *Ergonomics.* 41(4), 433–447, 1998.

Kvalseth, T.O. Human performance comparisons between digital pursuit and compensatory control. *Ergonomics.* 21(6), 419–425, 1978.

Lanzetta, T.M., Dember, W.N., Warm, J.S., and Berch, D.B. Effects of task type and stimulus heterogeneity on the event rate function in sustained attention. *Human Factors.* 29(6), 625–633, 1987.

Lee, M.D. and Fisk, A.D. Disruption and maintenance of skilled visual search as a function of degree of consistency. *Human Factors.* 35(2), 205–220, 1993.

Loeb, M., Noonan, T.K., Ash, D.W., and Holding, D.H. Limitations of the cognitive vigilance decrement. *Human Factors.* 29(6), 661–674, 1987.

Maddox, M.E. and Turpin, J.A. The effect of number ordering and orientation on marking speed and errors for mark-sensed labels. *Human Factors.* 28(4), 401–405, 1986.
Matthews, M.L., Lovasik, J.V., and Mertins, K. Visual performance and subjective discomfort in prolonged viewing of chromatic displays. *Human Factors.* 31(3), 259–271, 1989.
McCann, P.H. The effects of ambient noise on vigilance performance. *Human Factors.* 11(3), 251–256, 1969.
McLeod, R.W. and Griffin, M.J. Effects of duration and vibration on performance of a continuous manual control task. *Ergonomics.* 36(6), 645–659, 1993.
Meister, D. *Human factors testing and evaluation.* New York: Elsevier, 1986.
Mertens, H.W. and Collins, W.E. The effects of age, sleep deprivation, and altitude on complex performance. *Human Factors.* 28(5), 541–551, 1986.
Millar, K. and Watkinson, N. Recognition of words presented during general anesthesia. *Ergonomics.* 26(6), 585–594, 1983.
Moseley, M.J. and Griffin, M.J. Whole-body vibration and visual performance: An examination of spatial filtering and time-dependency. *Ergonomics.* 30(4), 613–626, 1987.
Mullin, J. and Corcoran, D.W.J. Interaction of task amplitude with circadian variation in auditory vigilance performance. *Ergonomics.* 20(2), 193–200, 1977.
O'Hanlon, J., Schmidt, A., and Baker, C.H. Sonar Doppler discrimination and the effect of a visual alertness indicator upon detection of auditory sonar signals in a sonar watch. *Human Factors.* 7(2), 1965.
Payne, D.G. and Lang, V.A. Visual monitoring with spatially versus temporally distributed displays. *Human Factors.* 33(4), 443–458, 1991.
Pepler, R.D. Warmth and performance: An investigation in the tropics. *Ergonomics.* 2(10), 63–88, 1958.
Pitrella, F.D. and Kruger, W. Design and validation of matching tests to form equal groups for tracking experiments. *Ergonomics.* 26(9), 833–845, 1983.
Raaijmaker, J.G.W. and Verduyn, W.W. Individual difference and the effects of an information aid in performance of a fault diagnosis task. *Ergonomics.* 39(7), 966–979, 1996.
Ralston, J.V., Pisoni, D.B., Lively, S.E., Greene, G.B., and Mullennix, J.W. Comprehension of synthetic speech produced by rule: Word monitoring and sentence-by-sentence listening tones. *Human Factors.* 33(4), 471–491, 1991.
Rosenberg, D.J. and Martin, G. Human performance evaluation of digitizer pucks for computer input of spatial information. *Human Factors.* 30(2), 231–235, 1988.
Ruffell-Smith, H.P. *A simulator study of the interaction of pilot workload with errors, vigilance, and decisions.* Moffett Field, CA: Ames Research Center, NASA TM 78432, January 1979.
Salame, P. and Baddeley, A. Noise, unattended speech and short-term memory. *Ergonomics.* 30(8), 1185–1194, 1987.
Smith, A.P. and Miles, C. The effects of lunch on cognitive vigilance tasks. *Ergonomics.* 29(10), 1251–1261, 1986.
Torle, G. Tracking performance under random acceleration: Effects of control dynamics. *Ergonomics.* 8(4), 481–486, 1965.
Tullis, T.S. Human performance evaluation of graphic and textual CRT displays of diagnostic data. *Proceedings of the Human Factors Society 24th Annual Meeting.* 310–316, 1980.
Tzelgov, J., Henik, A., Dinstein, I., and Rabany, J. Performance consequences of two types of stereo picture compression. *Human Factors.* 32(2), 173–182, 1990.
Van Orden, K.F., Benoit, S.L., and Osga, G.A. Effects of cold air stress on the performance of a command and control task. *Human Factors.* 38(1), 130–141, 1996.
Vermeulen, J. Effects of functionally or topographically presented process schemes on operator performance. *Human Factors.* 29(4), 383–394, 1987.
Vidulich, M.A. The Bedford Scale: Does it measure spare capacity. *Proceedings of the 6th International Symposium on Aviation Psychology.* 2, 1136–1141, 1991.

Wierwille, W.W., Rahimi, M., and Casali, J.G. Evaluation of 16 measures of mental workload using a simulated flight task emphasizing mediational activity. *Human Factors*. 27(5), 489–502, 1985.

Yeh, Y. and Silverstein, L. D. Spatial judgments with monoscopic and stereoscopic presentation of perspective displays. *Human Factors*. 34(5), 583–600, 1992.

Zaitzeff, L.P. Aircrew task loading in the Boeing multimission simulator. *Proceedings of Measurement of Aircrew Performance—The Flight Deck Workload and its Relation to Pilot Performance (AGARD-CP-56)*. Neuilly-sur-Seine, France: AGARD, 1969.

Zeiner, A.R. and Brecher, G.A. Reaction time with and without backscatter from intense pulsed light. *Aviation, Space, and Environmental Medicine*, 46(2), 125–127, 1975.

2.2 TIME

The second category of human performance measures is time. Measures in this category assume that tasks have a well-defined beginning and end so that the duration of task performance can be measured. Measures in this category include dichotic listening detection time (section 2.2.1), glance duration (section 2.2.2), lookpoint time (section 2.2.3), marking speed (section 2.2.4), movement time (section 2.2.5), RT (section 2.2.6), reading speed (section 2.2.7), search time (section 2.2.8), task load (section 2.2.9), and time to complete (section 2.2.10).

2.2.1 Dichotic Listening Detection Time

General description. Subjects are presented with auditory messages through headphones. Each message contains alphanumeric words. Different messages are presented simultaneously to each ear. Subjects must detect specific messages when they are presented in a specific ear.

Strengths and limitations. Gopher (1982) reported significant differences in scores between flight cadets who completed training and those who failed.

Data requirements. Auditory tapes are required as well as an audio system capable of displaying different messages to each ear. Finally, a system is required to record omissions, intrusions, and switching errors.

Thresholds. Low threshold is zero. High threshold was not stated.

SOURCE

Gopher, D. Selective attention test as a predictor of success in flight training. *Human Factors*. 24(2), 173–183, 1982.

2.2.2 Glance Duration

General description. The duration that a human visually samples a single scene is a glance. Glance duration has been used to evaluate controls, displays, and procedures.

Strengths and limitations. Glance duration has long been used to evaluate driver performance. In an early study, Mourant and Rockwell (1970) analyzed the glance behavior of eight drivers traveling at 50 mph (miles per hour) (80.47 kmph [kilometers per hour]) on an expressway. As the route became more familiar, drivers increased glances to the right edge marker and horizon. While following a car, drivers glanced more often at lane markers. Imbeau, Wierwille, Wolf, and Chun (1989) used time glancing at a display to evaluate instrument panel lighting in automobiles. Not unexpectedly, higher complexity of messages was associated with significantly longer (+0.05 s more) glance times. Similarly, Land (1993) reported that drivers' glance behavior varies throughout a curve: a search for cues during approach, into the curve during the bend, and target beyond the bend on exit.

Mourant and Donohue (1977) reported that novice and young experienced drivers made fewer glances to the left outside mirror than did mature drivers. Novice drivers also made more direct looks than glances in the mirrors prior to executing a maneuver. In addition, novice drivers have longer and more frequent glances at vehicles than at obstacles (Masuda, Nagata, Kureyama, and Sato, 1990). Glance duration increases as blood alcohol concentration level increases (Masuda, Nagata, Kureyama, and Sato, 1990).

There has been some nondriver research as well. For example, Fukuda (1992) reported a maximum of six characters could be recognized in a single glance.

Data requirements. Eye movements must be recorded to an accuracy of ±0.5° horizontal and ±1° vertical (Mourant and Rockwell, 1970).

Thresholds. Minimum glance duration 0.68 s (Mourant and Donohue, 1977). Maximum glance duration 1.17 s (Mourant and Donohue, 1977).

SOURCES

Fukuda, T. Visual capability to receive character information Part I: How many characters can we recognize at a glance? *Ergonomics*. 35(5), 617–627, 1992.

Imbeau, D., Wierwille, W.W., Wolf, L.D., and Chun, G.A. Effects of instrument panel luminance and chromaticity on reading performance and preference in simulated driving. *Human Factors*. 31(2), 147–160, 1989.

Land, M.F. Eye-head coordination during driving. *IEEE Systems, Man and Cybernetics Conference Proceedings*. 490–494, 1993.

Masuda, K., Nagata, M., Kureyama, H., and Sato, T.B. *Visual behavior of novice drivers as affected by traffic conflicts (SAE Paper 900141)*. Warrendale, P.A. Society of Automotive Engineers, 1990.

Mourant, R.R. and Donohue, R.J. Acquisition of indirect vision information by novice, experienced, and mature drivers. *Journal of Safety Research*. 9(1), 39–46, 1977.

Mourant, R.R. and Rockwell, T.H. Mapping eye movement patterns to the visual scene in driving: An exploratory study. *Human Factors*. 12(1), 81–87, 1970.

2.2.3 Lookpoint Time

General description. Lookpoint is "the current coordinates of where the pilot is looking during any one thirtieth of a second" (Harris, Glover, and Spady, 1986, p. 38). Lookpoint is usually analyzed by either real-time viewing of lookpoint superimposed on the instrument panel or examination of time histories.

Strengths and limitations. Real-time observation of lookpoint efficiently informs the researcher about scanning behavior as well as helps identify any calibration problems. Analysis of time histories of lookpoint provides information such as average dwell time, dwell percentage, dwell time, fixation, fixations per dwell, one- and two-way transitions, saccades, scans, transition, and transition rate. Such information is useful in (1) arranging instruments for optimum scanning, (2) assessing the time required to assimilate information from each display, (3) estimating the visual workload associated with each display, and task criticality (i.e., blink rate decreases as task criticality increases; Stern, Walrath, and Goldstein, 1984). Corkindale (1974) reported significant differences between working conditions in the percentage of time looking at a HUD. Spady (1977) reported different scanning behavior during approach between manual (73% of time on flight director, 13% on airspeed) and autopilot with manual throttle (50% on flight director, 13% on airspeed).

Limitations of lookpoint include the following: (1) inclusion of an oculometer into the workplace or simulator, (2) requirement for complex data-analysis software, (3) lack of consistency, (4) difficulty in interpreting the results, and (5) lack of sensitivity of average dwell time. The first limitation is being overcome by the development of miniaturized oculometers, the second by the availability of standardized software packages, the third by collecting enough data to establish a trend, the fourth by development of advanced analysis techniques, and the fifth by use of the dwell histogram.

Data requirements. The oculometer must be calibrated and its data continuously monitored to ensure that it is not out of track. Specialized data reduction and analysis software is required.

Thresholds. Not stated.

SOURCES

Corkindale K.G.G. A flight simulator study of missile control performance as a function of concurrent workload. *Proceedings of Simulation and Study of High Workload (AGARD-CP-146)*, 1974.

Harris, R.L., Glover, B.J., and Spady, A.A. *Analytical techniques of pilot scanning behavior and their application (NASA Technical Paper 2525).* Hampton, VA: NASA Langley, July 1986.

Spady, A.A. Airline pilot scanning behavior during approaches and landing in a Boeing 737 simulator. *Proceedings of Guidance and Control Design Considerations for Low Altitude and Terminal Area Flight (AGARD-CP-240)*, 1977.

Stern, J.A., Walrath, L.C., and Goldstein, R. The indigenous eye blink. *Psychophysiology.* 21(1), 22–23, 1984.

2.2.4 Marking Speed

General description. Maddox and Turpin (1986) used speed to evaluate performance using mark-sensed forms.

Strengths and limitations. Marking speed was not affected by number orientation (horizontal or vertical), number ordering (bottom-to-top or top-to-bottom), or handedness of users (Maddox and Turpin, 1986).

Data requirements. Start and stop times must be recorded.

Thresholds. The average marking speed was 4.74 s for a five-digit number.

SOURCE

Maddox, M.E. and Turpin, J.A. The effect of number ordering and orientation on marking speed and errors for mark-sensed labels. *Human Factors*. 28(4), 401–405, 1986.

2.2.5 Movement Time

General description. Arnaut and Greenstein (1990) provided three definitions of movement time. "Gross movement was defined as the time from the initial touch on the [control] … to when the cursor first entered the target. Fine adjustment was the time from the initial target entry to the final lift-off of the finger from the [control] …. Total movement time was the sum of these two measures" (p. 655).

Strengths and limitations. Arnaut and Greenstein (1990) reported significant increases in gross movements and significant decreases in fine adjustment times for larger (120 mm) than smaller (40 mm) touch tablets. Total movement times were significantly longer for the largest and smallest touch tablets than for the intermediate size tablets (60, 80, or 100 mm). For a trackball, gross and total movement times were significantly longer for a longer distance (160 or 200 mm) than the shortest distance (40 mm). In a second experiment, there were no significant differences in any of the movement times for the touch tablet with and without a stylus. All three measures were significantly different between different display amplitudes and target widths. "In general performance was better with the smaller display amplitudes and the larger display target widths" (p. 660).

Lin, Radwin, and Vanderheiden (1992) used movement time to assess a head-controlled computer input device. They reported the longest movement time for small-width targets (2.9 mm versus 8.1 or 23.5 mm), high gains (1.2 versus 0.15, 0.3, or 0.6), and large-movement amplitudes (61.7 mm versus 24.3 mm). Hancock and Caird (1993) reported that movement time increased as the shrink rate of a target increased. Further, movement time decreased as path length increased.

Hoffman and Sheikh (1994) reported that movement times increased as target height decreased from 200 to 1 mm and also as target width decreased from 40 to 10 mm. Li, Zhu, and Adams (1995) reported that movement time increased with distance moved but disproportionately increased at extreme height (1080 above seat reference point) and angle (90° ipsilaterally),

Data requirements. The time at both the beginning and the end of the movement must be recorded.

Thresholds. Movement times below 50 ms are very uncommon.

SOURCES

Arnaut, L.Y. and Greenstein, J.S. Is display/control gain a useful metric for optimizing an interface? *Human Factors*. 32(6), 651–663, 1990.

Hancock, P.A. and Caird, J.K. Experimental evaluation of a model of mental workload. *Human Factors*. 35(3), 413–419, 1993.

Hoffman, E.R. and Sheikh, I.H. Effect of varying target height in a Fitts' movement task. *Ergonomics*. 36(7), 1071–1088, 1994.

Li, S., Zhu, Z., and Adams, A.S. An exploratory study of arm-reach reaction time and eye-hand coordination. *Ergonomics*. 38(4), 637–650, 1995.

Lin, M.L., Radwin, R.G., and Vanderheiden, G.C. Gain effects on performance using a head-controlled computer input device. *Ergonomics*. 35(2), 159–175, 1992.

2.2.6 Reaction Time

General description. RT is the time elapsed between stimulus onset and response onset. The stimulus is usually a visually presented number requiring a manual key press but any stimulus and any input or output mode are possible.

Strengths and limitations. RT may measure the duration of mental-processing stages (Donders, 1969). RT is sensitive to physiological state such as fatigue, sleep deprivation, aging, brain damage, and drugs (Boer, Ruzius, Minpen, Bles, and Janssen, 1984; Frowein, 1981; Frowein, Gaillard, and Varey, 1981; Frowein, Reitsma, and Aquarius, 1981; Gaillard, Gruisen, and de Jong, 1986; Gaillard, Rozendaal, and Varey, 1983; Gaillard, Varey, and Ruzius, 1985; Logsdon, Hochhaus, Williams, Rundell, and Maxwell, 1984; Moraal, 1982; Sanders, Wijnen, and van Arkel, 1982; Steyvers, 1987).

Fowler, Elcombe, Kelso, and Portlier (1987) used RT for the first correct response to a five-choice visual RT task to examine the effects of hypoxia. The response time was greater at 82% arterial oxyhemoglobin saturation than at 84% or 86%. In a related study, Fowler, Mitchell, Bhatia, and Portlier (1989) reported increased RT as a function of inert gas narcosis.

RT is also sensitive to the effects of time. For example, Coury, Boulette, and Smith (1989) reported significant decreases in RT over time. Harris, Hancock, Arthur, and Caird (1995) reported significant decreases in response time as time on task increased. Pigeau, Angus, O'Neill, and Mack (1995) measured response time of air defense operators to detect incoming aircraft. There was a significant interaction between shift and time on task. Subjects working the midnight shift had longer RTs in the 60/60 min work/rest schedule than those in the evening shift. There was also a shift-by-zone significant interaction: longer RTs for midnight shift for northern regions (i.e., low air traffic). Finally, between the two sessions, RT increased during the midnight shift but decreased during the evening shift.

RT is a reliable measure (e.g., split-half reliabilities varying between 0.81 and 0.92, AGARD, 1989, p. 12). However, Vidulich (1991) reported a test-retest reliability of +0.39 of a visual choice RT task. However, Boer (1987) suggests that RT performance may require 2000 trials to stabilize. Further, Carter, Krause, and Harbeson (1986) reported that slope was less reliable than RT for a choice RT task.

RT has been used to evaluate auditory, tactile, visual, and vestibular stimuli. In an early study Swink (1966) compared RTs to a light, a buzzer, and an electropulse. RTs were shortest for the pulse.

2.2.6.1 Auditory Stimuli

Early work in the 1960s examined the effects of stimulus characteristics on RT. For example, Loeb and Schmidt (1963) measured the RT of eight subjects of randomly occurring auditory signals over eight 50 min sessions. In four sessions the signal was 10 dB above threshold. In the other four it was 60 dB above threshold. RT in the 10 dB sessions increased if the responses were merely acknowledged rather than the subjects being told the signal would be faster or slower. In a 6 month study, Warrick, Kibler, and

Topmiller (1965) measured the response time of five secretaries to a buzzer that was activated without warning once or twice a week. Response time decreased over time and with alerting. Simon (1966) reported significantly longer RTs with cross-stimulus response correspondence (i.e., responding to an auditory stimulus in the left ear with the right hand) than for same-side response. This was especially true for older subjects.

Later work focused on communication and feedback. For example, Payne and Lang (1991) reported shorter RTs for rapid communication than for conventional visual displays. In another communication experiment, RTs were significantly faster for target words in natural than in synthetic speech (Ralston, Pisoni, Lively, Greene, and Mullennix, 1991).

Akamatsu, MacKenzie, and Hasbroucq (1995) reported that there were no differences in response times associated with the type of feedback (normal, auditory, color, tactile, and combined) provided by a computer mouse.

Begault and Pittman (1996) used detection time to compare conventional versus 3-D (dimensional) audio warnings of aircraft traffic. Detection time was significantly shorter (500 ms) for the 3-D audio display. Haas and Casali (1995) reported that response time to an auditory stimulus decreased as perceived urgency increased and as pulse level increased.

Similar to other RTs to other stimuli, Horne and Gibbons (1991) reported a significant increase in RT with increases in alcohol dose.

2.2.6.2 Tactile Stimuli

In moving-base simulators, RT to cross-wind disturbances was significantly shorter when physical-motion cues were present than when they were not present (Wierwille, Casali, and Repa, 1983).

2.2.6.3 Visual Stimuli

The vast majority of the research using RT has been performed with visual stimuli. The research has examined the effects of task, environment, and subject variables. There have also been efforts to break RT into components and models.

Task Variables

Task variables examined using RT include target characteristics, display formats, task scheduling, and task type.

Target characteristics. In an early study, Baker, Morris, and Steedman (1960) reported increases in search time as the number of irrelevant items increased and as the difference in resolution of reference and form increased. RT increased as stimulus complexity (1, 2, or 4 vertical lines) increased and interstimulus interval (100, 300, 500, 700, 900, and 1,100 ms) decreased (Aykin, Czaja, and Drury, 1986). Mackie and Wylie (1994) reported decreased RT to sonar signals with performance feedback and increased signal rate. Hancock and Caird (1993) reported a significant decrease in RT as the length of a path from the cursor to the target increased.

In a field study, Cole, Johnston, Gibson, and Jacobs (1978) recorded the RTs of Air Traffic Controllers (ATC) to aircraft initial movement prior to takeoff. The independent variables were delay in takeoff, angle of observer's line of sight to the centerline of the runway, experience of the ATC personnel, speed of aircraft acceleration, and use of binoculars. RT was significantly related to each of these independent variables.

Yeh and Silverstein (1992) measured RT as subjects made spatial judgments pertaining to a simplified aircraft landing display. They reported significantly shorter RT for targets in the high front versus low back portions of the visual field of the display, for larger altitude separations between targets, for the 45° versus the 15° viewing orientation, and with the addition of binocular disparity. Kerstholt, Passenier, Houltuin, and Schuffel (1996) reported a significant increase in detection time with simultaneous targets. There was also a significant increase in detection time for subsequent targets (first versus second versus third) in two complex target conditions.

Display formats. Formats investigated included flashing, background, color, viewing angle, and type of symbology.

In an early study on flashing, Crawford (1962) reported that RTs were longer for a visual stimulus against a background of flashing lights than against a background of steady lights. However, Wolcott, Hanson, Foster, and Kay (1979) found no significant differences in choice RT with light flashes.

The effect of background has often been investigated. In an early study, Haines (1968) reported a significant increase in detection time of a spot light source with the introduction of a star background (either real or simulated) or a glare source. His subjects were 127 untrained observers in a planetarium. Thackray, Bailey, and Touchstone (1979) reported significant increases in RT to critical targets in the presence of 16 targets rather than 4 or 8 targets. Zeiner and Brecher (1975) reported significant increases in RTs to visual stimuli in the presence of backscatter from strobe lights.

Lee and Fisk (1993) reported faster RTs in a visual search task if the consistency of the stimuli remained 100% than if it did not (67%, 50%, or 33% consistent). Holahan, Culler, and Wilcox (1978) asked subjects to state "stop" or "go" when presented a photo of a traffic scene with or without a stop sign. RTs increased as the number of distracters increased and in the presence of some red in the distracters. In a more recent study, Ho, Scialfi, Caird, and Graw (2001) measured RT to traffic signs in visual scenes with varying amounts of clutter. There were significant effects of age (younger drivers had shorter RTs), clutter (clutter increased RTs), and presence of target (presence made RTs shorter).

Related to background are the effects of redundant cues. Simon and Overmeyer (1984) used RT to evaluate the effects of redundant visual cues. Simon, Peterson, and Wang (1988) in contrast examined the effects of redundant versus different cues on RT. Jubis (1990) used RT to evaluate display codes. She reported significantly faster RT for redundant color and shape ($\bar{x}$ = 2.5 s) and color ($\bar{x}$ = 2.3 s) coding than for partially redundant color ($\bar{x}$ = 2.8 s) or shape ($\bar{x}$ = 3.5 s). Perrott, Sadralodabai, Saberi, and Strybel (1991) reported significant decreases in RT to a visual target when spatially correlated sounds were presented with the visual targets. Also, MacDonald and Cole (1988) reported a significant decrease in RT for flight tasks in which relevant information was color-coded as compared to monochromatic displays.

Murray and Caldwell (1996) reported significantly longer RTs as the number (1, 2, 3) of displays to be monitored increased and number (1, 2, 3) of display figures increased. RTs to process independent, compared to redundant, images were longer.

Tzelgov, Henik, Dinstein, and Rabany (1990) used RT to compare two types of stereo picture compression. They reported significant task (faster for object decision than depth decision task), depth (faster for smaller depth differences), size (faster for no size difference between compared objects), and presentation effects. They also found numerous interactions.

Tullis (1980) reported significant difference on RT as a function of display format. Color graphics was associated with significantly shorter RTs than narrative text with structured text and black-and-white graphics in between. Thackray and Touchstone (1991) used detection times of secondary targets in a simulated ATC task to examine the effect of color versus flashing as a cue under low and high task load.

In an early study on viewing angle, Simon and Wolf (1963) reported increases in RT as viewing angle increased. Layout is related to viewing angle. Downing and Sanders (1987) reported longer response times in simulated control room emergencies with mirror-image panels than nonmirror-image panels.

The effects of type of symbology have been investigated in aviation using RT. For example, Remington and Williams (1986) used RT to measure the efficiency with which helicopter situation display symbols could be located and identified. RTs for numeric symbols were significantly shorter than for graphic symbols. Negative trials (target not present) averaged 120 ms longer than positive trials (target present); however, there were more errors on positive than on negative trials.

Taylor and Selcon (1990) compared response time for four display formats (HUD, Attitude Indicator, Aircraft Reference, and Command Indicator) in a simulated aircraft unusual attitude recovery. There were significantly longer response times for the Aircraft Reference display. Pigeau, Angus, O'Neill, and Mack (1995) measured response time of air defense operators to incoming aircraft. Significantly longer RTs occurred when the same geographical area was displayed in two rather than four zones.

Symbology has also been investigated in other domains. For example, Chapanis and Lindenbaum (1959) used response time to stove control-burner arrangements. They found a decrease in RT over trials 1 through 40 but not trials 41 through 81. They also found significantly shorter RTs with one of the four stove control-burner configurations. Buttigieg and Sanderson (1991) reported significant differences in RT between display formats. There were also significant decreases in RT over the 3 d of the experiment.

Coury, Boulette, and Smith (1989) reported significantly faster RTs to a digital display than to a configural or bar graph display. The task was classification. In a similar display study, Steiner and Camacho (1989) reported significant increases in RT as the number of bits of information increased, especially for alphanumeric representation as compared to icons. In addition, Imbeau, Wierwille, Wolf, and Chun (1989) used the time from stimulus presentation to a correct answer to evaluate driver reading performance. These authors reported significantly longer RTs for 7-arcmin (arc minute) characters (3.45 to 4.86 s) than for 25-arcmin characters (1.35 s).

In another driving study, Kline, Ghali, Kline, and Brown (1990) converted visibility distances of road signs to sight times in seconds, assuming a constant travel speed. This measure made differences between icon and text signs very evident. Also, McKnight and Shinar (1992) reported significantly shorter brake RT in following vehicles when the forward vehicle had center high-mounted stop lamps.

RT has also been useful in discriminating between display utility (Nataupsky and Crittenden, 1988). Further, Boehm-Davis, Holt, Koll, Yastrop, and Peters (1989) reported faster RT to database queries if the format of the database was compatible with the information sought. Finally, Perriment (1969) examined RT to bisensory stimuli, i.e., auditory and visual, and Morgerstern and Haskell (1971) reported significantly slower RTs for visual than for auditory stimuli.

Task scheduling. Adams, Humes, and Stenson (1962) reported a significant decrease in RT with a 3 h visual detection task but not over 9 d of task performance. In the same year Teichner (1962) reported that RT increased as the probability of detection during an initial session increased.

Task type. Dewar, Ells, and Mundy (1976) reported shorter RTs for classifying than identifying visually presented traffic signs. RTs were also shorter for warning than for regulatory signs and for text than symbol signs. Wierwille, Rahimi, and Casali (1985) reported that RT was significantly affected by the difficulty of a mathematical problem-solving task. They defined RT as the time from problem presentation to a correct response.

Koelega, Brinkman, Hendriks, and Verbaten (1989) reported no significant differences in RT between four types of visual vigilance tasks "1) the cognitive Continuous Performance Test (CPT), in which the target was the sequence of the letters AX; 2) a visual version of the cognitive Bakan task in which a target was defined as three successive odd but unequal digits; 3) an analogue of the Bakan using nondigital stimuli; and 4) a pure sensory task in which the critical signal was a change in brightness" (p. 46). Van Assen and Eijkman (1971) used RT to assess learning in a two-dimensional tracking task.

Fisk and Jones (1992) reported significant effects of search consistency (varying the ratio of consistent [all words were targets] to inconsistent words: 8:0, 6:2, 4:4, 2:6, and 0:8; shorter RTs with increased consistency) and practice (shorter RT from trial 1 to 12) on correct RT.

Elvers, Adapathya, Klauer, Kancler, and Dolan (1993) reported significant decreases in RT as a function of practice. This was especially true for a volume estimation compared to a distance estimation task.

Kimchi, Gopher, Rubin, and Raij (1993) reported significantly shorter RTs for a local-directed rather than a global-directed task in a focused than in a divided attention condition. Significantly shorter RTs were reported for the global-directed than the local-directed task but only in the divided-attention condition.

There have even been studies to assess the effect of how a task is performed. For example, Welford (1971) reported RTs were longer when made by the ring and middle fingers than when made by the index and little fingers.

Environment Variables

In an early environment study Pepler (1958) reported an increase in the number of missed signals as temperature increased from 67°F to 92°F. Using the same stimuli, Chiles (1958) reported no significant differences associated with temperature. However, Shvartz, Meroz, Mechtinger, and Birnfeld (1976) reported 30% longer RTs during exercise in temperate (23°C) and hot environments (30°C) than during rest. This reduced to 10% higher RT after 8 d of heat acclimation. Liebowitz, Abernathy,

Buskirk, Bar-or, and Hennessy (1972) reported no significant difference in reaction time to red lights in the central visual area while subjects were on a treadmill in a heat chamber with or without fluid replacement, although there was a decrease in RT to lights in the periphery, however, with practice. However, Enander (1987) reported no increase in moderate cold (+5°C).

Miles, Auburn, and Jones (1984) reported longer RTs in noise environments (95 dBC [decibel (C scale)]) versus quiet (65 dBC). Warner and Heimstra (1972) reported significant difference in RT as a function of noise (0, 80, 90, and 100 dB), task difficulty (8, 16, and 32 letter display), and target location (central and peripheral). Beh and Hirst (1999) reported shorter RTs to a stop-light display during music than during quiet. However, RT to peripheral visual targets was increased with high-intensity (i.e., 85 or 87 dBA) music.

Macintosh, Thomas, Olive, Chesner, and Knight (1988) reported significantly longer RTs for those individuals showing effects of acute mountain sickness at either 4790 or 5008 m altitude. The RT task was a visual three-choice task. In another hypoxia experiment, Fowler and Kelso (1992) reported increased RT for hypoxia than normoxia, especially for low-intensity (0.38 cd/m^2 [candela/square meter (cd/m^2)]) than high-intensity (1.57 cd/m^2) visual stimuli (males and female names). Fowler, White, Wright, and Ackles (1982), however, reported that hypoxia increased RT, especially with lower-luminance stimuli. Fowler, Paul, Porlier, Elcombe, and Taylor (1985) reported increased RT with hypoxia. Leifflen, Poquin, Savourey, Barraud, Raphel, and Bittel (1997) used the manikin choice RT task to evaluate the effects of hypoxia. Hypoxia had no significant effects on either the number of errors or the RT, even at 7000 m simulated altitude.

McCarthy, Corban, Legg, and Faris (1995) reported significantly slower reaction times in judging the orientation of visual targets at 7000 and 12000 ft relative to sea level. In a similar study, Linden, Nathoo, and Fowler (1996) reported significant increases in RT associated with hypoxia and with increased angle of rotation of a target. Gliner, Matsen-Twisdale, and Horvath (1979) report any significant differences in detecting 1 s light pulses associated with ozone detection (0.00, 0.25, 0.50, and 0.75 ppm [parts per million]) when the rate of signals to nonsignals was low (1 out of 30). There was a decrease in performance at 0.75 ppm when the signal-to-nonsignal ratio was increased to 1 to 6, however.

Mertens and Collins (1986) used RT to warning light changes in visual pointer position and successive presentation of targets to evaluate the effects of age (30 to 39 versus 60 to 69 years old), altitude (ground versus 3810 m), and sleep (permitted versus deprived). RT was standardized and then transformed so that better performance was associated with a higher score. Older persons had lower scores in the light task than younger persons. Sleep deprivation decreased performance on all three RT tasks; altitude did not significantly affect performance on either task. Heimstra, Fallesen, Kinsley, and Warner (1980) reported that deprivation of cigarette smoking had no effect on RT.

In an aviation-related study (Stewart and Clark, 1975), 13 airline pilots were exposed to rotary accelerations (0.5, 1.0, and 5.0 degrees per second^2) prior (0, 50, 90, 542, 926, 5503, and 9304 ms) to performance of a visual choice RT task. RT increased significantly with increases in each independent variable.

Finally, Krause (1982), on the basis of data from 15 navy enlisted men performing 50 trials on each of 15 d, recommended using at least 1000 practice trials prior to using RT to assess environmental effects.

Subject Variables

Age, gender, and experience have been investigated using RT.

Deupree and Simon (1963) reported longer RTs in the two-choice than in the single-choice condition. Older subjects (median age 75) moved faster in the single than in the choice RT condition.

Adam, Paas, Buekers, Wuyts, Spijkers, and Wallmeyer (1999) reported that males tended to have shorter RTs than females. In a simulated ATC task, Thackray, Touchstone, and Bailey (1978) reported a significant increase in RT over a 2 h session. There were, however, no significant differences between the 26 men and 26 women who participated in the experiment. Matthews and Ryan (1994) reported longer RTs to a line length comparison task in the premenstrual phase than in the intermenstrual phase. The subjects were female undergraduate students.

DeMaio, Parkinson, and Crosby (1978) asked instructor and student pilots to indicate if photos of an aircraft cockpit were correct. Instructors responded more quickly and more accurately.

Briggs and Goldberg (1995) used response time to evaluate armored tank recognition ability. There were significant differences between subjects, presentation time (shorter RTs as presentation time increased), view (flank view faster than frontal view), and model (M1 fastest, British Challenger slowest). Effects of component versus friend or foe were not significant.

Korteling (1990) compared visual choice RTs among diffuse brain injury patients, older males (61 to 73 years old), and younger males (21 to 43 years old). The RTs of the patients and older males were significantly longer than those of the younger males. RTs were significantly different as a function of response-stimulus interval (RSI; RSIs and RTs were 100, 723; 500, 698; and 1250 ms, 713 ms, respectively). There was also a significant RSI by stimulus sequence interaction, specifically, "interfering aftereffects of alternating stimuli decreased significantly with increasing RSI" (p. 99).

RT is not always sensitive to differences between subjects, however. For example, Leonard and Carpenter (1964) reported no correlation between a five-choice RT task and the A.H.4 Intelligence Task. There was, however, a significant correlation between RT and typewriter task performance measured as number of words typed in 5 min. Further, Park and Lee (1992) reported that RT in a computer-aided aptitude task did not predict performance of flight trainees.

Components

Taking a different approach, Jeeves (1961) broke total RT into component times to assess the effects of advanced warning. Hilgendorf (1966) reported that RT was a $\log_2 n$ function with n being the number of bits per stimulus and was linear even at high n's. Krantz, Silverstein, and Yeh (1992) went one step further and developed a mathematical model to predict RT as a function of spatial frequency, forward field of view/display luminance mismatch, and luminance contrast.

2.2.6.4 Vestibular Stimuli

Gundry (1978) asked subjects to detect roll motion while seated on a turntable. There were no significant differences in RT for left versus right roll except when subjects were given a visual cue. However, they detected rightward roll significantly faster (236 ms) than leftward roll (288 ms).

2.2.6.5 Related Measures

Related measures include detection time and recognition time. Detection time is defined as the onset of a target presentation until a correct detection is made. Recognition time is the time from onset of the target presentation until a due target is correctly recognized (Norman and Ehrlich, 1986). In an early study Johnston (1968) reported significant effects of horizontal resolution, shades of gray, and slant range on target recognition time. There were 12 subjects with 20/20 vision or better. The targets were models of three military vehicles (2½ ton cargo truck, 5 ton flatbed truck, and tank with 90 mm gun) on a terrain board viewed through a closed-circuit television system.

Bemis, Leeds, and Winer (1988) did not find any significant differences in threat detection times between conventional and perspective radar displays. Response time to select the interceptor nearest to the threat was significantly shorter with the perspective display.

Damos (1985) used a variant of RT, specifically, the average interval between correct responses (CRI). CRI includes the time to make incorrect responses. CRI was sensitive to variations in stimulus mode and trial.

Data requirements. Start time of the stimulus and response must be recorded to the nearest millisecond.

Thresholds. RTs below 50 ms are very uncommon.

SOURCES

Adam, J.J., Paas, F.G.W.C., Buekers, M.J., Wuyts, I.J., Spijkers, W.A.C., and Wallmeyer, P. Gender differences in choice reaction time: Evidence for differential strategies. *Ergonomics.* 42(2), 327–335, 1999.

Adams, J.A., Humes, J.M., and Stenson, H.H. Monitoring of complex visual displays: III. Effects of repeated session on human vigilance. *Human Factors.* 4(3), 149–158, 1962.

Akamatsu, M., MacKenzie, I.S., and Hasbroucq, T. A comparison of tactile, auditory, and visual feedback in a pointing task using a mouse-type device. *Ergonomics.* 38(4), 816–827, 1995.

Aykin, N., Czaja, S.J., and Drury, C.G. A simultaneous regression model for double stimulation tasks. *Human Factors.* 28(6), 633–643, 1986.

Baker, C.A., Morris, D.F., and Steedman, W.C. Target recognition on complex displays. *Human Factors.* 2(2), 51–61, 1960.

Begault, D.R. and Pittman, M.T. Three-dimensional audio versus head-down traffic alert and collision avoidance system displays. *International Journal of Aviation Psychology.* 6(1), 79–93, 1996.

Beh, H.C. and Hirst, R. Performance on driving-related tasks during music. *Ergonomics.* 42(8), 1087–1098, 1999.

Bemis, S.V., Leeds, J.L., and Winer, E.A. Operator performance as a function of type of display: Conventional versus perspective. *Human Factors.* 30(2), 163–169, 1988.

Boehm-Davis, D.A., Holt, R.W., Koll, M., Yastrop, G., and Peters, R. Effects of different data base formats on information retrieval. *Human Factors.* 31(5), 579–592, 1989.

Boer, L.C. *Psychological fitness of Leopard I-V crews after a 200-km drive (Report Number IZF 1987-30).* Soesterberg, Netherlands: TNO Institute for Perception, 1987.

Boer, L.C., Ruzius, M.H.B., Minpen, A.M., Bles, W., and Janssen, W.H. *Psychological fitness during a maneuver (Report Number IZF 1984-17).* Soesterberg, Netherlands: TNO Institute for Perception, 1984.

Briggs, R.W. and Goldberg, J.H. Battlefield recognition of armored vehicles. *Human Factors.* 37(3), 596–610, 1995.

Buttigieg, M.A. and Sanderson, P.M. Emergent features in visual display design for two types of failure detection tasks. *Human Factors.* 33(6), 631–651, 1991.

Carter, R.C., Krause, M., and Harbeson, M.M. Beware the reliability of slope scores for individuals. *Human Factors.* 28(6), 673–683, 1986.

Chapanis, A. and Lindenbaum, L.E. A reaction time study of four control-display linkages. *Human Factors.* 1(4), 1–7, 1959.

Chiles, W.D. Effects of elevated temperatures on performance of a complex mental task. *Ergonomics.* 2(1), 89–107, 1958.

Cole, B.L., Johnston, A.W., Gibson, A.J., and Jacobs, R.J. Visual detection of commencement of aircraft takeoff runs. *Aviation, Space, and Environmental Medicine.* 49(2), 395–405, 1978.

Coury, B.G., Boulette, M.D., and Smith, R.A. Effect of uncertainty and diagnosticity on classification of multidimensional data with integral and separable displays of system status. *Human Factors.* 31(5), 551–569, 1989.

Crawford, A. The perception of light signals: The effect of the number of irrelevant lights. *Ergonomics.* 5(3), 417–428, 1962.

Damos, D. The effect of asymmetric transfer and speech technology on dual-task performance. *Human Factors.* 27(4), 409–421, 1985.

DeMaio, J., Parkinson, S.R., and Crosby, J.V. A reaction time analysis of instrument scanning. *Human Factors.* 20(4), 467–471, 1978.

Deupree, R.H. and Simon, J.R. Reaction time and movement time as a function of age, stimulus duration, and task difficulty. *Ergonomics.* 6(4), 403–412, 1963.

Dewar, R.E., Ells, J.G., and Mundy, G. Reaction time as an index of traffic sign perception. *Human Factors.* 18(4), 381–392, 1976.

Donders, F.C. On the speed of mental processes. In Koster, W.G. (Ed.) *Attention and performance* (pp. 412–431). Amsterdam: North Holland, 1969.

Downing, J.V. and Sanders, M.S. The effects of panel arrangement and focus of attention on performance. *Human Factors.* 29(5), 551–562, 1987.

Elvers, G.C., Adapathya, R.S., Klauer, K.M., Kancler, D.E., and Dolan, N.J. Effects of task probability on integral and separable task performance. *Human Factors.* 35(4), 629–637, 1993.

Enander, A. Effects of moderate cold on performance of psychomotor and cognitive tasks. *Ergonomics.* 30(10), 1431–1445, 1987.

Fisk, A.D. and Jones, C.D. Global versus local consistency: Effects of degree of within-category consistency on performance and learning. *Human Factors.* 34(6), 693–705, 1992.

Fowler, B., Elcombe, D.D., Kelso, B., and Portlier, G. The thresholds for hypoxia effects on perceptual-motor performance. *Human Factors.* 29(1), 61–66, 1987.

Fowler, B. and Kelso, B. The effects of hypoxia on components of the human event-related potential and relationship to reaction time. *Aviation, Space, and Environmental Medicine.* 63(6), 510–516, 1992.

Fowler, B., Mitchell, I., Bhatia, M., and Portlier, G. Narcosis has additive rather than interactive effects on discrimination reaction time. *Human Factors*. 31(5), 571–578, 1989.

Fowler, B., Paul, M., Porlier, G., Elcombe, D.D., and Taylor, M. A re-evaluation of the minimum altitude at which hypoxic performance decrements can be detected. *Ergonomics*. 28(5), 781–791, 1985.

Fowler, B., White, P.L., Wright, G.R., and Ackles, K.N. The effects of hypoxia on serial response time. *Ergonomics*. 25(3), 189–201, 1982.

Frowein, H.W. *Selective drug effects on information processing*. Dissertatie, Katholieke Hogeschool, Tilburg, 1981.

Frowein, H.W., Gaillard, A.W.K., and Varey, C.A. EP components, visual processing stages, and the effect of a barbiturate. *Biological Psychology*. 13, 239–249, 1981.

Frowein, H.W., Reitsma, D., and Aquarius, C. Effects of two counteractivity stresses on the reaction process. In Long, J. and Baddeley, A.D. (Eds.) *Attention and performance*. Hillsdale, NJ: Erlbaum, 1981.

Gaillard, A.W.K., Gruisen, A., and de Jong, R. *The influence of loratadine (sch 29851) on human performance (Report Number IZF 1986-C-19)*. Soesterberg, Netherlands: TNO Institute for Perception, 1986.

Gaillard, A.W.K., Rozendaal, A.H., and Varey, C.A. *The effects of marginal vitamin-deficiency on mental performance (Report Number IZF 1983-29)*. Soesterberg, Netherlands: TNO Institute for Perception, 1983.

Gaillard, A.W.K., Varey, C.A., and Ruzius, M.H.B. *Marginal vitamin deficiency and mental performance (Report Number 1ZF 1985-22)*. Soesterberg, Netherlands: TNO Institute for Perception, 1985.

Gliner, J.A., Matsen-Twisdale, J.A., and Horvath, S.M. Auditory and visual sustained attention during ozone exposure. *Aviation, Space, and Environmental Medicine*. 50(9), 906–910, 1979.

Gundry, A.J. Experiments on the detection of roll motion. *Aviation, Space, and Environmental Medicine*. 49(5), 657–664, 1978.

Haas, E.C., and Casali, J.G. Perceived urgency of and response time to multi-tone and frequency modulated warning signals in broadband noise. *Ergonomics*. 38(11), 2313–2326, 1995.

Haines, R.F. Detection time to a point source of light appearing on a star field background with and without a glare source present. *Human Factors*. 10(5), 523–530, 1968.

Hancock, P.A. and Caird, J.K. Experimental evaluation of a model of mental workload. *Human Factors*. 35(3), 413–429, 1993.

Harris, W.C., Hancock, P.A., Arthur, E.J., and Caird, J.K. Performance, workload, and fatigue changes associated with automation. *International Journal of Aviation Psychology*. 5(2), 169–185, 1995.

Heimstra, N.W., Fallesen, J.J., Kinsley, S.A., and Warner, N.W. The effects of deprivation of cigarette smoking on psychomotor performance. *Ergonomics*. 23(11), 1047–1055, 1980.

Hilgendorf, L. Information input and response time. *Ergonomics*. 9(1), 31–38, 1966.

Ho, G., Scialfi, C.T., Caird, J.K., and Graw, T. Visual search for traffic signs: The effects of clutter, luminance, and aging. *Human Factors*. 43(2), 194–207, 2001.

Holahan, C.J., Culler, R.E., and Wilcox, B.L. Effects of visual detection on reaction time in a simulated traffic environment. *Human Factors*. 20(4), 409–413, 1978.

Horne, J.A. and Gibbons, H. Effects of vigilance performance and sleepiness of alcohol given in the early afternoon ("post lunch") vs. early evening. *Ergonomics*. 34(1), 67–77, 1991.

Imbeau, D., Wierwille, W.W., Wolf, L.D., and Chun, G.A. Effects of instrument panel luminance and chromaticity on reading performance and preference in simulated driving. *Human Factors*. 31(2), 147–160, 1989.

Jeeves, M.A. Changes in performance at a serial-reaction task under conditions of advance and delay of information. *Ergonomics*. 4(4), 327–338, 1961.

Johnston, D.M. Target recognition on TV as a function of horizontal resolution and shades of gray. *Human Factors*. 10(3), 201–210, 1968.

Jubis, R.M. Coding effects on performance in a process control task with uniparameter and multiparameter displays. *Human Factors*. 32(3), 287–297, 1990.

Kerstholt, J.H., Passenier, P.O., Houltuin, K., and Schuffel, H. The effect of a priori probability and complexity on decision making in a supervisory task. *Human Factors*. 38(1), 65–78, 1996.

Kimchi, R., Gopher, D., Rubin, Y., and Raij, D. Performance under dichoptic versus binocular viewing conditions: Effects of attention and task requirements. *Human Factors*. 35(1), 35–56, 1993.

Kline, T.J.B., Ghali, L.M., Kline, D., and Brown, S. Visibility distance of highway signs among young, middle-aged, and older observers: Icons are better than text. *Human Factors*. 32(5), 609–619, 1990.

Koelega, H.S., Brinkman, J., Hendriks, L. and Verbaten, M.N. Processing demands, effort, and individual differences in four different vigilance tasks. *Human Factors*. 31(1), 45–62, 1989.

Korteling, J.E. Perception-response speed and driving capabilities of brain-damaged and older drivers. *Human Factors*. 32(1), 95–108, 1990.

Krantz, J.H., Silverstein, L.D., and Yeh, Y. Visibility of transmissive liquid crystal displays under dynamic lighting conditions. *Human Factors*. 34(5), 615–632, 1992.

Krause, M. Repeated measures on a choice reaction time task. *Proceedings of the Human Factors Society 26th Annual Meeting,* 359–363, 1982.

Lee, M.D. and Fisk, A.D. Disruption and maintenance of skilled visual search as a function of degree of consistency. *Human Factors*. 35(2), 205–220, 1993.

Leifflen, D., Poquin, D., Savoourey, G., Barraud, P., Raphel, C., and Bittel, J. Cognitive performance during short acclimation to severe hypoxia. *Aviation, Space, and Environmental Medicine*. 68(11), 993–997, 1997.

Leonard, J.A. and Carpenter, A. On the correlation between a serial choice reaction task and subsequent achievement at typewriting. *Ergonomics*. 7(2), 197–204, 1964.

Liebowitz, H.W., Abernathy, C.N., Buskirk, E.R., Bar-or, O., and Hennessy, R.T. The effect of heat stress on reaction time to centrally and peripherally presented stimuli. *Human Factors*. 14(2), 155–160, 1972.

Linden, A.E., Nathoo, A., and Fowler, B. An AFM investigation of the effects of acute hypoxia on mental rotation. *Ergonomics*. 39(2), 278–284, 1996.

Loeb, M. and Schmidt, E.A. A comparison of the effects of different kinds of information in maintaining efficiency on an auditory monitoring task. *Ergonomics*. 6(1), 75–82, 1963.

Logsdon, R., Hochhaus, L., Williams, H.L., Rundell, O.H., and Maxwell, D. Secobarbital and perceptual processing. *Acta Psychologica*. 55, 179–193, 1984.

MacDonald, W.A. and Cole, B.L. Evaluating the role of colour in a flight information cockpit display. *Ergonomics*. 31(1), 13–37, 1988.

Macintosh, J.H., Thomas, D.J., Olive, J.E., Chesner, I.M. and Knight, R.J.E. The effect of altitude on tests of reaction time and alertness. *Aviation, Space, and Environmental Medicine*. 59(3), 246–248, 1988.

Mackie, R.R. and Wylie, C.D. Countering loss of vigilance in sonar watch standing using signal injection and performance feedback. *Ergonomics*. 37(7), 1157–1184, 1994.

Matthews, G. and Ryan, H. The expression of the "pre-menstrual" syndrome in measures of mood and sustained attention. *Ergonomics*. 37(8), 1407–1417, 1994.

McCarthy, D., Corban, R., Legg, S., and Faris, J. Effects of mild hypoxia on perceptual-motor performance: A signal-detection approach. *Ergonomics*. 38(10), 1979–1992, 1995.

McKnight, A.J. and Shinar, D. Brake reaction time to center high-mounted stop lamps on vans and trucks. *Human Factors*. 34(2), 205–213, 1992.

Mertens, H.W. and Collins, W.E. The effects of age, sleep deprivation, and altitude on complex performance. *Human Factors.* 28(5), 541–551, 1986.

Miles, C., Auburn, T.C., and Jones, D.M. Effects of loud noise and signal probability on visual vigilance. *Ergonomics.* 27(8), 855–862, 1984.

Moraal, J. *Age and information processing: an application of Sternberg's additive factor method (Report Number IZF 1982-18).* Soesterberg, Netherlands: TNO Institute for Perception, 1982.

Morgerstern, F.S. and Haskell, S.H. Disruptive reaction times in single and multiple response units. *Ergonomics.* 14(2), 219–230, 1971.

Murray, S.A. and Caldwell, B.S. Human performance and control of multiple systems. *Human Factors.* 38(2), 323–329, 1996.

Nataupsky, M. and Crittenden, L. Stereo 3-D and non-stereo presentations of a computer-generated pictorial primary flight display with pathway augmentation. *Proceedings of the 9th AIAA/IEEE Digital Avionics Systems Conference*, 1988.

Norman, J. and Ehrlich, S. Visual accommodation and virtual image displays: Target detection and recognition. *Human Factors.* 28(2), 135–151, 1986.

Park, K.S. and Lee, S.W. A computer-aided aptitude test for predicting flight performance of trainees. *Human Factors.* 34(2), 189–204, 1992.

Payne, D.G. and Lang, V.A. Visual monitoring with spatially versus temporally distributed displays. *Human Factors.* 33(4), 443–458, 1991.

Pepler, R.D. Warmth and performance: An investigation in the tropics. *Ergonomics.* 2(1), 63–88, 1958.

Perriment, A.D. The effect of signal characteristics on reaction time using bisensory stimulation. *Ergonomics.* 12(1), 71–78, 1969.

Perrott, D.R., Sadralodabai, T., Saberi, K., and Strybel, T.Z. Aurally aided visual search in the central visual field: Effects of visual load and visual enhancement of the target. *Human Factors.* 33(4), 389–400, 1991.

Pigeau, R.A., Angus, R.G., O'Neill, P., and Mack, I. Vigilance latencies to aircraft detection among NORAD surveillance operators. *Human Factors.* 37(3), 622–634, 1995.

Ralston, J.V., Pisoni, D.B., Lively, S.E., Greene, B.G., and Mullennix, J.W. Comprehension of synthetic speech produced by rule: Word monitoring and sentence-by-sentence listening times. *Human Factors.* 33(4), 471–491, 1991.

Remington, R. and Williams, D. On the selection and evaluation of visual display symbology: Factors influencing search and identification times. *Human Factors.* 28(4), 407–420, 1986.

Sanders, A.F., Wijnen, J.I.C., and van Arkel, A.E. An additive factor analysis of the effects of sleep-loss on reaction processes. *Acta Psychologica.* 51, 41–59, 1982.

Shvartz, E., Meroz, A., Mechtinger, A., and Birnfeld, H. Simple reaction time during exercise, heat exposure, and heat acclimation. *Aviation, Space, and Environmental Medicine.* 47(11), 1168–1170, 1976.

Simon, J.R. Choice reaction time as a function of auditory S-R correspondence, age, and sex. *Ergonomics.* 10(6), 659–664, 1967.

Simon, J.R. and Overmeyer, S.P. The effect of redundant cues on retrieval time. *Human Factors.* 26(3), 315–321, 1984.

Simon, J.R., Peterson, K.D., and Wang, J.H. Same-different reaction time to stimuli presented simultaneously to separate cerebral hemispheres. *Ergonomics.* 31(12), 1837–1846, 1988.

Simon, J.R. and Wolf, J.D. Choice reaction time as a function of angular stimulus-response correspondence and age. *Ergonomics.* 6(1), 99–106, 1963.

Steiner, B.A. and Camacho, M.J. Situation awareness: Icons vs. alphanumerics. *Proceedings of the Human Factors Society 33rd Annual Meeting.* 1, 28–32, 1989.

Stewart, J.D. and Clark, B. Choice reaction time to visual motion during prolonged rotary motion in airline pilots. *Aviation, Space, and Environmental Medicine*, 46(6), 767–771, 1975.

Steyvers, F.J.J.M. The influence of sleep deprivation and knowledge of results on perceptual encoding. *Acta Psychologica*. 66, 173–178, 1987.

Swink, J.R. Intersensory comparisons of reaction time using an electro-pulse tactile stimulus. *Human Factors*. 8(2), 143–146, 1966.

Taylor, R.M. and Selcon, S.J. Cognitive quality and situational awareness with advanced aircraft attitude displays. *Proceedings of the Human Factors Society 34th Annual Meeting*. 1, 26–30, 1990.

Teichner, W.H. Probability of detection and speed of response in simple monitoring. *Human Factors*. 4(4), 181–186, 1962.

Thackray, R.I., Bailey, J.P., and Touchstone, R.M. The effect of increased monitoring load on vigilance performance using a simulated radar display. *Ergonomics*. 22(5), 529–539, 1979.

Thackray, R.I. and Touchstone, R.M. Effects of monitoring under high and low task load on detection of flashing and coloured radar targets. *Ergonomics*. 34(8), 1065–1081, 1991.

Thackray, R.I., Touchstone, R.M., and Bailey, J.P. Comparison of the vigilance performance of men and women using a simulated radar task. *Aviation, Space, and Environmental Medicine*. 49(10), 1215–1218, 1978.

Tullis, T.S. Human performance evaluation of graphic and textual CRT displays of diagnostic data. *Proceedings of the Human Factors Society 24th Annual Meeting,* 310–316. 1980.

Tzelgov, J., Henik, A., Dinstein, I., and Rabany, J. Performance consequence of two types of stereo picture compression. *Human Factors*. 32(2), 173–182, 1990.

Van Assen, A. and Eijkman, E.G. Reaction time and performance in learning a two dimensional compensatory tracking task. *Ergonomics*. 13(9), 707–717, 1971.

Vidulich, M.A. The Bedford Scale: Does it measure spare capacity? *Proceedings of the 6th International Symposium on Aviation Psychology*. 2, 1136–1141, 1991.

Warner, H.D. and Heimstra, N.W. Effects of noise intensity on visual target-detection performance. *Human Factors*. 14(2), 181–185, 1972.

Warrick, M.J., Kibler, A.W., and Topmiller, D.A. Response time to unexpected stimuli. *Human Factors*. 7(1), 81–86, 1965.

Welford, A.T. What is the basis of choice reaction-time? *Ergonomics*. 14(6), 679–693, 1971.

Wierwille, W.W., Casali, J.G., and Repa, B.S. Driver steering reaction time to abrupt-onset crosswinds, as measured in a moving-base driving simulator. *Human Factors*. 25(1), 103–116, 1983.

Wierwille, W.W., Rahimi, M., and Casali, J.G. Evaluation of 16 measures of mental workload using a simulated flight task emphasizing mediational activity. *Human Factors*. 27(5), 489–502, 1985.

Wolcott, J. H., Hanson, C.A., Foster, W.D., and Kay, T. Correlation of choice reaction time performance with biorhythmic criticality and cycle phase. *Aviation, Space, and Environmental Medicine*, 50(1), 34–39, 1979.

Yeh, Y. and Silverstein, L.D. Spatial judgments with monoscopic and stereoscopic presentation of perspective displays. *Human Factors*. 34(5), 583–600, 1992.

Zeiner, A.R. and Brecher, G.A. Reaction time with and without backscatter from intense pulsed light. *Aviation, Space, and Environmental Medicine*, 46(2), 125–127, 1975.

2.2.7 Reading Speed

General description. Reading speed is the number of words read divided by the reading time interval. Reading speed is typically measured in words per minute.

Strengths and limitations. In an early study, Seminar (1960) measured reading speed for tactually presented letters. He reported an average speed of 5.5 s for two letters and 25.5 s for seven letters. The subjects were three men and three women.

Cushman (1986) reported that reading speeds tend to be slower for negative than for positive images. Because there may be a speed–accuracy trade-off, Cushman (1986) also calculated overall reading performance (reading speed × percentage of reading comprehension questions answered correctly). In the same year, Gould and Grischkowsky (1986) reported that reading speed decreases as visual angles increase over 24.3°. In a follow-on study, Gould, Alfaro, Barnes, Finn, Grischkowsky, and Minuto (1987) reported in a series of 10 experiments that reading speed is slower from CRT displays than from paper. In a similar study, Jorna and Snyder (1991) reported equivalent reading speeds for hard copy and soft copy displays if the image qualities are similar.

Gould, Alfaro, Finn, Haupt, and Minuto (1987) concluded on the basis of six studies that reading speed was equivalent on paper and CRT if the CRT displays contained "character fonts that resemble those on paper (rather than dot matrix fonts, for example), that have a polarity of dark characters on a light background, that are anti-aliased (i.e., contain grey level), and that are shown on displays with relatively high resolution (e.g., 1000 × 800)." (p. 497).

A year later, Chen, Chan, and Tsoi (1988) used the average reading rate, in words per minute (wpm), to evaluate the effects of window size (20 versus 40 character) and jump length (i.e., number of characters that a message is advanced horizontally) of a visual display. These authors reported that reading rate was significantly less for one-jump (90–91 wpm) than for five- (128 wpm) and nine-jump (139–144) conditions. Reading rate was not significantly affected by window size, however.

Campbell, Marchetti, and Mewhort (1981) reported increased reading speed for justified rather than unjustified text. Moseley and Griffin (1986) reported increased reading time and reading error in vibration of the display, the subject, or both.

Lachman (1989) used the inverse of reading time, i.e., reading rate to evaluate the effect of presenting definitions concurrently with text on a CRT display. There was a significantly higher reading rate for the first 14 screens read than for the second 14 screens.

Data requirements. The number of words and the duration of the reading interval must be recorded.

Thresholds. Cushman (1986) reported the following average words/minute: paper = 218; positive image microfiche = 210; negative image microfiche = 199; positive image, negative contrast VDT (Video Display Terminal) = 216; and negative image, positive contrast VDT = 209.

SOURCES

Campbell, A.J., Marchetti, F.M., and Mewhort, D.J.K. Reading speed and text production: A note on right-justification techniques. *Ergonomics*. 24(8), 633–640, 1981.

Chen, H., Chan, K., and Tsoi, K. Reading self-paced moving text on a computer display. *Human Factors*. 30(3), 285–291, 1988.

Cushman, W.H. Reading from microfiche, a VDT, and the printed page: Subjective fatigue and performance. *Human Factors*. 28(1), 63–73, 1986.

Gould, J.D., Alfaro, L., Barnes, V., Finn, R., Grischkowsky, N., and Minuto, A. Reading is slower from CRT displays than from paper. Attempts to isolate a single-variable explanation. *Human Factors*. 29(3), 269–299, 1987.

Gould, J.D., Alfaro, L., Finn, R., Haupt, B., and Minuto, A. Reading from CRT displays can be as fast as reading from paper. *Human Factors*. 29(5), 497–517, 1987.

Gould, J.D. and Grischkowsky, N. Does visual angle of a line of characters affect reading speed. *Human Factors*. 28(2), 165–173, 1986.

Jorna, G.C. and Snyder, H.L. Image quality determines differences in reading performance and perceived image quality with CRT and hard-copy displays. *Human Factors*. 33(4), 459–469, 1991.

Lachman, R. Comprehension aids for on-line reading of expository text. *Human Factors*. 31(1), 1–15, 1989.

Moseley, M.J. and Griffin, M.J. Effects of display vibration and whole-body vibration on visual performance. *Ergonomics*. 29(8), 977–983, 1986.

Seminar, J.L. Accuracy and speed of tactual reading: An exploratory study. *Ergonomics*. 3(1), 62–67, 1960.

2.2.8 Search Time

General description. Search time is the length of time for a user to retrieve the desired information from a database. Lee and MacGregor (1985) provided the following definition:

$$st = r\,(at + k + c),$$

where

st = search time
r = total number of index pages accessed in retrieving a given item
a = number of alternatives per page
t = time required to read one alternative
k = key-press time
c = computer response time (pp. 158, 159).

Matthews (1986) defined search time as the length of time for a subject to locate and indicate the position of a target.

Strengths and limitations. Search time has been used to evaluate displays, clutter, and time on task.

Fisher, Coury, Tengs, and Duffy (1989) used search time to evaluate the effect of highlighting on visual displays. Search time was significantly longer when the probability that the target was highlighted was low (0.25) rather than high (0.75). This is similar to an earlier study by Monk (1976) in which target uncertainty increased search time by 9.5%.

In another display study, Harpster, Freivalds, Shulman, and Leibowitz (1989) reported significantly longer search times using a low resolution/addressability ratio (RAR) than using a high RAR or hard copy. Vartabedian (1971) reported a significant increase in search time for lowercase than for uppercase words. Matthews, Lovasik, and Mertins (1989) reported significantly longer search times for green on black displays (7.71 s) than red on black displays (7.14 s). Hollands, Parker, McFadden, and

Boothby (2002) reported significantly longer search times for diamond shapes rather than square shapes on a CRT. There was no difference, however, on an LCD (Liquid Crystal Display). There were, however, longer search times on the LCD for red and blue symbols than for white symbols. There were no color effects on the CRT.

Erickson (1964) reported few significant correlations between search time and peripheral visual acuity. There was a significant interaction between target shape and number of objects in a display: rings were found in significantly shorter times than blobs when there were fewer objects in a display (16 versus 48).

For clutter effects, Brown and Monk (1975) reported longer visual search times as the number of nontargets increased and as the randomness of the location of the nontargets increased. Matthews (1986) reported that search time in the current trial was significantly increased if the visual load of the previous trial was high. Nagy and Sanchez (1992) used mean log search time to investigate the effects of luminance and chromaticity differences between targets and distractors. "Results showed that mean search time increased linearly with the number of distractors if the luminance difference between target and distractors was small but was roughly constant if the luminance difference was large" (p. 601).

Bednall (1992) reported shorter search times for spaced rather than nonspaced targets, targets with alternating lines rather than nonalternating lines, and insertions of blank lines. There was no effect of all-capital versus mixed-case letters.

Lee and MacGregor (1985) reported that their search time model was evaluated using a videotex information retrieval system. It was useful in evaluating menu design decisions.

Carter, Krause, and Harbeson (1986) reported that RT was more reliable than slope for this task.

For time on task, Lovasik, Matthews, and Kergoat (1989) reported significantly longer search times in the first half hour than in the remaining three and a half hours of a visual search task.

Data requirement. User search time can be applied to any computerized database in which the parameters a, c, k, r, and t can be measured.

Thresholds. Not stated.

SOURCES

Bednall, E.S. The effect of screen format on visual list search. *Ergonomics.* 35(4), 369–383, 1992.

Brown, B. and Monk, T.H. The effect of local target surround and whole background constraint on visual search times. *Human Factors.* 17(1), 81–88, 1975.

Carter, R.C., Krause, M., and Harbeson, M.M. Beware the reliability of slope scores for individuals. *Human Factors.* 28(6), 673–683, 1986.

Erickson, R.A. Relation between visual search time and peripheral visual acuity. *Human Factors.* 6(2), 165–178, 1964.

Fisher, D.L., Coury, B.G., Tengs, T.O., and Duffy, S.A. Minimizing the time to search visual displays: The role of highlighting. *Human Factors.* 31(2), 167–182, 1989.

Harpster, J.K. Freivalds, A., Shulman, G.L., and Leibowitz, H.W. Visual performance on CRT screens and hard-copy displays. *Human Factors.* 31(3), 247–257, 1989.

Hollands, J.G., Parker, H.A., McFadden, S., and Boothby, R. LCD versus CRT displays: A comparison of visual search performance for colored symbols. *Human Factors.* 44(2), 210–221, 2002.

Lee, E. and MacGregor, J. Minimizing user search time in menu retrieval systems. *Human Factors*. 27(2), 157–162, 1985.
Lovasik, J.V., Matthews, M.L., and Kergoat, H. Neural, optical, and search performance in prolonged viewing of chromatic displays. *Human Factors*. 31(3), 273–289, 1989.
Matthews, M.L. The influence of visual workload history on visual performance. *Human Factors*. 28(6), 623–632, 1986.
Matthews, M.L., Lovasik, J.V., and Mertins, K. Visual performance and subjective discomfort in prolonged viewing of chromatic displays. *Human Factors*. 31(3), 259–271, 1989.
Monk, T.H. Target uncertainty in applied visual search. *Human Factors*. 18(6), 607–612, 1976.
Nagy, A.L. and Sanchez, R.R. Chromaticity and luminance as coding dimensions in visual search. *Human Factors*. 34(5), 601–614, 1992.
Vartabedian, A.G. The effects of letter size, case, and generation method on CRT display search time. *Human Factors*. 13(4), 363–368, 1971.

2.2.9 Task Load

General description. Task load is the time required to perform a task divided by the time available to perform the task. Values above 1 indicate excessive task load.

Strengths and limitations. Task load is sensitive to workload in in-flight environments. For example, Geiselhart, Schiffler, and Ivey (1976) used task load to identify differences in workload between four types of refueling missions. Geiselhart, Koeteeuw, and Schiffler (1977) used task load to estimate the workload of KC-135 crews. Using this method, these researchers were able to quantify differences in task load between different types of missions and crew positions. Gunning and Manning (1980) calculated the percentage of time spent on each task for three crew members during an aerial refueling. They reported the following inactivity percentages by crew position: pilot, 5%; copilot, 45%; navigator, 65%. Task load was high during takeoff, air refueling, and landing.

Stone, Gulick, and Gabriel (1984), however, identified three problems with task load: "(1) It does not consider cognitive or mental activities. (2) It does not take into account variations associated with ability and experience or dynamic, adaptive behavior. (3) It cannot deal with simultaneous or continuous-tracking tasks" (p. 14).

Data requirements. Use of the task load method requires the following: (1) clear visual and auditory records of pilots in flight and (2) objective measurement criteria for identifying the starts and ends of tasks.

Thresholds. Not stated.

SOURCES

Geiselhart, R., Koeteeuw, R.I., and Schiffler, R.J. *A study of task loading using a four-man crew on a KC-135 aircraft (Giant Boom) (ASD-TR-76-33)*. Wright-Patterson Air Force Base, OH: Aeronautical Systems Division, April 1977.
Geiselhart, R., Schiffler, R.J., and Ivey, L.J. *A study of task loading using a three man crew on a KC-135 aircraft (ASD-TR-76-19)*. Wright-Patterson Air Force Base, OH: Aeronautical Systems Division, October 1976.
Gunning, D. and Manning, M. The measurement of aircrew task loading during operational flights. *Proceedings of the Human Factors Society 24th Annual Meeting* (pp. 249–252). Santa Monica, CA: Human Factors Society, 1980.

Stone, G., Gulick, R.K., and Gabriel, R.F. *Use of task/timeline analysis to assess crew workload (Douglas Paper 7592).* Longbeach, CA: Douglas Aircraft Company, 1984.

2.2.10 Time to Complete

General description. Time to complete is the duration from the subject's first input to the last response (Casali, Williges, and Dryden, 1990).

Strengths and limitations. Time to complete provides a measure of task difficulty but may be traded off for accuracy. Adelman, Cohen, Bresnick, Chinnis, and Laskey (1993) used time to complete an aircraft identification task to evaluate expert system interfaces and capabilities. They reported that operators took longer to examine aircraft with the screening rather than with the override interface.

For unmanned systems, Massimino and Sheridan (1994) reported no difference in task competition times between direct and video viewing during teleoperation.

In an editing task, Brand and Judd (1993) reported significant differences in editing time as a function of the angle of hard copy (316.75 s for 30°; 325.03 s for 0°; and 371.92 s for 90°).

Casali, Williges, and Dryden (1990) reported significant effects of speech recognition system accuracy and available vocabulary but not a significant age effect. There were also several interactions.

Frankish and Noyes (1990) used rate of data entry to evaluate data feedback techniques. They reported significantly higher rates for visual presentation and for visual feedback than for spoken feedback.

In early work, Burger, Knowles, and Wulfeck (1970) used experts to estimate the time it would take to perform tasks. The estimates were then tested against the real times. Although the correlation between the estimates and the real times was high (+0.98), estimates of the minimum performance times were higher than the actual times and varied widely between judges. The tasks were throw to toggle switch, turn rotary switch to a specified value, push toggle, observe and record data, and adjust dial.

In an unusual study, Troutwine and O'Neal (1981) reported that time was judged shorter on an interesting rather than a boring task, but only for subjects with volition. For subjects without volition there was no significant difference in time estimation between the two types of tasks.

Three studies were completed looking at individual differences. In the first, Hartley, Lyons, and Dunne (1987) used completion times to assess the effect of menstrual cycle. Times were slower on verbal reasoning involving complex sentences during ovulation than during menstruation and premenstruation. In the second study, Richter and Salvendy (1995) reported that introverted users of a computer program performed faster with interfaces that were perceived as introverted than with interfaces that were perceived as extroverted. Finally, Best, Littleton, Gramopadhye, and Tyrrell (1996) reported that persons with very near dark convergence positions before a visual inspection task performed significantly faster than persons with far dark convergence positions. Their subjects were 38 university students with a mean age of 20.6 years.

Data requirements. The start and end of the task must be well-defined.

Thresholds. Minimum time is 30 ms.

SOURCES

Adelman, L., Cohen, M.S., Bresnick, T.A., Chinnis, J.O., and Laskey, K.B. Real-time expert system interfaces, cognitive processes, and task performance: An empirical assessment. *Human Factors* 35(2), 243–261, 1993.

Best, P.S., Littleton, M.H., Gramopadhye, A.K., and Tyrrell, R.A. Relations between individual differences in oculomotor resting states and visual inspection performance. *Human Factors*. 39(1), 35–40, 1996.

Brand, J.L and Judd, K. Angle of hard copy and text-editing performance. *Human Factors*. 35(1), 57–70, 1993.

Burger, W.J., Knowles, W.B., and Wulfeck, J.W. Validity of expert judgments of performance time. *Human Factors*, 12(5), 503–510, 1970.

Casali, S.P., Williges, B.H., and Dryden, R.D. Effects of recognition accuracy and vocabulary size of a speech recognition system on task performance and user acceptance. *Human Factors*. 32(2), 183–196, 1990.

Frankish, C. and Noyes, J. Sources of human error in data entry tasks using speech input. *Human Factors*. 32(6), 697–716, 1990.

Hartley, L.R., Lyons, D., and Dunne, M. Memory and menstrual cycle. *Ergonomics*. 30(1), 111–120, 1987.

Massimino, J.J. and Sheridan, T.B. Teleoperator performance with varying force and visual feedback. *Human Factors*. 36(1), 145–157, 1994.

Richter, L.A. and Salvendy, G. Effects of personality and task strength on performance in computerized tasks. *Ergonomics*. 38(2), 281–291, 1995.

Troutwine, R. and O'Neal, E.C. Volition, performance of a boring task and time estimation. *Perceptual and Motor Skills*. 52, 865–866, 1981.

2.3 TASK BATTERIES

The third category is task battery. Task batteries are collections of two or more tasks performed in series or in parallel to measure range of abilities or effects. These batteries assume that human abilities vary across types of tasks or are differentially affected by independent variables. Examples include AGARD's Standardized Tests for Research with Environmental Stressors (STRES) Battery (section 2.3.1), the Armed Forces Qualification Test (section 2.3.2), Deutsch and Malmborg Measurement Instrument Matrix (section 2.3.3), Performance Evaluation Tests for Environmental Research (PETER) (section 2.3.4), the Work and Fatigue Test Battery (section 2.3.5), and the Unified Tri-Services Cognitive Performance Assessment Battery (UTCPAB) (section 2.3.6).

2.3.1 AGARD's Standardized Tests for Research with Environmental Stressors Battery

General description. The STRES Battery is made up of seven tests:

1. Reaction time
2. Mathematical processing
3. Memory search

4. Spatial processing
5. Unstable tracking
6. Grammatical reasoning
7. Dual task performance of Tests 3 and 5

Strengths and limitations. Tests were selected for the STRES Battery based on the following criteria: "(1) preliminary evidence of reliability, validity, and sensitivity, (2) documented history of application to assessment of a range of stressor effects, (3) short duration (maximum of three minutes per trial block), (4) language-independence, (5) sound basis in [Human Performance Theory] HPT, [and] (6) ability to be implemented on simple and easily-available computer systems" (AGARD, 1989, p. 7).

Data requirements. Each STRES Battery test has been programmed for computer administration. The order of presentation is fixed, as presented earlier. Standardized instructions are used as well as a standardized data file format. Test stimuli must be presented in white on a dark background. The joystick must have 30° lateral travel from the vertical position, friction not greater than 50 g, linear relationship between angular rotation and lateral movement, and 8-bit resolution.

Thresholds. Not stated.

SOURCE

AGARD. *Human performance assessment methods (AGARD-AG-308).* Neuilly-sur-Seine, France, AGARD, June 1989.

2.3.2 Armed Forces Qualification Test

General description. The Armed Forces Qualification Test measures mechanical and mathematical aptitude. Scores from the test are used to place army recruits into a military occupational specialty.

Strengths and limitations. To minimize attrition due to poor placement, supplementary aptitude tests are required. Further, the results are affected by the "substitutability with compensation" principle. This principle states, "a relatively high ability in one area makes up for a low level in another so that observed performance equals that predicted from a linear combination of two predictor measures" (Uhlaner, 1972, p. 206).

Data requirements. Cognitive–noncognitive variance should be considered when evaluating the test scores.

Thresholds. Not stated.

SOURCE

Uhlaner, J.E. Human performance effectiveness and the systems measurement bed. *Journal of Applied Psychology.* 56(3), 202–210, 1972.

2.3.3 Deutsch and Malmborg Measurement Instrument Matrix

General description. The measurement instrument matrix consists of activities that must be performed by an organization along the vertical axis and the metrics used to evaluate the performance of those activities along the horizontal axis. A value of one is placed in every cell in which the metric is an appropriate measure of the activity; otherwise, a zero is inserted (Deutsch and Malmborg, 1982).

Strengths and limitation. This method handles the complexity and interaction of activities performed by organizations. It has been used to assess the impact of information overload on decision-making effectiveness. The measurement instrument matrix should be analyzed in conjunction with an objective matrix. The objective matrix lists the set of activities along the horizontal axis and the set of objectives along the vertical axis.

Data requirement. Reliable and valid metrics are required for each activity performed.

Threshold. Zero is the lower limit; one, the upper limit.

SOURCE

Deutsch, S.J. and Malmborg, C.J. The design of organizational performance measures for human decision making, Part I: Description of the design methodology. *IEEE Transactions on Systems, Man, and Cybernetics.* SMC-12 (3), 344–352, 1982.

2.3.4 Performance Evaluation Tests for Environmental Research

General description. The PETER test battery is made up of 26 tests: (1) aiming, (2) arithmetic, (3) associative memory, (4) Atari air combat maneuvering, (5) Atari anti-aircraft, (6) choice RT: 1-choice, (7) choice RT: 4-choice, (8) code substitution, (9) flexibility of closure, (10) grammatical reasoning, (11) graphic and phonemic analysis, (12) letter classification: name, (13) letter classification: category, (14) manikin, (15) Minnesota rate of manipulation, (16) pattern comparison, (17) perceptual speed, (18) search for typos in prose, (19) spoke control, (20) Sternberg item recognition: positive set 1, (21) Sternberg item recognition: positive set 4, (22) Stroop, (23) tracking: critical, (24) tracking: dual critical, (25) visual contrast sensitivity, and (26) word fluency (Kennedy, 1985). The process for test selection is described by Carter, Kennedy, and Bittner (1980). A tabular summary of the individual tests is presented by Kennedy, Carter, and Bittner (1980).

Strengths and limitations. Tests were selected for the PETER battery on the following criteria: (1) administration time, (2) total stabilization time, and (3) reliability. Kennedy and Bittner (1978) measured performance of 19 Navy enlisted men on 10 tasks over a 15 d period. Eight tasks had significant learning effects over days (grammatical reasoning, code substitution, Stroop, arithmetic, Neisser letter search, critical tracking task, subcritical two-dimensional compensatory tracking, and Spoke test); two did not (time estimation and complex counting). Two tasks (time estimation and Stroop) had low reliabilities. In another study using time estimation and tracking, Bohnen and Gaillard (1994) reported that time estimation was not affected by sleep loss, whereas tracking was. Seales, Kennedy, and Bittner (1979) reported that performance on a paper and pencil arithmetic test was stable after 9 d and had constant variance throughout 15 d of testing. Their subjects were 18 navy enlisted men. McCauley, Kennedy, and Bittner (1979) examined time estimation performance. Their subjects were 19 navy enlisted men. Forty trials per day were performed over 15 d. There were no significant differences in performance over time. McCafferty, Bittner, and Carter (1980) reported performance on an auditory forward digit span task was stable after 4 d with no significant differences in variance over 12 d. Their subjects were nine navy enlisted men. Guignard, Bittner, Einbender, and Kennedy (1980) reported unstable speed and error measures for the Landolt C reading test over 12 d. Their subjects were eight navy enlisted men. Carter, Kennedy, Bittner, and Krause (1980) reported that response time for the Sternberg task was stable after the fourth day in a 15 d experiment. The slope, however, was unreliable over time. The subjects were 21 navy enlisted men. Harbeson, Krause, and Kennedy (1980) reported acceptable reliability in two (interference susceptibility and free recall) tests and unacceptable reliability in two other memory tests (running recognition and list differentiation). Their subjects were 23 navy enlisted men who performed the tasks over 15 consecutive days.

Performance on individual tests (e.g., navigation plotting: Wiker, Kennedy, and Pepper, 1983; vertical addition, grammatical reasoning, perceptual speed, flexibility of closure: Bittner, Carter, Krause, Kennedy, and Harbeson, 1983) was compared in the laboratory and at sea. Bittner, Carter, Kennedy, Harbeson, and Krause (1984) summarized the results of all 112 tests studied for possible inclusion in PETER by placing each task into one of the following categories: good, good but redundant, ugly (flawed), and bad.

Data requirements. Each PETER test has been programmed for a NEC PC 8201A.

Thresholds. Not stated.

SOURCES

Bittner, A.C., Carter, R.C., Kennedy, R.S., Harbeson, M.M., and Krause, M. Performance Evaluation Tests for Environmental Research (PETER): The good, bad, and ugly. *Proceedings of the Human Factors Society 28th Annual Meeting.* 1, 11–15, 1984.

Bittner, A.C., Carter, R.C., Krause, M., Kennedy, R.S., and Harbeson, M.M. Performance Evaluation Tests for Environmental Research (PETER): Moran and computer batteries. *Aviation, Space, and Environmental Medicine.* 54(10), 923–928, 1983.

Bohnen, H.G.M. and Gaillard, A.W.K. The effects of sleep loss in a combined tracking and time estimation task. *Ergonomics.* 37(6), 1021–1030, 1994.

Carter, R.C., Kennedy, R.S., and Bittner, A.C. *Selection of performance evaluation test for environmental research.* 320–324, 1980.

Carter, R.C., Kennedy, R.S., Bittner, A.C., and Krause, M. *Item recognition as a Performance Evaluation Test for Environmental Research.* 340–343, 1980.

Guignard, J.C., Bittner, A.C., Einbender, S.W., and Kennedy, R.S. Performance Evaluation tests for Environmental Research (PETER): Landolt C reading test. *Proceedings of the Human Factors Society 24th Annual Meeting.* 335–339, 1980.

Harbeson, M.M., Krause, M., and Kennedy, R.S. Comparison of memory tests for environmental research. *Proceedings of the Human Factors Society 24th Annual Meeting.* 349–353, 1980.

Kennedy, R.S. *A portable battery for objective, non-obtrusive measures of human performance (NASA-CR-171868).* Pasadena, CA: Jet Propulsion Laboratory, 1985.

Kennedy, R.S. and Bittner, A.C. Progress in the analysis of a Performance Evaluation test for Environmental research (PETER). *Proceedings of the Human Factors Society 22nd Annual Meeting.* 29–35, 1978.
Kennedy, R.S., Carter, R.C., and Bittner, A.C. A catalogue of Performance Evaluation Tests for Environmental Research. *Proceedings of the Human Factors Society 24th Annual Meeting.* 344–348, 1980.
McCafferty, D.B., Bittner, A.C., and Carter, R.C. Performance Evaluation Test for Environmental Research (PETER): Auditory digit span. *Proceedings of the Human Factors Society 24th Annual Meeting.* 330–334, 1980.
McCauley, M.E., Kennedy, R.S., and Bittner, R.S. Development of Performance Evaluation Tests for Environmental Research (PETER): Time estimation test. *Proceedings of the Human Factors Society 23rd Annual Meeting.* 513–517, 1979.
Seales, D.M., Kennedy, R.S., and Bittner, A.C. Development of Performance Evaluation Test for Environmental Research (PETER): Arithmetic computation. *Proceedings of the Human Factors Society 23rd Annual Meeting.* 508–512, 1979.
Wiker, S.F., Kennedy, R.S., and Pepper, R.L. Development of Performance Evaluation Tests for Environmental Research (PETER): Navigation plotting. *Aviation, Space, and Environmental Medicine.* 1983, 54(2), 144–149.

2.3.5 Simulated Work and Fatigue Test Battery

General description. The Simulated Work and Fatigue Test Battery was developed by the National Institute for Occupational Safety and Health to assess the effects of fatigue in the workplace. The simulated work is a data entry task. The fatigue test battery has 11 tasks: grammatical reasoning, digit addition, time estimation, simple auditory RT, choice RT, two-point auditory discrimination, response alternation performance (tapping), hand steadiness, the Stanford sleepiness scale, the Neuropsychiatric Research Unit (NPRU) Mood Scale adjective checklist, and oral temperature. Two tasks, grammatical reasoning and simple RT, are also performed in dual-task mode.

Strengths and limitations. The Simulated Work and Fatigue Test Battery is portable, brief, easy to administer, and requires little training of the subject. Rosa and Colligan (1988) used the battery to evaluate the effect of fatigue on performance. All tasks showed significant fatigue effects except data entry, time production, and two-point auditory discrimination.

Data requirements. The microcomputer provides all task stimuli, records all data, and scores all tasks.

Thresholds. Not stated.

SOURCE

Rosa, R.R. and Colligan, M.J. Long workdays versus rest days: Assessing fatigue and alertness with a portable performance battery. *Human Factors.* 30(3), 305–317, 1988.

2.3.6 Unified Tri-Services Cognitive Performance Assessment Battery

General description. The UTCPAB is made up of 25 tests: (1) linguistic processing, (2) grammatical reasoning (traditional), (3) grammatical reasoning (symbolic), (4) two-column addition, (5) mathematical processing, (6) continuous recognition,

(7) four-choice serial RT, (8) alphanumeric visual vigilance, (9) memory search, (10) spatial processing, (11) matrix rotation, (12) manikin, (13) pattern comparison (simultaneous), (14) pattern comparison (successive), (15) visual scanning, (16) code substitution, (17) visual probability monitoring, (18) time wall, (19) interval production, (20) Stroop, (21) dichotic listening, (22) unstable tracking, (23) Sternberg-tracking combination, (24) matching to sample, and (25) item order.

Strengths and limitations. Tests were selected for the UTCPAB based on the following criteria: (1) used in at least one Department of Defense laboratory, (2) proven validity, (3) relevance to military performance, and (4) sensitivity to hostile environments and sustained operations (Perez, Masline, Ramsey, and Urban, 1987).

A nine-test version, the Walter Reed Performance Assessment Battery was used to assess the effects of altitude. Using this shortened version, Crowley, Wesensten, Kamimori, Devine, Iwanyk, and Balkin (1992) reported decrements on three tasks due to altitude effects: code substitution, Stroop, and logical reasoning.

Using another shortened version, the Criterion Task Set (which includes probability monitoring, unstable tracking, continuous recall, grammatical reasoning, linguistic processing, mathematical processing, memory search, spatial processing, and interval production), Chelen, Ahmed, Kabrisky, and Rogers (1993) reported no significant effects on performance of any of these tasks of phenytoin serum (motion sickness therapy) levels.

Rogers, Spencer, Stone, and Nicholson (1989) used the same approach to select tests to evaluate the effects of a 1 h nap and caffeine on performance. The tests were sustained attention, auditory vigilance and tracking, complex vigilance, two-letter cancellation, digit symbol substitution, logic, short-term memory, and visual vigilance. There were significant effects on sustained attention, auditory vigilance and tracking, visual vigilance, and complex vigilance, but not on short-term memory.

Using still another task battery, Simple Portable Aviation Relevant Test battery and Answer-scoring System (SPARTANS), Stokes, Belger, Banich, and Bernadine (1994) reported no significant effects of aspartame. The tasks included in SPARTANS were: maze tracing, hidden figures recognition, hidden figures rotation, visual number, scheduling, Sternberg, first-order pursuit tracking, dual task of Sternberg and tracking, minefield, and Stroop.

In another task set, Paul and Fraser (1994) reported that mild acute hypoxia had no effect on the ability to learn new tasks. In this case, the tasks were Manikin, choice RT, and logical reasoning. In fact, performance on the first two tasks improved over time. In yet another task set (simple RT, four-choice RT, tracking, visual search, and visual analog series) Cherry, Johnston, Venables, and Waldron (1983) reported decreases in performance of the tracking and visual search task as alcohol level increased. Toluene, a rubber solvent, had no effect on any of the tasks.

In still another grouping of tasks, Beh and McLaughlin (1991) investigated the effects of desynchronization on airline crews. The tasks used were grammatical reasoning, horizontal addition, vertical addition, letter cancellation, and card sorting. There were no significant effects on the number of errors. The control group, however, completed more items on the grammatical reasoning and vertical addition tasks than the desynchronized flight crews. There was a significant group by test period interaction for the horizontal addition task and for the card-sorting task.

Rosa and Bonnet (1993) applied a slightly different set to evaluate the effect of 8 and 12 h rotating shifts on performance. Tests included mental arithmetic, dual task of grammatical reasoning and auditory RT, simple auditory RT, and hand steadiness. For mental arithmetic, the most correct answers occurred on the 8 h evening shift and fewest errors at 24 h. For grammatical reasoning RT the first day on 8 h shift was significantly longer than the fourth day, which was a 12 h shift. There was also 9% more errors on 12 h shifts. Dual RT was fastest at 24:00 hours. For simple RT percent misses, there were significantly more misses on the fourth day (i.e., the first day of 12 h shift) than any of the 8 h shifts (day, evening, or night). Hand steadiness was 0.5% greater per day over the 5 d test.

In another test battery, the Automated Performance Test System (APTS), Kennedy, Turnage, Wilkes, and Dunlap (1993) reported that eight of the nine tests showed significant effects of alcohol (up to 0.15% blood alcohol concentration). The tests were preferred hand tapping, nonpreferred hand tapping, grammatical reasoning, mathematical processing, code substitution, pattern comparison, manikin, short-term memory, and four-choice RT. Grammatical reasoning did not show a significant alcohol effect. Kennedy, Dunlap, Ritters, and Chavez (1996) compared APTS with its successor Delta. There were essentially no differences in task performance associated with hardware or software differences between the two systems.

Salame (1993) identified a major defect in the grammatical reasoning test that was part of the Standardized Tests for Research and Environmental Stressors (STRES). The defect was in the pattern of answers: if there was one match, the answer was always "the same" and if there were no or two matches, the answer was always "different."

Bonnet and Arand (1994) used addition, vigilance, and logical reasoning tasks to assess the effects of naps and caffeine. The researchers concluded that naps and caffeine resulted in near-baseline performance over a 24 h period without sleep. Their subjects were 18- to 30-year-old males.

The utility of test batteries is demonstrated by research into the effects of shifts in work–rest cycles reported by Porcu, Bellatreccia, Ferrara, and Casagrande (1998). They reported no effects on a digit symbol substitution task or on the Deux Barrages task (a paper-and-pencil task requiring marking two target symbols embedded among similar symbols), but a significant decrement on a letter cancellation task.

Data requirements. Each UTCPAB test has been programmed for computer administration. Standardized instructions are used as well as a standardized data file format.

Thresholds. Not stated.

SOURCES

Beh, H.C. and McLaughlin, P.J. Mental performance of air crew following layovers on transzonal flights. *Ergonomics*. 34(2), 123–135, 1991.

Bonnet, M.H. and Arand, D.L. The use of prophylactic naps and caffeine to maintain performance during a continuous operation. *Ergonomics*. 37(6), 1009–1020, 1994.

Chelen, W., Ahmed, N., Kabrisky, M., and Rogers, S. Computerized task battery assessment of cognitive and performance effects of acute phenytoin motion sickness therapy. *Aviation, Space, and Environmental Medicine*. 64(3), 201–205, 1993.

Cherry, N., Johnston, J.D., Venables, H., and Waldron, H.A. The effects of tooulene and alcohol on psychomotor performance. *Ergonomics*. 26(11), 1081–1087, 1983.
Crowley, J.S., Wesensten, N., Kamimori, G., Devine, J., Iwanyk, E., and Balkin, T. *Aviation, Space, and Environmental Medicine*. 63(8), 696–701, 1992.
Kennedy, R.S., Dunlap, W.P., Ritters, A.D., and Chavez, L.M. Comparison of a performance test battery implements on different hardware and software: APTS versus DELTA. *Ergonomics*. 39(8), 1005–1016, 1996.
Kennedy, R.S., Turnage, J.J., Wilkes, R.L., and Dunlap, W.P. Effects of graded dosages of alcohol on nine computerized repeated-measures tests. *Ergonomics*. 36(10), 1195–1222, 1993.
Paul, M. A. and Fraser, W.D. Performance during mild acute hypoxia. *Aviation, Space, and Environmental Medicine*. 65(10), 891–899, 1994.
Perez, W.A., Masline, P.J., Ramsey, E.G., and Urban, K.E. *Unified tri-services cognitive performance assessment battery: Review and methodology (AAMRL-TR-87-007)*. Wright-Patterson Air Force Base, OH: Armstrong Aerospace Medical Research Laboratory, March 1987.
Porcu, S., Bellatreccia, A., Ferrara, M., and Casagrande, M. Sleepiness, alertness and performance during a laboratory simulation of an acute shift of the wake-sleep cycle. *Ergonomics*. 41(8), 1192–1262, 1998.
Rogers, A.S., Spencer, M.B., Stone, B.M., and Nicholson, A.N. The influence of a 1 h nap on performance overnight. *Ergonomics*. 32(10), 1193–1205, 1989.
Rosa, R.R. and Bonnet, M.H. Performance and alertness on 8 h and 12 h rotating shifts at a natural gas utility. *Ergonomics*. 36(10), 1177–1193, 1993.
Salame, P. The AGARD grammatical reasoning task: A defect and proposed solutions. *Ergonomics*. 36(12), 1457–1464, 1993.
Stokes, A.F., Belger, A., Banich, M.T., and Bernadine, E. Effects of alcohol and chronic aspartame ingestion upon performance in aviation relevant cognitive tasks. *Aviation, Space, and Environmental Medicine*. 65, 7–15, 1994.

2.4 DOMAIN-SPECIFIC MEASURES

The fourth category of human performance measures is domain-specific measures, which assess abilities to perform a family of related tasks. These measures assume that abilities and effects vary across segments of a mission or with the use of different controllers. Examples in this category are aircraft parameters (section 2.4.1), Boyett and Conn's White-Collar Performance Measures (section 2.4.2), Charlton's Measures of Human Performance in Space Control Systems (section 2.4.3), driving parameters (section 2.4.4), Eastman Kodak Company Measures for Handling Task (section 2.4.5), and Haworth–Newman Avionics Display Readability Scale (section 2.4.6).

2.4.1 Aircraft Parameters

General description. Aircrew performance is often estimated from parameters describing aircraft state. These parameters include airspeed, altitude, bank angle, descent rate, glide slope, localizer, pitch rate, roll rate, and yaw rate. Measures derived from these include root-mean-square values, minimums and maximums, correlations between two or more of these parameters, and deviations between actual and assigned values.

Strengths and limitations. One universal measure is number of errors. For example, Hardy and Parasuraman (1997) developed an error list based on component pilot activities (see table 1).

Not all universal aircraft parameters are sensitive to nonflying stressors. For example, Wierwille, Rahimi, and Casali (1985) reported that neither pitch nor roll

TABLE 1

Component Abilities of Commercial Airline Pilot Performance Determined by Frequency of Errors Extracted from Accident Reports, Critical Incidents, and Flight Checks

	Frequency of Errors			
Component Ability	**Accidents**	**Incidents**	**Flight Checks**	**Total**
Establishing and maintaining angle of glide, rate of descent, and gliding speed on approach to landing	47	41	11	99
Operating controls and switches	15	44	33	92
Navigating and orienting	4	39	19	62
Maintaining safe airspeed and attitude, recovering from stalls and spins	11	28	18	57
Following instrument flight procedures and observing instrument flight regulations	5	27	13	45
Carrying out cockpit procedures and routines	7	31	4	42
Establishing and maintaining alignment with runway on approach or takeoff climb	3	31	5	39
Attending, remaining alert, maintaining lookout	14	23	1	38
Utilizing and applying essential pilot information	0	19	18	37
Reading, checking, and observing instruments, dials, and gauges	1	26	7	34
Preparing and planning of flight	2	27	3	32
Judging type of landing or recovering from missed or poor landing	1	23	8	32
Breaking angle of glide on landing	1	25	5	31
Obtaining and utilizing instructions and information from control personnel	3	21	0	24
Reacting in an organized manner to unusual or emergency situations	0	17	7	24
Operating plane safely on ground	7	15	1	23
Flying with precision and accuracy	0	7	15	22
Operating and attending to radio	0	7	10	17
Handling of controls smoothly and with coordination	0	6	8	14
Preventing plane from experiencing undue stress	0	5	7	12
Taking safety precautions	2	5	4	11

Note: Based on "The Airline Pilot's Job," (Gordon, 1949).

Source: From Hardy, D.J. and Parasuraman, R. Cognition and flight performance in older pilots. *Journal of Experimental Psychology.* 3(4), 313–348, 1997. With permission.

high-pass mean-square scores from a primary simulated flight task were sensitive to changes in the difficulty of a secondary mathematical problem-solving task.

Researchers have also examined aircraft parameters sensitive to phase of flight (takeoff, climb, cruise, approach, and landing), task (air-to-air combat), and flight mode (hover).

2.4.1.1 Takeoff and Climb Measures

Cohen (1977), using a flight simulator, reported increased airspeed within the first 60 s after being launched from the deck of an aircraft carrier. Airspeed was also significantly different between three types of flight displays, as were variance of airspeed, vertical speed, altitude, angle of attack, pitch attitude, and frequency of pitch adjustments.

2.4.1.2 Cruise Measures

In an in-flight helicopter study, Berger (1977) reported significant differences in airspeed, altitude, bank angle, descent rate, glide slope, localizer, pitch rate, roll rate, and yaw rate and some combination of Visual Meteorological Conditions (VMC), Instrument Meteorological Conditions (IMC) with fixed sensor, IMC with stabilization sensor, and IMC with a sensor looking ahead through turns.

North, Stackhouse, and Graffunder (1979) reported that: (1) pitch error, heading error, roll acceleration, pitch acceleration, speed error, and yaw position were sensitive to differences in display configurations, (2) pitch error, roll error, heading error, roll acceleration, pitch acceleration, yaw acceleration, speed error, pitch position, roll position, yaw position, power setting, altitude error, and cross-track error were sensitive to differences in winds, and (3) heading error, roll acceleration, pitch acceleration, yaw acceleration, speed error, roll position, yaw position, altitude error, and cross-track error were sensitive to motion cues.

Bortolussi and Vidulich (1991) calculated a figure of merit (FOM) for simulated flight from rudder standard deviation (SD), elevator SD, aileron SD, altitude SD, mean altitude, airspeed SD, mean airspeed, heading SD, and mean heading. Only the airspeed and altitude FOMs showed significant differences between scenarios. The primary measures for these variables were also significant as well as aileron SD and elevator SD.

Janowsky, Meacham, Blaine, Schoor, and Bozzetti (1976) reported significant decrements in the following performance measures after pilots smoked 0.9 mg/kg (milligram per kilogram) of marijuana: number of major and minor errors, altitude deviations, heading deviations, and radio navigation errors. The data were collected from ten male pilots who had smoked marijuana socially. The aircraft simulator was a general aviation model used for instrument flight training.

2.4.1.3 Approach and Landing Measures

Morello (1977), using a B-737 aircraft, reported differences in localizer, lateral, and glide slope deviations during three-nautical-mile and close-in approaches between a baseline and an integrated display format. However, neither localizer rmse nor glide slope rmse was sensitive to variations in pitch-stability level, wind-gust disturbance, or crosswind direction and velocity (Wierwille and Connor, 1983).

Brictson (1969) reported differences between night and day carrier landings for the following aircraft parameters: altitude error from glide slope, bolster rate, wire arrestments, percentage of unsuccessful approaches, and probability of successful recovery. There were no significant day–night differences for lateral error from centerline, sink speed, and final approach airspeed.

Kraft and Elworth (1969) reported significant differences in generated altitude during final approach due to individual differences, city slope, and lighting. Also related to individual differences, Billings, Gerke, and Wick (1975) compared the performance of five experienced pilots in a Cessna 172 and a Link Singer General Aviation Trainer 1 during ILS (Instrument Landing System) approaches in IMC. Pitch, roll, and airspeed errors were twice as large in the aircraft as in the simulator. Further, there were improvements in performance in the simulator over time but not in the aircraft.

Swaroop and Ashworth (1978) reported significantly higher glide slope intercepts and flight path elevation angles with than without a diamond marking on the runway. Touchdown distance was significantly smaller without the diamond for research pilots and with the diamond for general aviation pilots. Also examining the effect of displays, Lewis and Mertens (1979) reported significant differences in rmse deviation on glide path approach angle between four different displays.

In another display study, Lintern, Kaul, and Collyer (1984) reported significant differences between conventional and modified Fresnel Lens Optical Landing System displays for glide slope rmse and descent rate at touchdown but not for localizer rmse. From the same research facility, Lintern and Koonce (1991) reported significant differences in vertical angular glide slope errors between varying scenes, display magnifications, runway sizes, and start points.

Gaidai and Mel'nikov (1985) developed an integral criterion to evaluate landing performance:

$$I(t) = \frac{1}{t_z} \int_0^{t_z} \sum_{j=1}^{K} a_i(t_i) \left[\frac{Y_i(t_i) - m_{yi}}{s_{yi}} \right]^2 dt$$

where

I = integral criteria
t = time
t_z = integration time
$Ka_j(t_i)$ = weighting coefficient for the path parameter Y_i at instant i
$Y_j(t_i)$ = instantaneous value of parameter Y_j at instant i
m_{Yj} = programmed value of the path parameter
s_{Yj} = standard deviation for the integral deviations:

$$s_{y_j} = \frac{1}{n_{j=1}} \sqrt[n]{\frac{1}{t_z} \int_0^{t_z} [Y_j(t_i) m_{yi}]^2 dt}$$

I(t) provides a multivariate assessment of pilot performance, but is difficult to calculate.

2.4.1.4 Air Combat Measures

Kelly (1988) reviewed approaches to automated aircrew performance measurement during air-to-air combat. He concluded that measures must include positional advantage or disadvantage, control manipulation, and management of kinetic and potential energy.

Barfield, Rosenberg, and Furness (1995) examined the effect of frame of reference, field of view (FOV), and eye point elevation on performance of a simulated air-to-ground targeting task. There was a significant frame-of-reference effect. Specifically, the pilot's eye display was associated with lower flight path rmse; faster target lock-on time, and faster target acquisition time than the God's eye display. There was also a significant effect of FOV: lower rmse for 30° and 90° FOV than for 60° FOV, fastest time to lockout target for 30° FOV, and fastest target acquisition time. Finally, eye point elevation also resulted in significant differences: lower rmse, faster lock-on times, and faster target acquisition times for 60° than for 30° elevation.

Kruk, Regan, and Beverley (1983) compared the performance of 12 experienced fighter pilots, 12 undergraduate training instructor pilots, and 12 student pilots on three tasks (formation, low-level, and landing) in a ground simulator. Only one parameter was sensitive to flight experience: students performed significantly worse than instructors in the distance of first correction to the runway during the landing task. The measures were: time spent in correct formation, percentage of bomb strikes within 36 m of target center, time in missile tracking, number of times shot down, number of crashes, altitude variability, heading variability, release height variability, and gravity load at release.

2.4.1.5 Hover Measures

Moreland and Barnes (1969) derived a measure of helicopter pilot performance using the following equation:

100 – (absolute airspeed error + absolute altitude error + absolute heading error + absolute change in torque)

This measure decreased when cockpit temperature increased above 85°F, was better in light-to-moderate turbulence than no turbulence, and was sensitive to basic piloting techniques. However, it was not affected by either clothing or equipment configurations.

Richard and Parrish (1984) used a vector combination of errors (VCE), to estimate hover performance. VCE was calculated as follows:

$$VCE = (x^2 + y^2 + z^2)^{1/2}$$

where x, y, and z refer to the x, y, and z axis errors. The authors argue that VCE is a good summary measure because it discriminated trends in the data.

Data requirements. Not stated.

Thresholds. Not stated.

SOURCES

Barfield, W., Rosenberg, C., and Furness, T.A. Situation awareness as a function of frame of reference, computer-graphics eye point elevation, and geometric field of view. *International Journal of Aviation Psychology*. 5(3), 233–256, 1995.

Berger, I.R. Flight performance and pilot workload in helicopter flight under simulated IMC employing a forward looking sensor. *Proceedings of Guidance and Control Design Considerations for Low-Altitude and Terminal-Area Flight (AGARD-CP-240)*. Neuilly-sur-Seine, France: AGARD, 1977.

Billings, C.E., Gerke, R.J., and Wick, R.L. Comparisons of pilot performance in simulated and actual flight. *Aviation, Space, and Environmental Medicine*, 46(3), 304–308, 1975.

Bortolussi, M.R. and Vidulich, M.A. An evaluation of strategic behaviors in a high fidelity simulated flight task: Comparing primary performance to a figure of merit. *Proceedings of the 6th International Symposium on Aviation Psychology*. 2, 1101–1106, 1991.

Brictson, C.A. Operational measures of pilot performance during final approach to carrier landing. *Proceedings of Measurement of Aircrew Performance—The Flight Deck Workload and Its Relation to Pilot Performance (AGARD-CP-56)*. Neuilly-sur-Seine, France: AGARD, 1969.

Cohen, M.M. Disorienting effects of aircraft catapult launching: III. Cockpit displays and piloting performance. *Aviation, Space, and Environmental Medicine,* 48(9), 797–804, 1977.

Gaidai, B.V. and Mel'nikov, E.V. Choosing an objective criterion for piloting performance in research on pilot training on aircraft and simulators. *Cybermetrics and Computing Technology*. 3, 162–169, 1985.

Hardy, D.J. and Parasuraman, R. Cognition and flight performance in older pilots. *Journal of Experimental Psychology*. 3(4), 313–348, 1997.

Janowsky, D.S., Meacham, M.P., Blaine, J.D., Schoor, M., and Bozzetti, L.P. Simulated flying performance after marihuana intoxication. *Aviation, Space, and Environmental Medicine,* 47(2), 124–128, 1976.

Kelly, M.J. Performance measurement during simulated air-to-air combat. *Human Factors*. 30(4), 495–506, 1988.

Kraft, C.L. and Elworth, C.L. Flight deck work and night visual approach. *Proceedings of Measurement of Aircrew Performance—The Flight Deck Workload and its Relation to Pilot Performance (AGARD-CP-56)*. Neuilly-sur-Seine, France: AGARD, 1969.

Kruk, R., Regan, D., and Beverley, K.I. Flying performance on the advanced simulator for pilot training and laboratory tests of vision. *Human Factors*. 25(4), 457–466, 1983.

Lewis, M.F. and Mertens, H.W. Pilot performance during simulated approaches and landings made with various computer-generated visual glidepath indicators. *Aviation, Space, and Environmental Medicine*. 50(10), 991–1002, 1979.

Lintern, G., Kaul, C.E., and Collyer, S.C. Glide slope descent-rate cuing to aid carrier landings. *Human Factors*. 26(6), 667–675, 1984.

Lintern, G. and Koonce, J.M. Display magnification for simulated landing approaches. *International Journal of Aviation Psychology*. 1(1), 59–72, 1991.

Moreland, S. and Barnes, J.A. Exploratory study of pilot performance during high ambient temperatures/humidity. *Proceedings of Measurement of Aircrew Performance (AGARD-CP-56)*. Neuilly-sur-Seine, France: AGARD, 1969.

Morello, S.A. Recent flight test results using an electronic display format on the NASA B-737. *Proceedings of Guidance and Control Design Considerations for Low-Altitude and Terminal Area Flight (AGARD-CP-240)*. Neuilly-sur-Seine, France: AGARD, 1977.

North, R.A., Stackhouse, S.P., and Graffunder, K. *Performance, physiological, and oculometer evaluation of VTOL landing displays (NASA-CR-3171)*. Hampton, VA: NASA Langley Research Center, 1979.

Richard, G.L., and Parrish, R.V. Pilot differences and motion cuing effects on simulated helicopter hover. *Human Factors*. 26(3), 249–256, 1984.

Swaroop, R. and Ashworth, G.R. *An analysis of flight data from aircraft landings with and without the aid of a painted diamond on the same runway (NASA-CR-143849)*. Edwards Air Force Base, CA: NASA Dryden Research Center, 1978.

Wierwille, W.W. and Connor, S.A. Sensitivity of twenty measures of pilot mental workload in a simulated ILS task. *Proceedings of the Annual Conference on Manual Control (18th)*, 150–162, 1983.

Wierwille, W.W., Rahimi, M., and Casali, J.G., Evaluation of 16 measures of mental workload using a simulated flight task emphasizing mediational activity. *Human Factors*. 27(5), 489–502, 1985.

2.4.1.6 Standard Rate Turn

General description. A Standard Rate Turn (SRT) is a turn of 3° compass heading per second. It is taught in both rotary and fixed-wing aircraft.

Strengths and limitations. Chapman, Temme, and Still (2001) used mean and standard deviations from assigned altitude, airspeed, and heading to evaluate direction of turn (left versus right), degree of turn (180° versus 360°), and segment (roll in, roll out, first 30, and last 30). There were significant main effects for all three independent variables. Degree-of-turn error and altitude error were significantly greater for right than left turns and for 360° versus 180° turns. Altitude error was greater for right than left turns. Airspeed error was greater for 180° than for 360° turns and for first 30 than for roll out. Degree-of-turn error standard deviation was significantly greater for roll out than roll in. Altitude error standard deviation was significantly greater for left than right turns, 180° than 360° turns, and first 30 rather than last 30 segments. Airspeed error standard deviation was significantly greater for left than right turns and for 180° than for 360° turns.

Data requirements. The variables are aircraft- and mission-specific.

Thresholds. Stated for TH-57 helicopter.

SOURCE

Chapman, F., Temme, L.A., and Still, D.L. The performance of the standard rate turn (SRT) by student Naval helicopter pilots. *Aviation, Space, and Environmental Medicine*. 72(4), 343–351, 2001.

2.4.1.7 Landing Performance Score

General description. Landing Performance Score (LPS) is a score derived from the multiple regression of the following variables: number of landings per pilot, log book scorings, environmental data (weather, sea state, etc.), aircraft data (type and configuration), carrier data (ship size, visual landing aids, accident rate, etc.), boarding and bolster rate, intervals between landings, mission type and duration, and flying cycle workload estimate. LPS was developed for navy carrier landings (Brictson, 1977).

Strengths and limitations. LPS distinguished between night and day carrier landings (Brictson, 1974).

Data requirements. LPS requires the use of regression techniques.

Thresholds. Not stated.

SOURCES

Brictson, C.A. Pilot landing performance under high workload conditions. In A.N. Nickolson (Ed.) *Simulation and study of high workload operations (AGARD-CP-146)*. Neuilly-sur-Scinc, Francc: AGARD, 1974.

Brictson, C.A. Methods to assess pilot workload and other temporal indicators of pilot performances effectiveness. In Auffret, R. (Ed.) *Advisory Group for Aerospace Research and Development (AGARD) Conference Proceedings Number 217, AGARD-CP-217, B9-7 to B9-10*, 1977.

2.4.1.8 Control Input Activity

General description. Corwin, Sandry-Garza, Biferno, Boucek, Logan, Jonsson, and Metalis (1989) used control input activity for the wheel (aileron) and column (elevator) as a measure of flight-path control. Griffith, Gros, and Uphaus (1984) defined control reversal rate as "the total number of control reversals in each controller axis divided by the interval elapsed time" (p. 993). They computed separate control reversal rates for steady-state and maneuver transition intervals in a ground-based flight simulator.

Strengths and limitations. Corwin et al. (1989) reported that control input activity was a reliable and valid measure. Griffith et al. (1984) were unable to compute control reversal rates for throttle and rudder pedal activity because of minimal control inputs. Further, there were no significant differences between display configurations for pitch-axis control reversal rate. There were, however, significant differences between the same display configurations in the roll-axis control reversal rate.

Wierwille, Rahimi, and Casali (1985) used the total number of elevator, aileron, and rudder inputs during simulated flight. This measure was not sensitive to the difficulty of a mathematical problem-solving task performed during simulated flight.

Data requirements. Both control reversals and time must be simultaneously recorded.

Thresholds. Not available.

SOURCES

Corwin, W.H., Sandry-Garza, D.L., Biferno, M.H., Boucek, G.P., Logan, A.L., Jonsson, J.E., and Metalis, S.A. *Assessment of crew workload measurement methods, techniques and procedures. Volume 1—Process, methods, and results* (WRDC-TR-89-7006). Wright-Patterson Air Force Base, OH, 1989.

Griffith, P.W., Gros, P.S., and Uphaus, J.A. Evaluation of pilot performance and workload as a function of input data rate and update frame rate on a dot-matrix graphics display. *Proceedings of the National Aerospace and Electronics Conference*. 988–995, 1984.

Wierwille, W.W., Rahimi, M., and Casali, J.G., Evaluation of 16 measures of mental workload using a simulated flight task emphasizing mediational activity. *Human Factors*. 27(5), 489–502, 1985.

2.4.1.9 Composite Scores

General description. Many composite scores have been developed. One of the first was the Pilot Performance Index. In consultation with subject matter experts, Stein

TABLE 2
Pilot Performance Index Variable List

Takeoff	Initial Approach
Pitch angle	Heading
Climb	Manifold left
Heading	Manifold right
Airspeed	Bank angle
En route	Final Approach
Altitude	Heading
Pitch angle	Gear position
Heading	Flap position
Course deviation indicator	Course deviation indicator
Omni bearing sensor	
Descent	
Heading	
Airspeed	
Bank angle	
Course deviation indicator	
Omni bearing sensor	

Source: From Stein, E.S. *The measurement of pilot performance: A master-journeyman approach (DOT/FAA/CT-83/15).* Atlantic City, NJ: Federal Aviation Administration Technical Center; May 1984, p. 20. With permission.

(1984) developed a list of performance variables and associated performance criteria for an air transport mission. The list was subsequently reduced by eliminating those performance measures that did not distinguish experienced from novice pilots (see table 2). This collection of performance measures was called the Pilot Performance Index (PPI). Several similar composite scores exist and are described in the strengths and limitations section.

Strengths and limitations. The PPI provides objective estimates of performance and can distinguish between experienced and novice pilots. It does not measure the amount of effort being applied by the pilot, however. In an early study, Simmonds (1960) compared the performance of 17 pilots with varying levels of experience on turning while changing altitude tasks. Simmonds reported that the effect of experience was greater on consistency of performance than on accuracy of performance.

Stave (1979) used deviations from assigned flight parameters to assess the effects of vibration on pilot performance. The data were collected in a helicopter simulator. Measures included navigation error (average distance off course from takeoff to ILS), ILS score (deviation from glide slope altitude, degrees of heading, and airspeed), hover score (average altitude error plus average distance off course), hover time (duration from crossing load marker to load/unload), and load position (distance between load and helicopter when tensioned or released). There was a tendency for performance to improve with increased vibration, which was interpreted as compensation by the pilots.

Leirer, Yesavage, and Morrow (1989) also used deviations from ideal flight as a measure of pilot performance. They reported decrements in a fixed-base flight simulator in older pilots (18 to 29 versus 30 to 48 years old), in turbulence, and after ingestion of marijuana. In a follow-on study, Leirer, Yesavage, and Morrow (1991) reported significant effects of marijuana on pilot performance 24 h after dosing. In another drug study, Izraeli, Avgar, Almog, Shochat, Tochner, Tamir, and Ribak (1990) used deviations in airspeed, true heading, altitude, vertical velocity and bank to evaluate the effects of pyridostigmine bromide. No significant effects were reported from a 30 mg dose. In yet another drug study, Caldwell, Stephens, Carter, and Jones (1992) reported significant effects of 4 mg of atropine sulfate on pilot performance in a helicopter simulator. Significant deviations occurred in the following flight parameters: heading, vertical ascent heading during hover, and airspeed control. There were significant interactive effects of atropine with flight and maneuver on vertical speed control and of atropine with flight on bank angle. Ross, Yeazel, and Chau (1992) also used pilot performance deviations to examine the effects of a drug, but the drug was alcohol. They reported significant effects but only during high-workload conditions (e.g., turbulence, crosswind, wind shear). Pilot performance measures included aircraft control errors (e.g., deviation from assigned altitude, heading, and bank), navigation errors, and communication errors. In a similar study, Morrow, Yesavage, Leirer, Dolhert, Taylor, and Tinklenberg (1993) reported reduced performance associated with 0.10% BAL (blood alcohol level) on pilot performance in a flight simulator. These authors used a summary of the deviations from ideal performance as the dependent variable.

Wildzunas, Barron, and Wiley (1996) examined the effect of time delay on pilot performance in a full-motion flight simulator. They calculated a composite performance score to reflect the degree to which the subject maintained the required standards for all maneuvers (takeoff, standard rate turns, straight and level flight, descending turn, approach, hover, hover turn, nap-of-the-earth flight, formation flight, and pinnacle landing). They reported significant decrements associated with the 400 and 533 ms visual display.

Paul (1996) had Canadian Forces pilots perform an instrumented departure, turn, and instrument landing in a simulator after G-Loss of Consciousness (G-LOC) induced in a centrifuge. They calculated rmse scores on 11 flight parameters—again using deviations from target values. Only 1 of the 29 pilots had a significant decrement in performance after G-LOC.

Reardon, Fraser, and Omer (1998) examined flight performance in a helicopter simulator while nine military pilots were exposed to either 70°F or 100°F temperatures while wearing standard or chemical protective uniforms. Composite scores were derived from deviations from assigned aircraft states during hover, hover turn, straight and level, left climbing turn, left descending turn, right standard rate turn, contour, and nap-of-the-earth flight. Wearing of the chemical protective suit was associated with a significant decrement in performance.

Data requirements. The variables are aircraft and mission specific.

Thresholds. Not stated for any of the composite scores.

SOURCES

Caldwell, J.A., Stephens, R.L., Carter, D.J., and Jones, H.D. Effects of 2 mg and 4 mg atropine sulfate on the performance of U.S. Army helicopter pilots. *Aviation, Space, and Environmental Medicine.* 63(10), 857–864, 1992.

Izraeli, S., Avgar, D., Almog, S., Shochat, I., Tochner, Z., Tamir, A., and Ribak, J. The effect of repeated doses of 30 mg pyridostigmine bromide on pilot performance in an A-4 flight simulator. *Aviation, Space, and Environmental Medicine.* 61(5), 430–432, 1990.

Leirer, V.O., Yesavage, J.A., and Morrow, D.G. Marijuana, aging, and task difficulty effects on pilot performance. *Aviation, Space, and Environmental Medicine.* 60(12), 1145–1152, 1989.

Leirer, V.O., Yesavage, J.A., and Morrow, D.G. Marijuana carry-over effects on aircraft pilot performance. *Aviation, Space, and Environmental Medicine.* 62(3), 221–227, 1991.

Morrow, D., Yesavage, J., Leirer, V., Dohlert, N., Taylor, J., and Tinklenberg, J. The time-course of alcohol impairment of general aviation pilot performance in a Frasca 141 Simulator. *Aviation, Space, and Environmental Medicine.* 64(8), 697–705, 1993.

Paul, M. A. Instrument flying performance after G-Induced Loss of Consciousness. *Aviation, Space, and Environmental Medicine.* 67(11), 1028–1033, 1996.

Reardon, M.J., Fraser, E.B., and Omer, J.M. Flight performance effects of thermal stress and two aviator uniforms in a UH-60 helicopter simulator. *Aviation, Space, and Environmental Medicine.* 69(9), 569–576, 1998.

Ross, L.E., Yeazel, L.M., and Chau, A.W. Pilot performance with blood alcohol concentrations below 0.04%. *Aviation, Space, and Environmental Medicine.* 63(11), 951–956, 1992.

Simmonds, D.C.V. An investigation of pilot skill in an instrument flying task. *Ergonomics.* 3(3), 249–254, 1960.

Stave, A.M. The influence of low frequency vibration on pilot performance (as measured in a fixed base simulator). *Ergonomics.* 22(7), 823–835, 1979.

Stein, E.S. *The measurement of pilot performance: A master-journeyman approach (DOT/FAA/CT-83/15).* Atlantic City, NJ: Federal Aviation Administration Technical Center, May 1984.

Wildzunas, R.M., Barron, T.L., and Wiley, R.W. Visual display delay effects on pilot performance. *Aviation, Space, and Environmental Medicine.* 67(3), 214–221, 1996.

2.4.2 Boyett and Conn's White-Collar Performance Measures

General description. Boyett and Conn (1988) identified lists of performance measures with which to evaluate personnel in white-collar, professional, knowledge-worker organizations. These lists are categorized by function being performed, i.e., engineering, production planning, purchasing, and management information systems. The lists are provided in table 3.

Strengths and limitations. These measures are among the few developed for white-collar tasks and were developed in accordance with the following guidelines: (1) "involve white-collar employees in developing their own measures" (Boyett and Conn, 1988, p. 210), (2) "measure results, not activities" (p. 210), (3) "use group or team-based measures" (p. 211), and (4) "use a family of indicators" (p. 211).

Data requirements. Not stated.

Thresholds. Not available.

TABLE 3
White-Collar Measures in Various Functions

Engineering

- Percent new or in-place equipment/tooling performing as designed
- Percent machines/tooling capable of performing within established specifications
- Percent operations with current detailed process/method sheets
- Percent work run on specified tooling
- Number bills of material errors per employee
- Percent engineering change orders per drawing issued
- Percent material specification changes per specifications issued
- Percent engineering change requests to drawings issued based on design or material changes due to drawing/specification errors
- Percent documents (drawings, specifications, process sheets, etc.) issued on time

Production Planning or Scheduling

- Percent deviation actual/planned schedule
- Percent on-time shipments
- Percent utilization manufacturing facilities
- Percent overtime attributed to production scheduling
- Percent earned on assets employed
- Number, pounds, or dollars delayed orders
- Percent back orders
- Percent on-time submission master production plan
- Hours/time lost waiting on materials
- Number of days receipt of work orders prior to scheduled work
- Percent turnover of parts and material (annualized)

Purchasing

- Dollar purchases made
- Percent purchases handled by purchasing department
- Dollar purchases by major type
- Percent purchases/dollar sales volume
- Percent "rush" purchases
- Percent orders exception to lowest bid
- Percent orders shipped "most economical"
- Percent orders shipped "most expeditious"
- Percent orders transportation allowance verified
- Percent orders price variance from original requisition
- Percent orders "cash discount" or "early payment discount"
- Percent major vendors–annual price comparison completed
- Percent purchases–corporate guidelines met
- Elapsed time–purchase to deliver
- Percent purchases under long-term or "master contract"
- Dollar adjustment obtained/dollar value "defective" or "reject"

TABLE 3 (Continued)
White-Collar Measures in Various Functions

- Purchasing costs/purchase dollars
- Purchasing costs/number purchases
- Dollar value rejects/dollar purchases
- Percent shortages

Management Information Systems

- Number of data entry/programming errors per employee
- Percent reports issued on time
- Data processing costs as percentage of sales
- Number of reruns
- Total data processing cost per transaction
- Percent target dates met
- Average response time to problem reports
- Number data entry errors by type
- Percent off-peak jobs completed by 8:00 am
- Percent end-user available (prime) online
- Percent online 3 s response time
- Percent print turnaround in 1 hr or less
- Percent prime shift precision available
- Percent uninterrupted power supply available
- Percent security equipment available
- Score on user satisfaction survey
- Percent applications on time
- Percent applications on budget
- Correction costs of programming errors
- Number programming errors
- Percent time on maintenance
- Percent time on development
- Percent budget maintenance
- Percent budget development

Source: From Boyett and Conn (1988) p. 214.

SOURCE

Boyett, J.H. and Conn, H.P. Developing white-collar performance measures. *National Productivity Review.* Summer, 209–218, 1988.

2.4.3 Charlton's Measures of Human Performance in Space Control Systems

General description. Charlton's (1992) measures to predict human performance in space control systems are divided into three phases (prepass, contact execution, and contact termination) and three crew positions (ground controller, mission controller, and planner analyst). The measures by phase and crew position are: (1) prepass phase, ground-controller time to complete readiness tests and errors during configuration; (2) contact-execution phase, mission-controller track termination time and commanding time; (3) contact-execution phase, planner-analyst communication duration; (4) contact-termination phase, planner-analyst communication duration; and (5) contact-termination phase, ground-controller deconfiguration time, time to return resources, and time to log off system.

Strengths and limitations. The measures were evaluated in a series of three experiments using both civilian and air force satellite crews.

Data requirements. Questionnaires are used as well as computer-based scoring sheets.

Thresholds. Not stated.

SOURCE

Charlton, S.G. Establishing human factors criteria for space control systems. *Human Factors.* 34, 485–501, 1992.

2.4.4 Driving Parameters

General description. Driving parameters include measures of driver behavior (e.g., average brake RT, brake pedal errors, control light response time, number of brake responses, perception-response time, speed, and steering wheel reversals) as well as measures of total system performance (e.g., time to complete a driving task, tracking error) and observational measures (e.g., vehicles passing, use of traffic lights).

Strengths and limitations. Most researchers use several driving parameters in a single study. For example, Popp and Faerber (1993) used speed, lateral distance, longitudinal distance to the leading car, acceleration in all axes, steering angle, heading angle and frequency of head movements of subjects driving a straight road in a moving-base simulator to evaluate four feedback messages that a voice command was received. There were no significant differences between these dependent variables as a function of type of feedback message.

2.4.4.1 Average Brake RT

In an early study, Johansson and Rumar (1971) measured the brake RT of 321 drivers in an anticipated event on the road. Five of these were exposed to the same condition two more times. The median brake RT was 0.9 s, with 25% of the drivers having brake RTs longer than 1.2 s.

In an unusual study, Richter and Hyman (1974) compared the brake RT times for three types of controllers: foot, hand, and trigger. The RT was shortest for the hand controls. In a similar study, Snyder (1976) measured the time to brake for three accelerator-brake pedal distances. He reported longer times for a 6.35 cm (centimeter) lateral and 5.08 cm vertical separation than for 10.16 or 15.24 cm lateral separation with no vertical separation.

Also investigating the effects of brake placement, Morrison, Swope, and Malcomb (1986) found significant differences in movement time from the accelerator to the brake as a function of brake placement (lower than the accelerator resulted in shorter movement times) and gender (women had longer movement times except when the brake was below the accelerator).

Sivak, Post, Olson, and Donohoe (1980) reported higher percentages of responding to brake lights on both single (54.8%) and dual (53.2%) high-mounted brake lights than conventional brake lights (31.4%). However, there was no difference in mean RT (1.39, 1.30, and 1.38 s, respectively). Luoma, Flannagan, Sivak, Aoki, and Traube (1997) reported that yellow turn signals had shorter brake RTs than red turn signals.

In a similar study, Sivak, Flannagan, Sato, Traube, and Aoki (1994) used RT to compare neon, LED (light emitting diode), and fast incandescent brake lamps. There were significant differences between the lamps, which were lit in the subject's peripheral vision while the subject was performing a tracking task.

Also using a secondary task, Drory (1985) reported significant differences in average brake RT associated with different types of secondary tasks. It was not affected by the amount of rest drivers received prior to the simulated driving task.

In a comparison of laboratory, stationary, and on-road driving, Korteling (1990) used the RT of correct responses and error percentages. Older drivers (61 to 73 years old) and brain-injured patients had significantly longer RTs than younger drivers (21 to 43 years old). RTs were significantly longer in on-road driving than in the laboratory. There was a significant effect of inter-stimulus interval (ISI). Specifically, the shortest ISI was associated with the longest RT. Patients made significantly more errors than older or younger drivers.

In another study from the lead author, Korteling (1994) used four measures to assess platoon car-following performance: (1) brake RT, (2) the correlation between the speeds of the two cars, (3) time to obtain the maximum correlation, and (4) the maximum correlation. Brake RT and delay time were significantly longer for the patients than for the older or younger drivers. Brake RT was significantly longest when the driving speed of the lead car was varied and the road was winding. Both correlation measures were significantly lower for the patients and older drivers than for the younger drivers.

In another lead car study, Brookhuis, de Waard, and Mulder (1994) used RT to speed variations in response to the speed of a leading car as a measure of driver attention. Also using a lead car, Schweitzer, Apter, Ben-David, Liebermann, and Parush (1995) measured total braking times of 51 drivers 21 to 30 years of age in response to a lead vehicle. There was no significant difference between 60 and 80 kmph speeds. There were, however, significant differences between 6 and 12 m following distance (shorter for 6 m) as well as significant differences between naive, partial knowledge, and full knowledge of the future actions of the lead vehicle (shortest for full knowledge).

Szlyk, Seiple, and Viana (1995) reported a significant increase in braking response time as age increased. The data were collected in a simulator. Age or visual impairment had no significant effects on the variability of brake pressure.

2.4.4.2 Brake Pedal Errors

Rogers and Wierwille (1988) used the type and frequency of pedal actuation errors to evaluate alternative brake pedal designs. They reported 297 errors over 72 h of testing. Serious errors occurred when the wrong or both pedals were depressed. Catch errors occurred when a pedal interfered with a foot movement. If the interference was minimal, the error was categorized as a scuff. Instructional errors occurred when the subject failed to perform the task as instructed.

Vernoy and Tomerlin (1989) used pedal error, "hitting the accelerator pedal when instructed to depress the brake pedal" (p. 369), to evaluate misperception of the centerline. There was no significant difference in pedal error between eight types of automobile used in the evaluation. There was a significant difference in deviation from the center line between the eight automobiles, however.

2.4.4.3 Control Light Response Time

Drory (1985) reported significant differences in control light response time associated with various types of secondary tasks. It was not affected by the amount of rest drivers received prior to the simulated driving task. Summala (1981) placed a small lamp which, however, was not a control light on roads and measured the RT of drivers for initiating an avoidance maneuver. RTs were about 2.5 s with an upper safe limit projected at 3 s.

2.4.4.4 Number of Brake Responses

Drory (1985) reported significant differences in the number of brake responses used to evaluate the effects of various types of secondary tasks. The type of task did not affect the number of brake responses nor did the amount of rest drivers received prior to the simulated driving task.

Attwood and Williams (1980) reported significant increase in brake reversal rate between 3.5 and 4.0 kg under alcohol than under either cannabis or cannabis and alcohol conditions. The data were collected from eight male drivers on a closed course in an instrumented vehicle.

2.4.4.5 Perception-Response Time

Olson and Sivak (1986) measured perception-response time from the first sighting of an obstacle until the accelerator was released and the driver contacted the brake. Their data were collected in an instrumented vehicle driven on a two-lane rural road.

2.4.4.6 Speed

Svenson (1976) reported that subjects overestimated speed of slower segments if faster segments were included. Their subjects made their estimates from model trains but the results were generalized to driving. Looking at roadway enhancements, Steyvers and de Waard (2000) reported significant increase in speed on roads with edge lines than in the absence of edge lines.

Attwood and Williams (1980) reported significant increases in both mean and median speed when drivers were under the influence of cannabis and a significant decrease when drivers were under the influence of both alcohol and cannabis. The data were collected from eight male drivers on a closed course in an instrumented vehicle. In the same year Wilson and Anderson (1980) compared driving speeds on a closed course and in traffic using radial or cross-ply tires. They reported that mean speed on the track and on the road were not significantly correlated. There was a significant correlation between speed on the test track and age of the driver, however. There were also significant increases in speed over trials and with the radial versus the cross-ply tire.

In a simulator study, Szlyk, Seiple, and Viana (1995) reported a significant decrease in speed as age increased. Using both a simulator and an old road vehicle, Reed and Green (1999) reported a significant increase in lateral speed when dialing a telephone while driving on the road or in a simulator. Steyvers and de Waard (2000) reported significant increase in speed on roads with edge lines than in their absence.

Applying a series of speed measures, Shinar and Stiebel (1986) used speed, speed above or below the speed limit, speed reduction (((original speed – speed at site 1)/original speed) × 100) and speed resumption (((speed at site 2 – speed limit)/speed limit) × 100) to evaluate the effectiveness of the presence of police cars in reducing speeding.

2.4.4.7 Steering Wheel Reversals

Hicks and Wierwille (1979) reported that steering reversals were sensitive to workload (i.e., gusts at the front of a driving simulator).

Drory (1985) reported significant differences between steering wheel reversals associated with various types of secondary tasks. It was not affected by the amount of rest drivers received prior to the simulated driving task, however. Similarly, Horne and Baumber (1991) reported that neither time of day nor alcohol had any effect on lateral corrective steering movements.

Frank, Casali, and Wierwille (1988) used the number of large steering reversals (greater than 5°), number of small steering reversals (less than 5°), and yaw standard deviation ("angle in the horizontal plane between the simulated vehicle longitudinal axis and the instantaneous roadway tangent" [p. 206]) to evaluate the effects of motion system transport delay and visual system transport delay. All three measures of driver performance were significantly related to transport delay.

2.4.4.8 Time

Gawron, Baum, and Perel (1986) used the time to complete a double-lane change as well as the number of pylons struck to evaluate side impact padding thickness (0, 7.5, 10 cm), direction of the lane change (left/right, right/left), and replication (1 to 12). Subjects took longer to perform the left/right than the right/left lane change. Time to complete the lane change increased as padding thickness increased (0 cm = 3.175 s, 7.5 cm = 3.222 s, and 10 cm = 3.224 s). There were no significant effects on the number of pylons struck.

Sidaway, Fairweather, Sekiya, and McNitt-Gray (1996) asked subjects to estimate time to collision after viewing videotapes of accidents. Subjects consistently underestimated the time. However, as velocity increased, the time estimate was more accurate.

Roge (1996) used time to line crossing (TLC) as a measuring of steering quality. He reported significantly better steering control among subjects who were frame independent. Others have used TLC to enhance preview-predictor models of human driving performance. In these models TLC equals the time for the vehicle to reach either edge of the driving lane. It is calculated from lateral lane position, the heading angle, vehicle speed, and commanded steering angle (Godthelp, Milgram, and Blaauw, 1984).

Godthelp, Milgram, and Blaauw (1984) evaluated TLC in an instrumented car driven on an unused straight, four-lane highway by six male drivers at six different speeds (20, 40, 60, 80, 100, and 120 kmph) with and without a visor. Based on the results, the authors argued that TLC was a good measure of open-loop driving performance. Godthelp (1986) reported, based on field study data, that TLC described anticipatory steering action during curve driving. In a later study, Godthelp (1988) used TLC to assess the effects of speed on driving strategies from error neglecting to error correcting at lane edges.

Finnegan and Green (1990) reviewed five studies in which TLC was measured. For these studies, the authors concluded that 6.6 s are required for the visual search with a single lane change and 1.5 s to complete the lane change.

Lamble, Laakso, and Summala (1999) used a similar measure, time to collision, to evaluate locations of in-vehicle displays and controls. Also using time to collision, van Winsum and Heino (1996) reported that time-to-collision information was used to initiate and control braking. Their data were also collected in a driving simulator. Their subjects were 54 male drivers with an average age of 29 years.

With yet another measure, Reed and Green (1999) reported that lane keeping was less precise in a simulator than on the road. Lane keeping was also less precise when dialing a telephone during driving in either the simulator or on the road.

2.4.4.9 Tracking Error

In an early study, Soliday (1975) measured lane position maintenance of 12 male drivers on the road. Drivers tended to stay in the center of the lane and oscillated around the center of two-lane roads more than four-lane roads. In another early study, Hicks and Wierwille (1979) reported that yaw deviation and lateral deviation were sensitive to workload (i.e., gusts at the front of a driving simulator).

Also using a driving simulator, Drory (1985) reported significant differences in tracking error associated with various types of secondary tasks. It was not affected by the amount of rest drivers received prior to the simulated driving task. Using a similar measure, Godthelp and Kappler (1988) reported larger standard deviations in lateral position when drivers were wearing safety goggles than when they were not. This measure also increased as speed increased.

Imbeau, Wierwille, Wolf, and Chun (1989) reported that if drivers failed to respond to a display reading task, the variance of lane deviation decreased. The data were collected in a driving simulator. Also using deviation, Korteling (1994) reported no significant differences in the standard deviation of lateral position between young (21–34) and old (65–74 years) drivers in a driving simulator. However, older drivers had significantly larger longitudinal standard deviation in a car-following task. There was also a decrement associated with fatigue in steering performance but not in car-following performance. In a related study, Summala, Nieminen, and Punto (1996) reported worse lane-keeping performance for novice drivers than for experienced drivers when the display was near the periphery rather than in the middle console of a driving simulation.

In another measure of deviation, Szlyk, Seiple, and Viana (1995) reported a significant increase in the number of lane crossings as age increased. The data were collected in a simulator. Using still other measures, van Winsum (1996) reported that steering wheel angle increased as curve radii decreased. Steering error increased as steering wheel angle increased. The data were collected in a driving simulator.

Using other related measures, Heimstra, Fallesen, Kinsley, and Warner (1980) used time on target and frequency off target to assess the effects of cigarette deprivation on smokers' tracking performance. There were significant effects on both measures. Horne and Baumber (1991) reported significant effects of alcohol on average following distance in a driving simulator.

2.4.4.10 Observational Measures

Wooller (1972) observed four drivers negotiate an 11.72 mi (mile) route to determine the following measures: (1) travel time, (2) vehicles passing subject, (3) vehicles passed by subject, (4) lateral position changes, (5) use of traffic lights, and (6) number of vehicles. Patterns varied across subjects but were consistent within a subject.

Data requirements. An instrumented simulator or vehicle is required.

Thresholds. Total time varied between 0.1 and 1.8 s (Olson and Sivak, 1986).

SOURCES

Attwood, D.A., and Williams, R.D. Braking performance of drivers under the influence of alcohol and cannabis. *Proceedings of the Human Factors Society 24th Annual Meeting.* 134–138, 1980.

Brookhuis, K., de Waard, D., and Mulder, B. Measuring driving performance by car-following in traffic. *Ergonomics.* 37(3), 427–434, 1994.

Drory, A. Effects of rest and secondary task on simulated truck-driving task performance. *Human Factors.* 27(2), 201–207, 1985.

Finnegan, P. and Green, P. *The time to change lanes: A literature review (UMTRI-90-34).* Ann Arbor, MI: The University of Michigan Transportation Research Institute, September 1990.

Frank, L.H., Casali, J.G., and Wierwille, W.W. Effects of visual display and motion system delays on operator performance and uneasiness in a driving simulator. *Human Factors.* 30(2), 201–217, 1988.

Gawron, V.J., Baum, A.S., and Perel, M. Effects of side-impact padding on behavior performance. *Human Factors.* 28(6), 661–671, 1986.

Godthelp, H. Vehicle control during curve driving. *Human Factors.* 28(2), 211–221, 1986.

Godthelp, H. The limits of path error neglecting in straight lane driving. *Ergonomics.* 31(4), 609–619, 1988.

Godthelp, H. and Kappler, W.D. Effects of vehicle handling characteristics on driving strategy. *Human Factors*. 30(2), 219–229, 1988.
Godthelp, H., Milgram, P., and Blaauw, G.J. The development of a time-related measure to describe driving strategy. *Human Factors*. 26(3), 257–268, 1984.
Heimstra, N.W., Fallesen, J.J., Kinsley, S.A., and Warner, N.W. The effects of deprivation of cigarette smoking on psychomotor performance. *Ergonomics*. 23(11), 1047–1055, 1980.
Hicks, T.G. and Wierwille, W.W. Comparison of five mental workload assessment procedures in a moving-base driving simulator. *Human Factors*. 21(2), 129–143, 1979.
Horn, J.A. and Baumber, C.J. Time-of-day effects of alcohol intake on simulated driving performance in women. *Ergonomics*. 34(11), 1377–1383, 1991.
Imbeau, D., Wierwille, W.W., Wolf, L.D., and Chun, G.A. Effects of instrument panel luminance and chromaticity on reading performance and preference in simulated driving. *Human Factors*. 31(2), 147–160, 1989.
Johansson, G. and Rumar, K. Driver's brake reaction times. *Human Factors*. 13(1), 23–27, 1971.
Korteling, J.E. Perception-response speed and driving capabilities of brain-damaged and older drivers. *Human Factors*. 32(1), 95–108, 1990.
Korteling, J.E. Effects of aging, skill modification, and demand alternation on multiple task performance. *Human Factors*. 36(1), 27–43, 1994.
Lamble, D., Laakso, M., and Summala, H. Detection thresholds in car following situations and peripheral vision for positioning of visually demanding in-car displays. *Ergonomics*. 42(6), 807–815, 1999.
Luoma, J., Flannagan, M.J., Sivak, M., Aoki, M., and Traube, E.C. Effects of turn-signal colour on reaction time to brake signals. *Ergonomics*. 40(1), 62–68, 1997.
Morrison, R.W., Swope, J.G., and Malcomb, C.G. Movement time and brake pedal placement. *Human Factors*. 28(2), 241–246, 1986.
Olson, P.L. and Sivak, M. Perception-response time to unexpected roadway hazards. *Human Factors*. 28(1), 91–96, 1986.
Popp, M.M. and Faerber, B. Feedback modality for nontransparent driver control actions: Why not visually? In Gale, A.G., Brown, J.D, C.M. Haslegrave, H.W. Kruysse, and S.P. Taylor (Eds.). *Vision in vehicles—IV*. Amsterdam: North-Holland, 1993.
Reed, M.P. and Green, P. A. Comparison of driving performance on-road and in a low-cost simulator using a concurrent telephone dialing task. *Ergonomics*. 42(8), 1015–1037, 1999.
Richter, R.L. and Hyman, W.A. Research note: Driver's brake reaction times with adaptive controls. *Human Factors*. 16(1), 87–88, 1974.
Roge, J. Spatial reference frames and driver performance. *Ergonomics*. 39(9), 1134–1145, 1996.
Rogers, S.B. and Wierwille, W.W. The occurrence of accelerator and brake pedal actuation errors during simulated driving. *Human Factors*. 30(1), 71–81, 1988.
Schweitzer, N., Apter, Y., Ben-David, G., Lieberman, D.G., and Parush, A. A field study on braking responses during driving. II. Minimum driver braking times. *Ergonomics*. 38(9), 1903–1910, 1995.
Shinar, D. and Stiebel, J. The effectiveness of stationary versus moving police vehicles on compliance with speed limit. *Human Factors*. 28(3), 365–371, 1986.
Sidaway, B., Fairweather, M., Sekiya, H., and McNitt-Gray, J. Time-to-collision estimation in a simulated driving task. *Human Factors*. 38(1), 101–113, 1996.
Sivak, M., Flannagan, M.J., Sato, T., Traube, E.C., and Aoki, M. Reaction times to neon, LED, and fast incandescent brake lamps. *Ergonomics*. 37(6), 989–994, 1994.
Sivak, M., Post, D.V., Olson, P.L., and Donohoe, R.J. Brake responses of unsuspecting drivers to high-mounted brake lights. *Proceedings of the Human Factors Society*. 139–142, 1980.
Snyder, H.L. Braking movement time and accelerator-brake separation. *Human Factors*. 18(2), 201–204, 1976.
Soliday, S.M. Lane position maintenance by automobile drivers on two types of highway. *Ergonomics*. 18(2), 175–183, 1975.

Steyvers, F.J.J.M. and de Waard, D. Road-edge delineation in rural areas: Effects on driving behavior. *Ergonomics*. 43(2), 223–238, 2000.
Summala, H. Drivers' steering reaction to a light stimulus on a dark road. *Ergonomics*. 24(2), 125–131, 1981.
Summala, H., Nieminen, T., and Punto, M. Maintaining lane position with peripheral vision during in-vehicle tasks. *Human Factors*. 38(3), 442–451, 1996.
Svenson, O. Experience of mean speed related to speeds over parts of a trip. *Ergonomics*. 19(1), 11–20, 1976.
Szlyk, J.P., Seiple, W., and Viana, M. Relative effects of age and compromised vision on driving performance. *Human Factors*. 37(2), 430–436, 1995.
van Winsum, W. Speed choice and steering behavior in curve driving. *Human Factors*. 38(3), 434–441, 1996.
van Winsum, W. and Heino, A. Choice of time-headway in car-following and the role of time-to-collision information in braking. *Ergonomics*. 39(4), 579–592, 1996.
Vernoy, M.W. and Tomerlin, J. Pedal error and misperceived center line in eight different automobiles. *Human Factors*. 31(4), 369–375, 1989.
Wilson, W.T. and Anderson, J.M. The effects of tyre type on driving speed and presumed risk taking. *Ergonomics*. 23(3), 223–235, 1980.
Wooller, J. The measurement of driver performance. *Ergonomics*. 15(1), 81–87, 1972.

2.4.5 Eastman Kodak Company Measures for Handling Tasks

General description. The Eastman Kodak Company (1986) has developed a total of eight measures to assess human performance in repetitive assembly, packing, or handling tasks. These eight measures have been divided into the following: "(1) measures of productivity over the shift: total units per shift at different levels and durations of effort and/or exposure, units per hour compared to a standard, amount of time on arbitrary work breaks or secondary work, amount of waste, and work interruptions, distractions, and accidents and (2) quality of output: missed defects/communications, improper actions, and incomplete work" (p. 104).

Strengths and limitations. These measures are well-suited for repetitive overt tasks but may be inappropriate for maintenance or monitoring tasks.

Data requirements. Task must require observable behavior.

Thresholds. Not stated.

SOURCE

Eastman Kodak Company. *Ergonomic design for people at work*. New York: Van Nostrand Reinhold, 1986.

2.4.6 Haworth–Newman Avionics Display Readability Scale

General description. The Haworth–Newman Avionics Display Readability Scale (see figure 2) is based on the Cooper–Harper Rating Scale. As such, it has a three-level-deep branching that systematically leads to a rating of 1 (excellent) to 10 (major deficiencies).

Strengths and limitations. The scale is easy to use. It has been validated in a limited systematic degradation (i.e., masking) of avionics symbology (Chiappetti, 1994). Recommendations from that validation study include: (1) provide a more precise definition of readability, (2) validate the scale using better-trained subjects, (3) use more realistic displays, and (4) use display resolution, symbol luminance, and symbol size to improve readability.

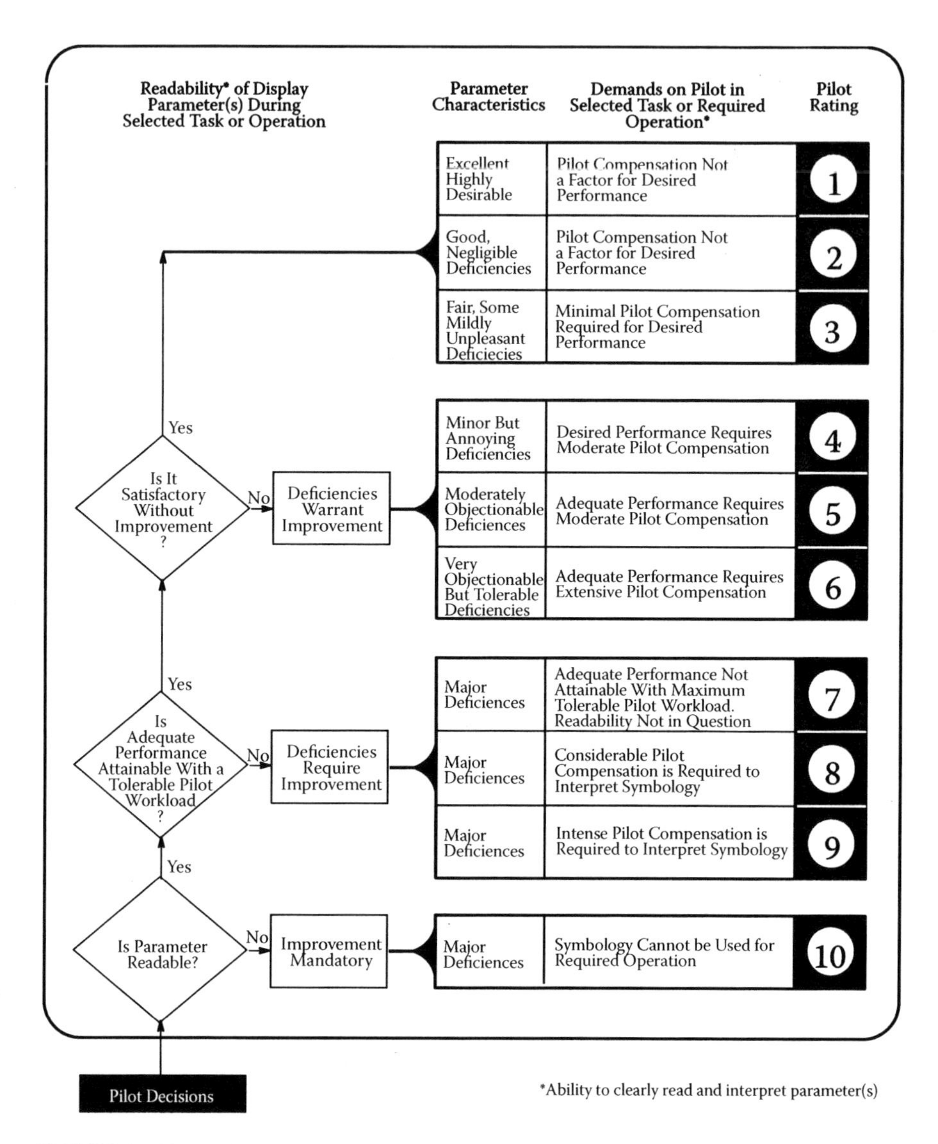

FIGURE 2 Haworth–Newman Display Readability Rating Scale. (From Haworth, 1993 cited in Chiappetti, C.F. *Evaluation of the Haworth–Newman avionics display readability scale*. Monterey, CA: Naval Postgraduate School Thesis, September 1994. With permission.)

Data requirements. Subjects must have a copy of the scale in front of them during rating.

Thresholds. 1 (excellent) to 10 (major deficiencies).

SOURCE

Chiappetti, C.F. *Evaluation of the Haworth–Newman avionics display readability scale*. Monterey, CA: Naval Postgraduate School Thesis, September 1994.

2.5 CRITICAL INCIDENT TECHNIQUE

The fifth category is critical incidents, which are typically used to assess worst-case performance.

General description. The Critical Incident Technique is a set of specifications for collecting data from observed behaviors. These specifications include the following:

(1) Persons to make the observations must have the following:
  (a) Knowledge concerning the activity
  (b) Relation to those observed
  (c) Training requirements
(2) Groups to be observed, including the following:
  (a) General description
  (b) Location
  (c) Persons
  (d) Times
  (e) Conditions
(3) Behaviors to be observed with an emphasis on the following:
  (a) General type of activity
  (b) Specific behaviors
  (c) Criteria of relevance to general aim
  (d) Criteria of importance to general aim (critical prints) (Flanagan, 1954, p. 339)

Strengths and limitations. The technique has been used successfully since 1947. It is extremely flexible, but can be applied only to observable performance activities.

Data requirements. The specifications listed in the general description paragraph must be applied.

Thresholds. Not stated.

SOURCE

Flanagan, J.C. The critical incident technique. *Psychological Bulletin.* 51(4), 327–358, 1954.

2.6 TEAM PERFORMANCE MEASURES

The final category of performance measures is team performance measures. These assess the abilities of two or more persons working in unison to accomplish a task or tasks. These measures assume that human performance varies when part of a team. There are literally hundreds of team performance measures. Most of them have been developed for production teams. Examples include the following: (1) defect percentage, (2) number of accidents, (3) difference between budgeted and actual costs, and (4) customer satisfaction.

There are many books to help develop measures for such teams (e.g., the book by Jones and Schilling [2000]). Jones and Schilling (2000) present eight principles for measuring team performance: capture the team strategy, align the strategy with the organization, stimulate problem solving that results in improved performance, use measurement to focus team meeting, measure the critical items, ensure that team members understand the measures, involve customers in development of the measures, and address the work of each member.

In another book, Heineman and Zeiss (2002) identify obstacles to measuring team performance in health-care settings: (1) many measures are proprietary and require contracting for training to use them; (2) measures are not reported in a standardized manner, hindering easy comparison between them; (3) many measures were developed for other businesses and are not relevant to health care; and (4) information on the measures is scattered across literature from different disciplines.

SOURCES

Heinemann, G.D. and Zeiss, A.M. *Team performance in health care assessment and development.* New York, New York: Kluwer Academic/Plenum Publishers, 2002.

Jones, S.D. and Schilling, D.J.M. *Measuring team performance.* San Francisco, California: Jossey-Bass, 2000.

2.6.1 Cicek, Koksal, and Ozdemirel's Team Performance Measurement Model

General description. The Team Performance Measurement Model has four categories of team performance measures: (1) structure, (2) process, (3) output, and (4) input. Using the behaviors of effective total quality management teams identified by Champion, Medsker, and Higgs (1993), O'Brien and Walley (1994), and Storey (1989), Cicek, Koksal, and Ozdemirel (2005) listed measures for each of the four categories. Structure measurement consists of team members rating their team on a scale of 0 to 100 on behaviors listed by Cicek (1997) relating to clear objectives, communication and conflict management, participation and relationships, good decision making and involvement, knowledge and skills, culture, motivation, and administration. Process measurement consists of data from critical operations and key intermediate results. Output measurements include customer satisfaction and objective measures based on customer expectations, e.g., service times. Input measures are quality of raw materials, information, equipment, services, and management support.

Strengths and limitations. The items on the questionnaires for behaviors were reliable in a hospital setting (Cicek, 1997). The model was applied in a neurological sciences team at a hospital and was reported to be effective (Cicek, Koksal, and Ozdemirel (2005).

Data requirements. List of behaviors from Cicek (1997), flow charts of process used by team being evaluated, and input and output data.

Thresholds. Not stated.

SOURCES

Champion, M.A., Medsker, G.J., and Higgs, A.C. Relations between workgroup characteristics and effectiveness: Implications for designing effective workgroups. *Personnel Psychology.* 46(3), 823–850, 1993.

Cicek, M.C. *A model for performance measurement of total quality teams and an application in a private hospital.* Master of Science Thesis, Middle East Technical University, Ankara, 1997.

Cicek, M.C., Koksal, G., and Ozdemirel, N.E. A team performance measurement model for continuous improvement. *Total Quality Management.* 16(3), 331–349, 2005.

O'Brien, P. and Walley, P. Total quality teamworking: What's different? *Total Quality Management.* 5(3), 151–160, 1994.

Storey, R. *C BACIE from team building.* British Association for Commercial and Industrial Education. 1989.

2.6.2 Command and Control Team Performance Measures

General description. Entin, Serfaty, Elliott, and Schiflett (2001) applied the following metrics to assess performance of four diverse teams on a command and control (C2) task: (1) value of enemy assets destroyed, (2) diminished value of own assets, (3) number of friendly assets not destroyed, (4) number of friendly assets lost to hostile fire, (5) number of friendly assets lost to fuel out, (6) number of hostile assets destroyed by friendly action, (7) kill ratio, (8) air refuelings completed, (9) number of transfers of resources in, (10) number of resources out, (11) number of e-mails sent, and (12) number of e-mails received.

Strengths and limitations. The results of the analyses were not presented.

Data requirements. Asset status and e-mail tally.

Thresholds. Not stated.

SOURCE

Entin, E.B., Serfaty, D., Elliott, L.R., and Schiflett, S. G. *DMT-RNet: An internet-based infrastructure for distributed multidisciplinary investigations of C2 performance.* Brooks Air Force Base, Texas: Air Force Research Laboratory, 2001.

2.6.3 Gradesheet

General description. The gradesheet contains 40 items that are rated by self, peers, and instructors. Items are related to air combat and include radar mechanics, tactics, tactical intercepts, communications, mutual support, and flight leadership. Criteria for rating each of these 40 items are the following: not applicable, unsafe, lack of ability or knowledge, limited proficiency, recognizes and corrects errors, correct, and unusually high ability.

Strengths and limitations. Krusmark, Schreiber, and Bennett (2004) used data from 148 F-16 pilots grouped into 32 teams performing air combat missions in groups of four aircraft. Ratings were made by seven F-16 subject matter experts using the gradesheet. The average interrater reliability was small, 0.42. There was very high internal consistency in the ratings, however; Cronbach's alpha = 98. Ratings were similar across multiple items. The authors performed a principal component analysis to identify the underlying variable accounting for the majority of the variance. One component explained 62.47% of the variance. The next highest component explained only 4.67%.

Data requirements. Ratings of 40 items included in the gradesheet.

Thresholds. Not stated.

SOURCE

Krusmark, M., Schreiber, B.T., and Bennett, W. *The effectiveness of a traditional gradesheet for measuring air combat team performance in simulated distributed mission operations (AFRL-HE-AZ-TR-2004-0090)*. Mesa, Arizona: Air Force Research Laboratory, May 2004.

2.6.4 Knowledge, Skills, and Ability

General description. Miller (2001) adapted the KSAs (Knowledge, Skills, and Ability) measured by the Teamwork Test. These KSAs are conflict resolution, collaborative problem solving, communication, goal setting and performance management, as well as planning and task management.

Strengths and limitations. Miller (2001) asked 176 undergraduate management majors to work in groups of three to five persons on an organization design task. Each participant completed the Teamwork Test. The team average score and variance were correlated with the group project grade as well as a self-report of satisfaction with the group. There was no significant relationship between team average score on the Teamwork Test and either task performance or team satisfaction. There was a trend for teams with high Teamwork Test variances and high Teamwork scores to have better task performance ($p = 0.07$).

Data requirements. Complete 35 item Teamwork Test.

Thresholds. Not stated.

SOURCE

Miller, D.L. Examining teamwork KSAs and team performance. *Small Group Research*. 32(6), 745–766, 2001.

2.6.5 Latent Semantic Analysis

General description. Latent Semantic Analysis (LSA) is the creation of a word by a document matrix in which cells are populated with the frequency of occurrence of that word in that document. Log entropy term weighting is then applied. Afterward, a singular value decomposition technique is used to identify the significant vectors. LSA does not consider word order or syntax.

Strengths and limitations. Dong, Hill, and Agogino (2004) applied LSA to documents produced by collaborative design teams over the 15 weeks of a graduate course in product design. There were eight teams. The Spearman rank correlation coefficient was calculated between the rankings of the semantic coherence and the ranking of the team performance by faculty members. There was a significant positive correlation between the two sets of rankings. The correlation was highest with documents only and lower with documents and e-mail included in the LSA.

Data requirements. Documents produced by the teams.

Thresholds. Not stated.

SOURCE

Dong, A., Hill, A.W., and Agogino, A.M. A document analysis method for characterizing design team performance. *Transactions of the ASME Journal of Mechanical Design*. 126(3), 378–385, 2004.

2.6.6 Load of the Bottleneck Worker

General description. Slomp and Molleman (2002) defined the load of the bottleneck worker as "the time in which all the work can be completed by the workers. The worker with the heaviest workload determines this load" (p. 1197). They used the following equation to calculate the load of the bottleneck worker (WB): WB = $\max_j \sum_i x_{ij}$ (p. 1200), where j is the worker index, i is the task index, and x_{ij} is the "time (in time units) assigned to worker j to perform task I" (p. 1197).

Strengths and limitations. WB is quantitative and has face validity. It can be used in discrete observable tasks with distant start and completion points. WB was significantly different as a function of training policy, absenteeism, fluctuation in demand, level of cross-training, and two of their interactions (training policy by level of cross-training and absenteeism by level of cross-training; Slomp and Molleman, 2002).

Data requirements. Number of workers, number of tasks, and time for task performance.

Thresholds. Not stated.

SOURCE

Slomp, J. and Molleman, E. Cross-training policies and team performance. *International Journal of Production Research.* 40(5), 1193–1219, 2002.

2.6.7 Nieva, Fleishman, and Rieck's Team Dimensions

General description. Nieva, Fleishman, and Rieck (1985) defined five measures of team performance: (1) matching number resources to task requirements, (2) response coordination, (3) activity pacing, (4) priority assignment among tasks, and (5) load balancing.

Strengths and limitations. The five measures are an excellent first step in developing measures of team performance, but specific metrics must be developed and tested.

Data requirements. The following group characteristics must be considered when using these measures of team performance: (1) group size, (2) group cohesiveness, (3) intra- and intergroup competition and cooperation, (4) communication, (5) standard communication nets, (6) homogeneity/heterogeneity in personality and attitudes, (7) homogeneity/heterogeneity in ability, (8) power distribution within the group, and (9) group training.

Thresholds. Not stated.

SOURCE

Nieva, V.F., Fleishman, E.A., and Rieck, A. *Team dimensions: Their identity, their measurement and their relationships. Research Note 85-12.* Alexandria, VA: Army Research Institute for the Behavioral and Social Sciences, January 1985.

2.6.8 Project Value Chain

General description. Bourgault, Lefebvre, Lefebvre, Pellerin, and Elia (2002) identified the need for performance measures to evaluate virtual organizations of distributed teams. They identified project value chains as critical in this measurement. They define these chains as "the process by which a series of activities are linked together for the purpose of creating value for a client" (p. 3).

Strengths and limitations. Project Value chains have two distinct advantages. First, they help the team identify "where their input generates values for the end user" (p. 3). Second, Project Value chains can be linked to other chains, including those outside the current team or organization, providing increased flexibility.

Data requirements. Process decomposition into activities and links among those activities.

Thresholds. Not stated.

SOURCE

Bourgault, M., Lefebvre, E., Lefebvre, L.A., Pellerin, R., and Elia, E. Discussion of metrics for distributed project management: Preliminary findings. *Proceedings of the 35th Hawaii International Conference on System Sciences*. 2002.

2.6.9 Team Communication

General description. Harville, Lopez, Elliott, and Barnes (2005) developed a set of communication codes (see figure 3) to assess team performance in a command and control (C2) task. The task was developed using a PC-based system emulating United States Air Force C2 tactical operations.

Svensson and Andersson (2006) used speech acts and communication problems to assess performance of fighter pilot teams. Speech acts were the following: (1) present activity information, (2) future or combat information, (3) tactics, (4) com-

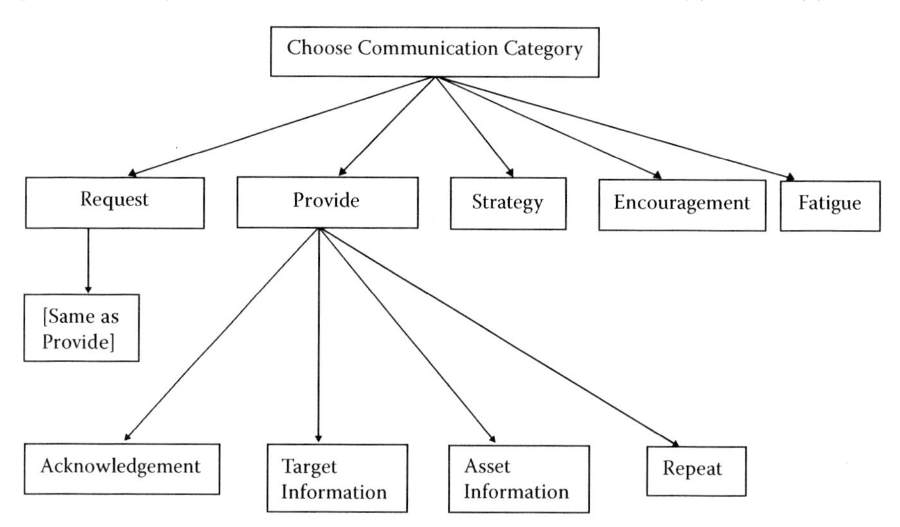

FIGURE 3 Communication Codes. (From Harville, D., Lopez, N., Elliott, L., and Barnes, C. *Team communication and performance during sustained working conditions (AFRL-HE-BR-TR-2005-0085)*. Brooks City Base, Texas: Air Force Research Laboratory, May 2005, p. 7. With permission.)

munication, (5) question, (6) others, such as practice or identifying issues. Communication problems were categorized as follows: (1) simultaneous, (2) unclear information, (3) uncodable, (4) unconventional, and (5) uncertain target.

Strengths and limitations. The communications codes were selected for relevance to team performance and distinctiveness. The codes were applied with 95% agreement among coders. Several of the codes were significantly affected by fatigue (i.e., comparison of first and sixth sessions). Total communication and total task communications were significantly less in session 6 than session 1. The same occurred for provide information, provide asset information, request asset information, and strategy. There was a significant increase in communication of fatigue state from first to sixth session.

Svensson and Andersson (2006) examined the performance of two teams of four pilots and one fighter controller performing aircraft escort or aircraft attack tasks in a simulator. Speech acts frequency was highest when a team was winning. Communication problems were highest when teams were tied.

Data requirements. Records of verbal communications.

Thresholds. Not stated.

SOURCES

Harville, D., Lopez, N., Elliott, L., and Barnes, C. *Team communication and performance during sustained working conditions (AFRL-HE-BR-TR-2005-0085).* Brooks City Base, Texas: Air Force Research Laboratory, May 2005.

Svensson, J. and Andersson, J. Speech acts, communication problems, and fighter pilot team performance. *Ergonomics*, 49(12, 13), 1226–1237, 2006.

2.6.10 Team Effectiveness Measure

General description. Kennedy (2002) developed a Team Effectiveness Measure (TEF) calculated from financial information. TEF is the "ratio of a team's total projects' benefits divided by total cost of implementing and maintain the project" (p. 28). Benefits include increased revenue (annualized) and incremental savings in material, labor, overhead, and other (also annualized). Costs include equipment, material, labor, utilities, and other (also annualized).

In the same year another team effectiveness measure was developed by Hexmoor and Beavers. The Hexmoor and Beavers (2002) measure was based on the following: (1) efficiency (how much resources are used to achieve goal), (2) influence (how team members affect the performance of other team members), (3) dependence (how much the ability of a team member is affected by the performance of other team members), and (4) redundancy (duplication by two or more team members). All measures are provided in percentages.

Strengths and limitations. TEF was designed to provide an objective measure to make comparisons across teams. Kennedy (2002) applied the measure when comparing 68 teams from two service and five manufacturing companies. Teams varied in size, stage of development, and type. TEF was applied by these diverse teams. A longitudinal study was recommended to further evaluate the utility of these measures of team performance.

The second measure of team effectiveness has not been validated as yet.

Data requirements. Annualized financial data including revenue, material, labor, utilizes, and other costs.

Thresholds. Not stated.

SOURCES

Hexmoor, H. and Beavers, G. Measuring team effectiveness. *Proceedings of the International Symposium on Artificial Intelligence and Applications International Conference Applied Infomatics.* 351–393, 2002.

Kennedy, F.A. *Team performance: Using financial measures to evaluate the influence of support systems on team performance.* University of North Texas: Dissertation, May 2002.

2.6.11 Team Knowledge Measures

General description. Cooke, Kiekel, Salas, Stout, Bowers, and Canon-Bowers (2003) identified four team knowledge measures: overall accuracy, positional accuracy, accuracy of knowledge of other team members' tasks, and intrateam similarity.

Strengths and limitations. Cooke, Kiekel, Salas, Stout, Bowers, and Canon-Bowers (2003) reported that hands-on cross-training of tasks of other team members resulted in significantly better team knowledge than exposure to a conceptual description of the tasks of other team members. The data were collected on 36 three-person teams of undergraduate students. Each team had a dedicated intelligence officer, navigation officer, and pilot. The task was a navy helicopter mission.

Data requirements. Knowledge of components of tasks performed by each member of a team.

Thresholds. Not stated.

SOURCE

Cooke, N.J., Kiekel, P.A., Salas, E., Stout, R., Bowers, C., and Canon-Bowers, J. Measuring team knowledge: A window to the cognitive underpinnings of team performance. *Group Dynamics: Theory, Research, and Practice.* 7(3), 179–199, 2003.

2.6.12 Temkin–Greener, Gross, Kunitz, and Mukamel Model of Team Performance

General description. The Temkin–Greener, Gross, Kunitz, and Mukamel Model of Team Performance is presented in figure 4. This model was developed in a long-term patient care application.

Strengths and limitations. The model was tested by survey of 26 Program of All-Inclusive Care for the Elderly (PACE) programs covering 1860 part-time and full-time employees. The survey included 9 items on leadership, 10 each on communication and conflict management, 6 on coordination, and 7 each on team cohesion and perceived unit effectiveness.

The response rate was 65%. The internal consistency of responses was +0.7. The authors argued that construct validity was demonstrated by team process variables

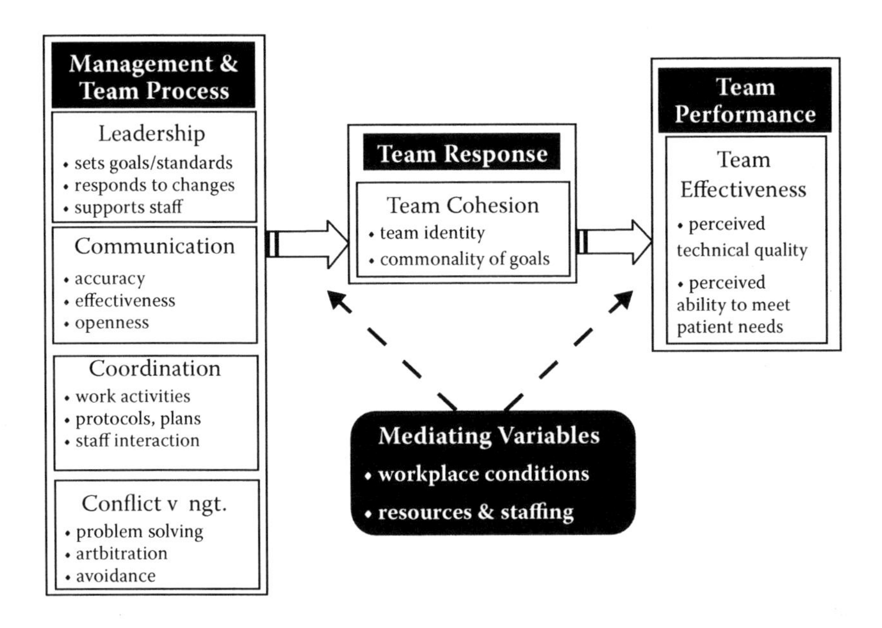

FIGURE 4 The Temkin-Greener, Gross, Kunitz, and Mukamel Model of Team Performance. (From Temkin-Greener, H., Gross, D., Kunitz, S.J., and Mukamel, D. Measuring interdisciplinary team performance in a long-term care setting. *Medical Care*. 42(5), 473, 2004. With permission.)

accounting for 55% of the team cohesion portion and 52% of the team effectiveness of the model. There were also statistically significant differences between professional and paraprofessionals.

Data requirements. Ratings by team members.

Thresholds. Not stated.

SOURCE

Temkin-Greener, H., Gross, D., Kunitz, S.J., and Mukamel, D. Measuring interdisciplinary team performance in a long-term care setting. *Medical Care*. 42(5), 472–481, 2004.

2.6.13 Uninhabited Aerial Vehicle Team Performance Score

General description. Cooke, Shope, and Kiekel (2001) identified the following measures for assessing performance of teams flying reconnaissance uninhabited aerial vehicles (UAVs): film and fuel used, number and type of photographic errors, route deviations, time spent in warning and alarm states, and waypoints visited. They generated a composite team score by subtracting from a starting team score of 1000 for film and fuel used, unphotographed targets, seconds in alarms state, and unvisited critical waypoints. Additional measures were related to communication and included number of communication episodes and length of each episode. Finally, the authors presented pairs of tasks and asked the subjects to rate the relatedness of concepts.

Strengths and limitations. Cooke, Shope, and Kiekel (2001) used the data from 11 teams of Air Force Reserve Officer Training Corps cadets for the first study and 18 teams of similar cadets for the second study. All teams performed a UAV reconnaissance task. Relatedness ratings were the best prediction of team performance.

Data requirements. Data needed include film and fuel used, unphotographed targets, seconds in alarms state, and unvisited critical waypoints. Also, ratings of concept relatedness for task-relevant concepts.

Thresholds. Not stated.

SOURCE

Cooke, N.J., Shope, S.M., and Kiekel, P.A. *Shared-knowledge and team performance: A cognitive engineering approach to measurement (AFRL-SB-BL-TR-01-0370).* Arlington, Virginia: Air Force Office of Scientific Research, March 29, 2001.

3 Human Workload

Workload has been defined as a set of task demands, as effort, and as activity or accomplishment (Gartner and Murphy 1979). The task demands (task load) are the goals to be achieved: the time allowed to perform the task, and the performance level to which the task is to be completed. The factors affecting the effort expended are the information and equipment provided the task environment, the subject's skills and experience, the strategies adopted, and the emotional response to the situation. These definitions provide a testable link between task load and workload. For example (paraphrased from an example given by Azad Madni, Vice President, Perceptronics, Woodland Hills, California, on results of an army study), the workload of a helicopter pilot in maintaining a constant hover may be 70 on a scale of 0 to 100. Given the task of maintaining a constant hover and targeting a tank, workload may again be 70. The discrepancy results from the pilot imposing a strict performance requirement on hover-only (no horizontal or vertical movement), but relaxing the performance requirement on hover (movement of a few feet) when targeting the tank to keep workload within a management level. These definitions enable task load and workload to be explained in real-world situations. These definitions also enable logically correct analysis of task load and workload.

Realistically speaking, workload can never exceed 100% (a person cannot do the impossible). Any theories or reported results that allow workload to exceed 100% are not realistic. However, as defined, task load may exceed 100%. An example is measuring "time required/time available." According to the proposed definition, this task load measurement may exceed 100% if the performance requirements are set too high (thereby increasing the time required) or the time available is set too low. In summary, workload cannot exceed 100%, even if task load does exceed 100%.

Workload has been measured as stand-alone performance (see section 3.1) or secondary task performance (section 3.2) or as subjective estimates (see section 3.3) or in digital simulation (see section 3.4). Physiological measures of workload have also been identified but are not discussed here. An excellent reference on physiological measures of workload is presented in Caldwell, Wilson, and Cetinguc (1994). Dissociation between workload and performance is discussed in section 3.5.

Guidelines for selecting the appropriate workload measure are given in Wierwille, Williges, and Schiflett (1979) and O'Donnell and Eggemeier (1986); for mental workload in Moray (1982). Wierwille and Eggemeier (1993) listed four aspects of measures that were critical: diagnosticity, global sensitivity, transferability, and implementation requirements. A general guide is presented in figure 5.

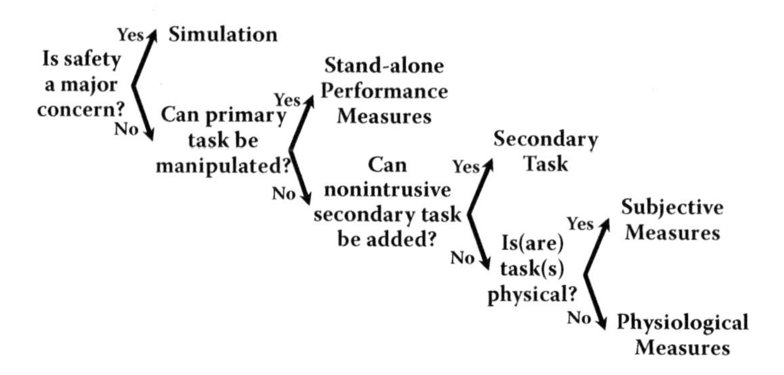

FIGURE 5 Guide for selecting a workload measure.

SOURCES

Caldwell, J.A., Wilson, G.F., and Cetinguc, M. *Psychophysiological Assessment Methods (AGARD-AR-324).* Neuilly-Sur-Seine, France: Advisory Group For Aerospace Research and Development, May 1994.

Gartner, W.B. and Murphy, M.R. Concepts of workload. In Hartman, B.O. and McKenzie, R.E. (Eds.) *Survey of methods to assess workload.* AGARD-AG-246, 1979.

Moray, N. Subjective mental workload. *Human Factors.* 24(1), 25–40, 1982.

O'Donnell, R.D. and Eggemeier, F.T. Workload assessment methodology. In Boff, K.R., Kaufman, L., and Thomas, J.P. (Eds.) *Handbook of perception and human performance.* New York, NY: John Wiley and Sons, 1986.

Wierwille, W.W., and Eggemeier, F.T. Recommendations for mental workload measurement in a test and evaluation environment. *Human Factors.* 35(2), 263–281, 1993.

Wierwille, W.W., Williges, R.C., and Schiflett, S.G. Aircrew workload assessment techniques. In Hartman, B.O. and McKenzie, R.E. (Eds.) *Survey of methods to assess workload.* AGARD-AG-246, 1979.

3.1 STAND-ALONE PERFORMANCE MEASURES OF WORKLOAD

Performance has been used to measure workload. These measures assume that as workload increases, the additional processing requirements will degrade performance. O'Donnell and Eggemeier (1986) identified four problems associated with using performance as a measure of workload: (1) underload may enhance performance, (2) overload may result in a floor effect, (3) confounding effects of information-processing strategy, training, or experience, and (4) measures are task specific and cannot be generalized to other tasks. Meshkati, Hancock, and Rahimi (1990) stated that multiple task measures are required when the task is complex or multidimensional. In addition, task measures may be intrusive and may be influenced by factors other than workload, for example, motivation and learning.

Stand-alone measures of performance include aircrew workload assessment (section 3.1.1), control movement per unit time (section 3.1.2), glance duration and frequency (section 3.1.3), load stress (section 3.1.4), observational workload area (section 3.1.5), rate of gain of information (3.1.6), relative condition efficiency (3.1.7), and speed stress (section 3.1.8).

SOURCES

Meshkati, N., Hancock, P.A., and Rahimi, M. Techniques in mental workload assessment. In Wilson, J.R., and Corlett, E.N. (Eds.) *Evaluation of human work: A practical ergonomics methodology.* New York: Taylor and Francis, 1990.

O'Donnell, R.D. and Eggemeier, F.T. Workload assessment methodology. In Boff, K.R., Kaufman, L., and Thomas, J.P. (Eds.) *Handbook of perception and human performance.* New York: Wiley and Sons, 1986.

3.1.1 Aircrew Workload Assessment System

General description. The Aircrew Workload Assessment System (AWAS) is a timeline analysis software developed by British Aerospace to predict workload. AWAS requires three inputs: (1) second-by-second description of pilot tasks during flight, (2) demands on each of Wicken's multiple resource theory-processing channels, and (3) effects of simultaneous demand on a single channel (Davies, Tomoszek, Hicks, and White [1995]).

Strengths and limitations. Davies et al. (1995) reported a correlation of +0.904 between AWAS workload prediction and errors in a secondary auditory discrimination task. The subjects were two experienced pilots flying a Sea Warrior Simulator.

Data requirements. Second-by-second timeline of pilot tasks, demands on each information-processing channel, and effect of simultaneous demand.

Thresholds. Not stated.

SOURCE

Davies, A.K., Tomoszek, A., Hicks, M.R., and White, J. AWAS (Aircrew Workload Assessment System): Issues of theory, implementation, and validation. In Fuller, R., Johnston, N., and McDonald, N. (Eds.) Human factors in aviation operations. *Proceedings of the 21st Conference of the European Association for Aviation Psychology (EAAP), Vol. 3,* Chapter 48, 1995.

3.1.2 Control Movements per Unit Time

General description. Control movements per unit time is the number of control inputs made summed over each control used by one operator divided by the unit of time over which the measurements were made.

Strengths and limitations. Wierwille and Connor (1983) stated that this measure was completely sensitive to workload. Their specific measure was the average count per second of inputs into the flight controls (ailerons, elevator, and rudder) in a moving-base, flight simulator. The workload manipulation was pitch stability, wind-gust disturbance, and crosswind direction and velocity.

Porterfield (1997), using a similar approach, evaluated the use of the duration of time that an en route air traffic controller was engaged in ground-to-air communications as a measure of workload. He reported a significant correlation (+0.88) between the duration and the Air Traffic Workload Input Technique (ATWIT), a workload rating based on Pilot Objective/Subjective Workload Assessment Technique (POSWAT).

Zeitlin (1995) developed a driver workload index based on brake actuations per minute plus the log of vehicle speed. This index was sensitive to differences in roadway (rural, city, and expressway).

Data requirements. Control movements must be well defined.

Thresholds. Not stated.

SOURCES

Porterfield, D.H. Evaluating controller communication time as a measure of workload. *International Journal of Aviation Psychology.* 7(2), 171–182, 1997.

Wierwille, W.W. and Connor, S.A. Evaluation of 20 workload measures using a psychomotor task in a moving-base aircraft simulator. *Human Factors.* 25(1), 1–16, 1983.

Zeitlin, L.R. Estimates of driver mental workload: A long term field trial of two subsidiary tasks. *Human Factors.* 37(3), 611–621, 1995.

3.1.3 Glance Duration and Frequency

General description. The duration and frequency of glances to visual displays have been used as measures of visual workload. The longer the durations and/or the greater the frequency of glances, the higher the visual workload.

Strengths and limitations. Fairclough, Ashby, and Parkes (1993) used glance duration to calculate the percentage of time that drivers looked at navigation information (a paper map versus an LCD text display), roadway ahead, rear-view mirror, dashboard, left-wing mirror, right-wing mirror, left window, and right window. Data were collected in an instrumented vehicle driven on British roads. The authors concluded that this "measure proved sensitive enough to (a) differentiate between the paper map and the LCD text display and (b) detect associated changes with regard to other areas of the visual scene" (p. 248). These authors warned, however, that reduction in glance durations might reflect the drivers' strategy to cope with the amount and legibility of the paper map.

These authors also used glance duration and frequency to compare two in-vehicle route guidance systems. The data were collected from 23 subjects driving an instrumented vehicle in Germany. The data indicate that "as glance frequency to the navigation display increases, the number of glances to the dashboard, rear-view mirror, and the left-wing mirror all show a significant decrease" (p. 251). Based on these results, the authors concluded that "glance duration appears to be more sensitive to the difficulty of information update." Glance frequency represents the amount of "visual checking behavior" (p. 251).

Wierwille (1993) concluded from a review of driver visual behavior that such behavior is "relatively consistent" (p. 278).

Data requirements. Record subject's eye position.

Threshold. Zero to infinity.

SOURCES

Fairclough, S.H., Ashby, M.C., and Parkes, A.M. In-vehicle displays, visual workload and visibility evaluation. In Gale, A.G., Brown, I.D., Haslegrave, C.M., Kruysse, H.W., and Taylor, S.P. (Eds.). *Vision in Vehicles—IV.* Amsterdam: North-Holland, 1993.

Wierwille, W.W. An initial model of visual sampling of in-car displays and controls. In Gale, A.G., Brown, I.D., Haslegrave, C.M., Kruysse, H.W., and Taylor, S.P. (Eds.) *Vision in Vehicles—IV.* Amsterdam: North-Holland, 1993.

3.1.4 Load Stress

General description. Load stress is the stress produced by increasing the number of signal sources that must be attended to during a task (Chiles and Alluisi, 1979).

Strengths and limitations. Load stress affects the number of errors made in task performance. Increasing load stress to measure operator workload may be difficult in nonlaboratory settings.

Data requirements. The signal sources must be unambiguously defined.

Thresholds. Not stated.

SOURCE

Chiles, W.D. and Alluisi, E.A. On the specification of operator or occupational workload with performance-measurement methods. *Human Factors.* 21(5), 515–528, 1979.

3.1.5 Observed Workload Area

General description. Laudeman and Palmer (1995) developed Observed Workload Area to measure workload in aircraft cockpits. The measure is not based on theory but, rather, on a logical connection between workload and task constraints. In their words:

> An objectively defined window of opportunity exists for each task in the cockpit. The observed workload of a task in the cockpit can be operationalized as a rating of maximum task importance supplied by a domain expert. Task importance increases during the task window of opportunity as a linear function of task importance versus time. When task windows of opportunity overlap, resulting in an overlap of task functions, the task functions can be combined in an additive manner to produce a composite function that includes the observed workload effects of two or more task functions. We called these composites of two or more task functions observed-workload functions. The dependent measure that we proposed to extract from the observed-workload function was the area under it that we called observed-workload area (pp. 188–190).

Strengths and limitations. Laudeman and Palmer (1995) reported a significant correlation between the first officers' workload-management ratings and the observed-workload area. This correlation was based on 18 two-person aircrews flying a high-fidelity aircraft simulator. Small observed-workload area was associated with high-workload management ratings. Higher-error-rate crews had higher observed-workload areas. The technique requires an expert to provide task importance ratings. It also requires well-defined beginnings and ends to tasks.

Thresholds. Not stated.

SOURCE

Laudeman, I.V. and Palmer, E.A. Quantitative measurement of observed workload in the analysis of aircrew performance. *International Journal of Aviation Psychology.* 5(2), 187–197, 1995.

3.1.6 Rate of Gain of Information

General description. This measure is based on Hick's law, which states that *RT* is a linear function of the amount of information transmitted, (Ht): $RT = a + B\ (H_t)$ (Chiles and Alluisi, 1979).

Strengths and limitations. Hick's law has been verified in a complete range of conditions. However, it is limited to only discrete tasks and, unless the task is part of the normal procedures, may be intrusive, especially in nonlaboratory settings.

Data requirements. Rate of gain of information is estimated from RT. Time is typically collected with either mechanical stop watches or software clocks. The first type of clock requires frequent (for example, before every trial) calibration; software clocks require a stable and constant source of power.

Thresholds. Not stated.

SOURCE

Chiles, W.D. and Alluisi, E.A. On the specification of operator or occupational workload with performance-measurement methods. *Human Factors.* 21(5), 515–528, 1979.

3.1.7 Relative Condition Efficiency

General description. Paas and van Merrienboer (1993) combined ratings of workload with task performance measures to calculate relative condition efficiency. Ratings varied from 1 (very, very low mental effort) to 9 (very, very high mental effort). Performance was measured as percent correct answers to test questions. Relative condition efficiency was calculated "as the perpendicular distance to the line that is assumed to represent an efficiency of zero" (p. 737).

Strengths and limitations. Efficiency scores were significantly different between work conditions.

Data requirements. Not stated.

Thresholds. Not stated.

SOURCE

Paas, F.G.W.C. and van Merrienboer, J.J.G. The efficiency of instructional conditions: An approach to combine mental effort and performance measures. *Human Factors.* 35(4), 737–743, 1993.

3.1.8 Speed Stress

General description. Speed stress is stress produced by increasing the rate of signal presentation from one or more signal sources.

Strengths and limitations. Speed stress affects the number of errors made as well as the time to complete tasks (Conrad, 1956; Knowles, Garvey, and Newlin, 1953). It may be difficult to impose speed stress on nonlaboratory tasks.

Data requirements. The task must include discrete signals whose presentation rate can be manipulated.

Thresholds. Not stated.

SOURCES

Conrad, R. The timing of signals in skill. *Journal of Experimental Psychology*. 51, 365–370, 1956.
Knowles, W.B., Garvey, W.D., and Newlin, E.P. The effect of speed and load on display-control relationships. *Journal of Experimental Psychology*. 46, 65–75, 1953.

3.1.9 Task Difficulty Index

General description. The Task Difficulty Index was developed by Wickens and Yeh (1985) to categorize the workload associated with typical laboratory tasks. The index has four dimensions:

1. Familiarity of stimuli:
   0 = letters
   1 = spatial dot patterns, tracking cursor
2. Number of concurrent tasks:
   0 = single
   1 = dual
3. Task difficulty:
   0 = memory set size 2
   1 = set size 4, second-order tracking, delayed recall
4. Resource competition:
   0 = no competition
   1 = competition for either modality of stimulus (visual, auditory) or central processing (spatial, verbal)" (Gopher and Braune, 1984).

The Task Difficulty Index is the sum of the scores on each of the four dimensions listed in the preceding text.

Strengths and limitations. Gopher and Braune (1984) reported a significant positive correlation (+0.93) between Task Difficulty Index and subjective measures of workload. Their data were based on responses from 55 male subjects performing 21 tasks, including Sternberg, hidden pattern, card rotation tracking, maze tracing, delayed digit recall, and dichotic listening.

Data requirements. This method requires the user to describe the tasks to be performed on the four dimensions given in the preceding text.

Thresholds. Values vary between 0 and 4.

SOURCES

Gopher, D. and Braune, R. On the psychophysics of workload: Why bother with subjective measures? *Human Factors*. 26(5), 519–532, 1984.
Wickens, C.D. and Yeh, Y. POCs and performance decrements: A reply to Kantowitz and Weldon. *Human Factors*, 27, 549–554, 1985.

3.1.10 Time Margin

General description. After a review of current in-flight workload measures, Gawron, Schiflett, and Miller (1989) identified five major deficiencies: (1) the subjective ratings showed wide individual differences well beyond those that could be attributed to experience and ability differences; (2) most of the measures were not comprehensive and assessed only a single dimension of workload; (3) many workload measures were intrusive in terms of requiring task responses or subjective ratings or the use of electrodes; (4) some measures were confusing to subjects under high stress, for example, the meanings of ratings would be forgotten in high-workload environments, so lower than actual values would be given by the pilot; and (5) subjects would misperceive the number of tasks to be performed and provide an erroneously low measure of workload. Gawron then returned to the purpose of workload measure: to identify potentially dangerous situations. Poor designs, inadequate procedures, poor training, or the proximity to catastrophic conditions could induce such situations. The most objective measure of danger in a situation is time until the aircraft is destroyed if control action is not taken. These times include time until impact, time until the aircraft is overstressed and breaks apart, and time until the fuel is depleted.

Strengths and limitations. The time-limit workload measure is quantitative, objective, directly related to performance, and can be tailored to any mission. For example, time until a surface-to-air missile destroys the aircraft is a good measure in air-to-ground penetration missions. In addition, the times can be easily computed from measures of aircraft performance. Finally, these times can be summed over intervals of any length to provide interval-by-interval workload comparisons.

Data requirements. This method is useful whenever aircraft performance data are available.

Thresholds. Minimum is 0 and maximum is infinity.

SOURCE

Gawron, V.J., Schiflett, S.G., and Miller, R.C. Measures of in-flight workload. In Jensen, R.S. (Ed.) *Aviation psychology.* London: Gower, 1989.

3.2 SECONDARY TASK MEASURES OF WORKLOAD

One of the most widely used techniques to measure workload is the secondary task. This technique requires an operator to perform the primary task within that task's specified requirements and to use any spare attention or capacity to perform a secondary task. The decrease in performance of the secondary task is operationally defined as a *measure of workload.*

The secondary-task technique has several advantages. First, it may provide a sensitive measure of operator capacity and may distinguish between alternative equipment configurations that are indistinguishable by single-task performance (Slocum, Williges, and Roscoe, 1971). Second, it may provide a sensitive index of task impairment due to stress. Third, it may provide a common metric for comparisons of different tasks.

The secondary-task technique may have one major disadvantage: intrusion on the performance of the primary task (Williges and Wierwille, 1979) and as Rolfe (1971) stated: "The final word, however, must be that the secondary task is no substitute for competent and comprehensive measurement of primary task performance"

(p. 146). Vidulich (1989a), however, concluded from two experiments that secondary tasks that do not intrude on primary-task performance are insensitive to primary-task difficulty. Vidulich (1989b) argued that added task sensitivity is directly linked to intrusiveness.

Further, subjects may use different strategies when performing secondary tasks. For example, Schneider and Detweiler (1988) identified seven compensatory activities that are associated with dual-task performance: "(1) shedding and delaying tasks and preloading buffers, (2) letting go of high-workload strategies, (3) utilizing non-competing resources, (4) multiplexing over time, (5) shortening transmissions, (6) converting interference from concurrent transmissions, and (7) chunking of transmissions" (p. 539). In addition, Meshkati, Hancock, and Rahimi (1990) recommend not using secondary task and subjective measures in the same experiment, because operators may include secondary-task performance as part of their subjective workload rating.

Further, Ogdon, Levine, and Eisner (1979) concluded from a survey of the literature that there was not a single best secondary task for measuring workload. In another review study, Damos (1993) analyzed the results of 14 studies in which single- and dual-task performances were evaluated. She concluded that "the effect sizes associated with both single- and multiple-task measures were both statistically different from 0.0, with the effect size for the multiple-task increases statistically greater than that of the corresponding single task measures. However, the corresponding predictive validities were low" (p. 615).

Poulton (1965) pointed out that comparing the results of performance tests that vary in sensitivity may be difficult. To address this concern, Colle, Amell, Ewry, and Jenkins (1988) developed the method of double trade-off curves to equate performance levels on different secondary tasks. In this method, "two different secondary tasks are each paired with the same primary tasks. A trade-off curve is obtained for each secondary task paired with the primary task" (p. 646).

To help researchers select secondary-task measures of workload, Knowles (1963) developed a comprehensive set of criteria for selecting a secondary task: (1) noninterference with the primary task, (2) ease of learning, (3) self-pacing, (4) continuous scoring, (5) compatibility with the primary task, (6) sensitivity, and (7) representativeness. In a similar vein, Fisk, Derrick, and Schneider (1983) developed three criteria, which they then tested in an experiment. The criteria were the following: (1) the secondary task must use the same resources as the primary task, (2) single- and dual-task performance must be maintained, and (3) the secondary task must require "controlled or effortful processing" (p. 230). Brown (1978) recommended that "the dual task method should be used for the study of individual difference in processing resources available to handle work-load" (p. 224). Also, Liu and Wickens (1987) reported that tasks using the same resources had increased workload more than tasks that did not.

Finally, Wetherell (1981) investigated seven secondary tasks (addition, verbal reasoning, attention, short-term memory, random digit generation, memory search, and white noise) with a primary driving task and concluded that none were "outstanding as a measure of mental workload." However, there was a significant gender difference, with degradations in the primary task occurring for female drivers.

SOURCES

Brown, I.D. Dual task methods of assessing work-load. *Ergonomics*. 21(3), 221–224, 1978.

Colle, H., Amell, J.R., Ewry, M.E., and Jenkins, M.L. Capacity equivalence curves: A double trade-off curve method for equating task performance. *Human Factors*. 30(5), 645–656, 1988.

Damos, D. Using meta-analysis to compare the predictive validity of single- and multiple-task measures of flight performance. *Human Factors*. 35(4), 615–628, 1993.

Fisk, A.D., Derrick, W.L., and Schneider, W. The assessment of workload: Dual task methodology. *Proceedings of the Human Factors Society 27th Annual Meeting*. 229–233, 1983.

Knowles, W.B. Operator loading tasks. *Human Factors*. 5, 151–161, 1963.

Liu, Y. and Wickens, C.D. The effect of processing code, response modality and task difficulty on dual task performance and subjective workload in a manual system. *Proceedings of the Human Factors Society 31st Annual Meeting*. 2, 847–851, 1987.

Meshkati, N., Hancock, P.A., and Rahimi, M. Techniques in mental workload assessment. In Wilson, J.R. and Corlett, E.N. (Eds.) *Evaluation of a human work: A practical ergonomics methodology*. New York: Taylor and Francis, 1990.

Ogdon, G.D., Levine, J.M., and Eisner, E.J. Measurement of workload by secondary tasks. *Human Factors*. 21(5), 529–548, 1979.

Poulton, E.C. On increasing the sensitivity of measures of performance. *Ergonomics*. 8(1), 69–76, 1965.

Rolfe, J.M. The secondary task as a measure of mental load. In Singleton, W.T., Fox, J.G., and Whitfield, D. (Eds.) *Measurement of man at work*. London: Taylor and Francis Ltd, 1971.

Schneider, W. and Detweiler, M. The role of practice in dual-task performance: Toward workload modeling in a connectionist/control architecture. *Human Factors*. 30(5), 539–566, 1988.

Slocum, G.K., Williges, B.H., and Roscoe, S.N. Meaningful shape coding for aircraft switch knobs. *Aviation Research Monographs*. 1(3), 27–40, 1971.

Vidulich, M.A. Objective measures of workload: Should a secondary task be secondary? *Proceedings of the 5th International Symposium on Aviation Psychology*. 802–807, 1989a.

Vidulich, M.A. Performance-based workload assessment: Allocation strategy and added task sensitivity. *Proceedings of the 3rd Annual Workshop on Space Operations, Automation, and Robotics (SOAR '89)*. 329–335, 1989b.

Wetherell, A. The efficacy of some auditory-vocal subsidiary tasks as measures of the mental load on male and female drivers. *Ergonomics*. 24(3), 197–214, 1981.

Williges, R.C. and Wierwille, W.W. Behavioral measures of aircrew mental workload. *Human Factors*. 21, 549–574, 1979.

3.2.1 Card-Sorting Secondary Task

General description. "The subject must sort playing cards by number, color, and/or suit." (Lysaght et al., 1989, p. 234).

Strengths and limitations. "Depending upon the requirements of the card sorting rule, the task can impose demands on perceptual and cognitive processes" (Lysaght et al., 1989, p. 234). Lysaght et al. (1989) state that dual-task pairing of a primary memory task with a secondary card-sorting task resulted in a decrease in performance in both tasks. Their statement is based on two experiments by Murdock (1965). Although used as a primary task, Courtney and Shou (1985) concluded that card sorting was a "rapid and simple means of estimating relative visual-lobe size" (p. 1319).

Data requirements. The experimenter must be able to record the number of cards sorted and the number of incorrect responses.

Thresholds. Not stated.

SOURCES

Courtney, A.J. and Shou, C.H. Simple measures of visual-lobe size and search performance. *Ergonomics.* 28(9), 1319–1331, 1985.

Lysaght, R.J., Hill, S.G., Dick, A.O., Plamondon, B.D., Linton, P.M., Wierwille, W.W., Zaklad, A.L., Bittner, A.C., and Wherry, R.J. *Operator Workload: Comprehensive review and evaluation of operator workload methodologies (Technical Report 851).* Alexandria, VA: Army Research Institute for the Behavioral and Social Sciences, June 1989.

Murdock, B.B. Effects of a subsidiary task on short-term memory. *British Journal of Psychology.* 56, 413–419, 1965.

3.2.2 Choice RT Secondary Task

General description. "The subject is presented with more than one stimulus and must generate a different response for each one" (Lysaght et al., 1989, p. 232).

Strengths and limitations. "Visual or auditory stimuli may be employed and the response mode is usually manual. It is theorized that choice RT imposes both central processing and response selection demands" (Lysaght et al., 1989, p. 232).

On the basis of 19 studies that included a choice RT secondary task, Lysaght et al. (1989) reported the following for dual-task pairings: performance of choice RT, problem solving, and flight simulation primary tasks remained stable; performance of tracking, choice RT, memory, monitoring, driving, and lexical decision primary tasks degraded; and tracking performance improved. Performance of the secondary task remained stable with tracking and driving primary tasks; and degraded with tracking, choice RT, memory, monitoring, problem-solving, flight simulation, driving, and lexical decision primary tasks (see table 4).

Hicks and Wierwille (1979) compared five measures of workload. They manipulated workload by increasing wind gust in a driving simulator. They reported that a secondary RT task was not as sensitive to wind gust as were steering reversals, yaw deviation, subjective opinion rating scales, and lateral deviations. Gawron (1982) reported longer RTs and lower percent correct scores when a four-choice RT task was performed simultaneously then sequentially.

Klapp, Kelly, and Netick (1987) asked subjects to perform a visual, zero-order, pursuit tracking task with the right hand while performing a two-choice, auditory reaction task with the left hand. In the dual-task condition, the tracking task was associated with hesitations lasting 333 ms or longer. Degradations in the tracking task were associated with enhancements of the RT task.

Data requirements. The experimenter must be able to record and calculate the following: mean RT for correct responses, mean (median) RT for incorrect responses, number of correct responses, and number of incorrect responses.

Thresholds. Not stated.

TABLE 4

References Listed by the Effect on Performance of Primary Tasks Paired with a Secondary Choice RT Task

	Primary Task			Secondary Task		
Type	**Stable**	**Degraded**	**Enhanced**	**Stable**	**Degraded**	**Enhanced**
Choice RT	Becker (1976) Ellis (1973)	Detweiler and Lundy (1995)* Gawron (1982)* Schvaneveldt (1969)		Hicks and Wierwille (1979)	Becker (1976) Detweiler and Lundy (1995)* Ellis (1973) Gawron (1982)* Schvaneveldt (1969)	
Driving	Kantowitz (1995)	Allen, Jex, McRuer, and DiMarco (1976) Brown, Tickner, and Simmonds (1969)		Allen et al. (1976) Drory (1985)*	Brown et al. (1969)	
Flight simulation	Bortolussi, Hart, and Shively (1987) Bortolussi, Kantowitz, and Hart (1986) Kantowitz, Hart and Bortolussi (1983)* Kantowitz, Hart, Bortolussi, Shively, and Kantowitz (1984)* Kantowitz, Bortolussi, and Hart (1987)				Bortolussi et al. (1987) Bortolussi et al. (1986)	
Lexical decision		Becker (1976)			Becker (1976)	

Memory		Logan (1970)		Logan (1970)	
Monitoring		Smith (1969)		Krol (1971) Smith (1969)	
Problem solving	Fisher (1975a) Fisher (1975b)			Fisher (1975a) Fisher (1975b)	
Tracking		Benson, Huddleston, and Rolfe (1965) Loeb and Jones (1978) Giroud, Laurencelle, and Proteau (1984) Israel, Chesney, Wickens, and Donchin (1980) Israel, Wickens, Chesney, and Donchin (1980) Klapp, Kelly, Battiste, and Dunbar (1984) Wempe and Baty (1968) Klapp, Kelly, and Netick (1987)*	Loeb and Jones (1978)	Benson et al. (1965) Damos (1978) Giroud et al. (1984) Israel et al. (1980) Klapp et al. (1984)	Klapp, Kelly, and Netick (1987)*

* Not included in Lysaght, R.J., Hill, S.G., Dick, A.O., Plamondon, B.D., Linton, P.M., Wierwille, W.W., Zaklad, A.L., Bittner, A.C., and Wherry, R.J. *Operator workload: Comprehensive review and evaluation of operator workload methodologies (Technical Report 851)*. Alexandria, VA: Army Research Institute for the Behavioral and Social Sciences; June 1989.

Source: From Lysaght, R.J., Hill, S.G., Dick, A.O., Plamondon, B.D., Linton, P.M., Wierwille, W.W., Zaklad, A.L., Bittner, A.C., and Wherry, R.J. *Operator workload: Comprehensive review and evaluation of operator workload methodologies (Technical Report 851)*. Alexandria, VA: Army Research Institute for the Behavioral and Social Sciences; June 1989, p. 246.

SOURCES

Allen, R.W., Jex, H.R., McRuer, D.T., and DiMarco, R.J. Alcohol effects on driving behavior and performance in a car simulator. *IEEE Transactions on Systems Man and Cybernetics*. SMC-5, 485–505, 1976.

Becker, C.A. Allocation of attention during visual word recognition. *Journal of Experimental Psychology: Human Perception and Performance*. 2, 556–566, 1976.

Benson, A.J., Huddleston, J.H.F., and Rolfe, J.M. A psychophysiological study of compensatory tracking on a digital display. *Human Factors*. 7, 457–472, 1965.

Bortolussi, M.R., Hart, S.G., and Shively, R.J. Measuring moment-to-moment pilot workload using synchronous presentations of secondary tasks in a motion-base trainer. *Proceedings of the 4th Symposium on Aviation Psychology*. Columbus, OH: Ohio State University, 1987. Also published in *Aviation, Space, and Environmental Medicine*. 60(2), 124–129, 1989.

Bortolussi, M.R., Kantowitz, B.H., and Hart, S.G. Measuring pilot workload in a motion base trainer. *Applied Ergonomics*. 17, 278–283, 1986.

Bortolussi, M.R., Kantowitz, B.H., and Hart, S.G. Measuring pilot workload in a motion base trainer: A comparison of four techniques. *Proceedings of the 3rd Symposium on Aviation Psychology*. 263–270, 1985.

Brown, I.D., Tickner, A.H., and Simmonds, D.C.V. Interference between concurrent tasks of driving and telephoning. *Journal of Applied Psychology*. 53, 419–424, 1969.

Damos, D. Residual attention as a predictor of pilot performance. *Human Factors*. 20, 435–440, 1978.

Detweiler, M. and Lundy, D.H. Effects of single-and dual-task practice an acquiring dual-task skill. *Human Factors*. 37(1), 193–211, 1995.

Drory, A. Effects of rest and secondary task on simulated truck-driving performance. *Human Factors*. 27(2), 201–207, 1985.

Ellis, J.E. Analysis of temporal and attentional aspects of movement control. *Journal of Experimental Psychology*. 99, 10–21, 1973.

Fisher, S. The microstructure of dual task interaction. 1. The patterning of main-task responses within secondary-task intervals. *Perception*. 4, 267–290, 1975a.

Fisher, S. The microstructure of dual task interaction. 2. The effect of task instructions on attentional allocation and a model of attentional-switching. *Perception*. 4, 459–474, 1975b.

Gawron, V.J. Performance effects of noise intensity, psychological set, and task type and complexity. *Human Factors*. 24(2), 225–243, 1982.

Giroud, Y., Laurencelle, L., and Proteau, L. On the nature of the probe reaction-time task to uncover the attentional demands of movement. *Journal of Motor Behavior*. 16, 442–459, 1984.

Hicks, T.G. and Wierwille, W.W. Comparison of five mental workload assessment procedures in a moving-base during simulator. *Human Factors*. 21, 129–143, 1979.

Israel, J.B., Chesney, G.L., Wickens, C.D., and Donchin, E. P300 and tracking difficulty: Evidence for multiple resources in dual-task performance. *Psychophysiology*. 17, 259–273, 1980.

Israel, J.B., Wickens, C.D., Chesney, G.L., and Donchin, E. The event related brain potential as an index of display-monitoring workload. *Human Factors*. 22, 211–224, 1980.

Kantowitz, B.H. Simulator evaluation of heavy-vehicle driver workload. *Proceedings of the Human Factors and Ergonomics Society 39th Annual Meeting*. 2: 1107–1111, 1995.

Kantowitz, B. H., Bortolussi, M.R., and Hart, S.G. Measuring pilot workload in a motion base simulator: III. Synchronous secondary task. *Proceedings of the Human Factors Society*. 2: 834–837, 1987.

Kantowitz, B.H., Hart, S.G., and Bortolussi, M.R. Measuring pilot workload in a moving-base simulator I. Asynchronous secondary choice-reaction time task. *Proceedings of the 27th Annual Meeting of the Human Factors Society*. Santa Monica, CA: Human Factors Society, 1983.

Kantowitz, B.H., Hart, S.G., Bortolussi, M.R., Shively, R.J., and Kantowitz, S.C. *Measuring pilot workload in a moving-base simulator II. Building levels of workload.* NASA 20th Annual Conference on Manual Control. 2, 373–396, 1984.

Klapp, S.T., Kelly, P.A., Battiste, V., and Dunbar, S. Types of tracking errors induced by concurrent secondary manual task. *Proceedings of the 20th Annual Conference on Manual Control* (pp. 299–304). Moffett Field, CA: Ames Research Center, 1984.

Klapp, S.T., Kelly, P.A., and Netick, A. Hesitations in continuous tracking induced by a concurrent discrete task. *Human Factors*. 29(3), 327–337, 1987.

Krol, J.P. Variations in ATC-workload as a function of variations in cockpit workload. *Ergonomics*. 14, 585–590, 1971.

Loeb, M. and Jones, P.D. Noise exposure, monitoring and tracking performance as a function of signal bias and task priority. *Ergonomics*. 21(4): 265–272, 1978.

Logan, G.D. On the use of a concurrent memory load to measure attention and automaticity. *Journal of Experimental Psychology: Human Perception and Performance*. 5, 189–207, 1970

Lysaght, R.J., Hill, S.G., Dick, A.O., Plamondon, B.D., Linton, P.M., Wierwille, W.W., Zaklad, A.L., Bittner, A.C., and Wherry, R.J. *Operator workload: Comprehensive review and evaluation of operator workload methodologies (Technical Report 851).* Alexandria, VA: Army Research Institute for the Behavioral and Social Sciences, June 1989.

Schvaneveldt, R.W. (1969). Effects of complexity in simultaneous reaction time tasks. *Journal of Experimental Psychology*. 81, 289–296, 1969.

Smith, M.C. Effect of varying channel capacity on stimulus detection and discrimination. *Journal of Experimental Psychology*. 82, 520–526, 1969.

Wempe, T.E. and Baty, D.L. Human information processing rates during certain multiaxis tracking tasks with a concurrent auditory task. *IEEE Transactions on Man-Machine Systems*. 9, 129–138, 1968.

3.2.3 Classification Secondary Task

General description. "The subject must judge whether symbol pairs are identical in form. For example, to match letters either on a physical level (AA) or on a name level (Aa)" (Lysaght et al., 1989, p. 233), or property (pepper is hot), or superset relation (an apple is a fruit). Cognitive processing requirements are discussed by Miller (1975).

Strengths and limitations. "Depending upon the requirements of the matching task, the task can impose demands on perceptual processes (physical match) and/or cognitive processes (name match or category match)" (Lysaght et al., 1989, p. 233).

Damos (1985) reported that percent correct scores for both single- and dual-task performance were affected only by trial and not by either behavior pattern or pacing condition. Correct RT scores, however, were significantly related to trial and pacing by behavior pattern in the single-task condition and trial, behavior pattern, trial by pacing, and trial by behavior pattern in the dual-task condition.

Mastroianni and Schopper (1986) reported that performance on a secondary auditory classification task (easiest version—tone low or high, middle version—previous tone low or high, hardest version—tone before the previous tone low or high) was degraded as the difficulty of the task increased or the amount of force needed on the primary pursuit tracking task increased. Performance on the primary task degraded when the secondary task was present compared to when it was absent.

Carter, Krause, and Harbeson (1986) reported that RT increased as the number of memory steps to verify a sentence increased. Slope was not a reliable measure of performance.

Beer, Gallaway, and Previc (1996) reported that performance on an aircraft classification task in singular task mode did not predict performance in dual-task mode.

Data requirements. The following data are used to assess performance of this task: mean RT for physical match, mean RT for category match, number of errors for physical match, and number of errors for category match (Lysaght et al., 1989, p. 236).

Thresholds. Kobus, Russotti, Schlichting, Haskell, Carpenter, and Wojtowicz (1986) reported the following times to correctly classify one of five targets: visual = 224.6 s, auditory = 189.6 s, and multimodal (i.e., both visual and auditory) = 212.7 s. These conditions were not significantly different.

SOURCES

Beer, M.A. Gallaway, R.A., and Previc, R.H. Do individuals' visual recognition thresholds predict performance on concurrent attitude control flight tasks? *International Journal of Aviation Psychology*. 6(3), 273–297, 1996.

Carter, R.C., Krause, M., and Harbeson, M.M. Beware the reliability of slope scores for individuals. *Human Factors*. 28(6), 673–683, 1986.

Damos, D. The relation between the Type A behavior pattern, pacing, and subjective workload under single- and dual-task conditions. *Human Factors*. 27(6), 675–680, 1985.

Kobus, D.A., Russotti, J., Schlichting, C., Haskell, G., Carpenter, S., and Wojtowicz, J. Multimodal detection and recognition performance of sonar operations. *Human Factors*. 28(1), 23–29, 1986.

Lysaght, R.J., Hill, S.G., Dick, A.O., Plamondon, B.D., Linton, P.M., Wierwille, W.W., Zaklad, A.L., Bittner, A.C., and Wherry, R.J. *Operator workload: Comprehensive review and evaluation of operator workload methodologies (Technical Report 851).* Alexandria, VA: Army Research Institute for the Behavioral and Social Sciences; June 1989.

Mastroianni, G.R. and Schopper, A.W. Degradation of force-loaded pursuit tracking performance in a dual-task paradigm. *Ergonomics*. 29(5), 639–647, 1986.

Miller, K. Processing capacity requirements for stimulus encoding. *Acta Psychologica*. 39, 393–410, 1975.

3.2.4 Cross-Adaptive Loading Secondary Task

General description. Cross-adaptive loading tasks are secondary tasks that the subject must perform only while primary-task performance meets or exceeds a previously established performance criterion (Kelly and Wargo, 1967).

Strengths and limitations. Cross-adaptive loading tasks are less likely to degrade performance on the primary task but are intrusive and, as such, difficult to use in nonlaboratory settings.

Data requirements. A well-defined, quantifiable criterion for primary-task performance as well as a method of monitoring this performance and cueing the subject on when to perform the cross-adaptive loading task are all required.

Thresholds. Dependent on type of primary and cross-adaptive loading tasks being used.

SOURCE

Kelly, C.R. and Wargo, M.J. Crossadaptive operator loading tasks. *Human Factors.* 9, 395–404, 1967.

3.2.5 Detection Secondary Task

General description. "The subject must detect a specific stimulus or event which may or may not be presented with alternative events. For example, to detect which of 4 lights is flickering. The subject is usually alerted by a warning signal (e.g., tone) before the occurrence of such events, therefore attention is required intermittently." (Lysaght et al., 1989, p. 233).

Strengths and limitations. "Such tasks are thought to impose demands on perceptual processes" (Lysaght et al., 1989, p. 233).

On the basis of a review of five studies in which detection was a secondary task, Lysaght et al. (1989) reported the following for dual-task pairings: performance of a primary classification task remained stable; and performance of tracking, memory, monitoring, and detection primary tasks degraded. In all cases, performance of the secondary detection task degraded (see table 5).

Data requirements. The following data are calculated for this task: mean RT for correct detections and number of correct detections.

Thresholds. Not stated.

TABLE 5
References Listed by the Effect on Performance of Primary Tasks Paired with a Secondary Detection Task

	Primary Task			Secondary Task		
Type	Stable	Degraded	Enhanced	Stable	Degraded	Enhanced
Detection		Wickens, Mountford, and Schreiner (1981)			Wickens et al. (1981)	
Driving					Verwey (2000)*	
Memory					Shulman and Greenberg (1971)	
Tracking		Wickens et al. (1981)			Wickens et al. (1981)	

* Not included in Lysaght, R.J., Hill, S.G., Dick, A.O., Plamondon, B.D., Linton, P.M., Wierwille, W.W., Zaklad, A.L., Bittner, A.C., and Wherry, R.J. *Operator workload: Comprehensive review and evaluation of operator workload methodologies (Technical Report 851).* Alexandria, VA: Army Research Institute for the Behavioral and Social Sciences; June 1989.

Source: From Lysaght, R.J., Hill, S.G., Dick, A.O., Plamondon, B.D., Linton, P.M., Wierwille, W.W., Zaklad, A.L., Bittner, A.C., and Wherry, R.J. *Operator workload: Comprehensive review and evaluation of operator workload methodologies (Technical Report 851).* Alexandria, VA: Army Research Institute for the Behavioral and Social Sciences; June 1989, p. 246.

SOURCES

Lysaght, R.J., Hill, S.G., Dick, A.O., Plamondon, B.D., Linton, P.M., Wierwille, W.W., Zaklad, A.L., Bittner, A.C., and Wherry, R.J. *Operator workload: Comprehensive review and evaluation of operator workload methodologies (Technical Report 851).* Alexandria, VA: Army Research Institute for the Behavioral and Social Sciences; June 1989.

Shulman, H.G. and Greenberg, S.N. Perceptual deficit due to division of attention between memory and perception. *Journal of Experimental Psychology*, 88, 171–176, 1971.

Verwey, W.B. On-line driver workload estimation. Effects of road situation and age on secondary task measures. *Ergonomics*. 43(2), 187–209, 2000.

Wickens, C.D., Mountford, S.J., and Schreiner, W. Multiple resources, task-hemispheric integrity, and individual differences in time sharing. *Human Factors*. 23, 211–229, 1981.

3.2.6 Distraction Secondary Task

General description. "The subject performs a task which is executed in a fairly automatic way such as counting aloud" (Lysaght et al., 1989, p. 233).

Strengths and limitations. "Such a task is intended to distract the subject to prevent the rehearsal of information that may be needed for the primary task" (Lysaght et al., 1989, p. 233).

Based on one study, Lysaght et al. (1989) reported degraded performance on a memory primary task when paired with a distraction secondary task. Drory (1985) reported significantly shorter brake RTs and fewer steering wheel reversals when a secondary distraction task (i.e., state the last two digits of the current odometer reading) was paired with a basic driving task in a simulator. The secondary distraction task had no effects on tracking error, number of brake responses, or control light response.

Zeitlin (1995) used two auditory secondary tasks (delayed digit recall and random digit generation) while driving on road. Performance on both tasks degraded as traffic density and average speed increased.

Data requirements. Not stated.

Thresholds. Not stated.

SOURCES

Drory, A. Effects of rest and secondary task on simulated truck-driving task performance. *Human Factors*. 27(2), 201–207, 1985.

Lysaght, R.J., Hill, S.G., Dick, A.O., Plamondon, B.D., Linton, P.M., Wierwille, W.W., Zaklad, A.L., Bittner, A.C., and Wherry, R.J. *Operator workload: Comprehensive review and evaluation of operator workload methodologies (Technical Report 851).* Alexandria, VA: Army Research Institute for the Behavioral and Social Sciences; June 1989.

Zeitlin, L.R. Estimates of driver mental workload: A long-term field trial of two subsidiary tasks. *Human Factors*. 37(3), 611–621, 1995.

3.2.7 Driving Secondary Task

General description. "The subject operates a driving simulator or actual motor vehicle" (Lysaght et al., 1989, p. 232).

Strengths and limitations. This "task involves complex psychomotor skills" (Lysaght et al., 1989, p. 232). Johnson and Haygood (1984) varied the difficulty of a primary simulated driving task by varying the road width. The secondary task was a visual choice RT task. Tracking score was highest when the difficulty of the primary task was adapted as a function of primary task performance. It was lowest when the difficulty was fixed.

Brouwer, Waterink, van Wolffelaar, and Rothengatten (1991) reported that older adults (mean age 64.4) were significantly worse than younger adults (mean age 26.1) in dual-task performance of compensatory lane tracking with a timed, self-paced visual analysis task.

Korteling (1994) did not find a significant difference in steering performance between single task (steering) versus dual task (addition of car-following task) between young (21 to 34) and old (65- to 74-year-old) drivers. There was, however, 24% performance deterioration in car-following performance with the addition of a steering task.

Data requirements. "The experimenter should be able to record the following: total time to complete a trial, number of acceleration rate changes, number of gear changes, number of footbrake operations, number of steering reversals, number of obstacles hit, high pass-steering deviation, yaw deviation, and lateral deviation" (Lysaght et al., 1989, p. 235).

Thresholds. Not stated.

SOURCES

Brouwer, W.H., Waterink, W., van Wolffelaar, P.C., and Rothengatten, T. Divided attention in experienced young and older drivers: Lane tracking and visual analysis in a dynamic driving simulator. *Human Factors.* 33(5), 573–582, 1991.

Johnson, D.F. and Haygood, R.C. The use of secondary tasks in adaptive training. *Human Factors.* 26(1), 105–108, 1984.

Korteling, J.E. Effects of aging, skill modification, and demand alternation on multiple-task performance. *Human Factors.* 36(1), 27–43, 1994.

Lysaght, R.J., Hill, S.G., Dick, A.O., Plamondon, B.D., Linton, P.M., Wierwille, W.W., Zaklad, A.L., Bittner, A.C., and Wherry, R.J. *Operator workload: Comprehensive review and evaluation of operator workload methodologies (Technical Report 851).* Alexandria, VA: Army Research Institute for the Behavioral and Social Sciences; June 1989.

3.2.8 Identification/Shadowing Secondary Task

General description. "The subject identifies changing symbols (digits and/or letters) that appear on a visual display by writing or verbalizing, or repeating a spoken passage as it occurs" (Lysaght et al., 1989, p. 233).

Strengths and limitations. "Such tasks are thought to impose demands on perceptual processes (i.e., attention)" (Lysaght et al., 1989, p. 233). Wierwille and Connor (1983), however, reported that a digit-shadowing task was not sensitive to variations in workload. Their task was control on moving-base aircraft simulator. Workload was varied by manipulating pitch-stability level, wind-gust disturbance level, and crosswind velocity and direction.

Savage, Wierwille, and Cordes (1978) evaluated the sensitivity of four dependent measures to workload manipulations (1, 2, 3, or 4 m) to a primary monitoring task. The number of random digits spoken on the secondary task was the most sensitive to workload. The longest consecutive string of spoken digits and the number of triplets spoken were also significantly affected by workload. The longest interval between spoken responses, however, was not sensitive to workload manipulations of the primary task.

Based on nine studies with an identification secondary task, Lysaght et al. (1989) reported that performance remained stable for an identification primary task and degraded for tracking, memory, detection, driving, and spatial transformation primary tasks. Performance of the identification secondary task remained stable for tracking and identification primary tasks and degraded for monitoring, detection, driving, and spatial transformation primary tasks (see table 6).

Data requirements. The following data are used for this task: number of words correct per minute, number of digits spoken, mean time interval between spoken digits, and number of errors of omission (Lysaght et al., 1989, p. 236).

Thresholds. Not stated.

TABLE 6

References Listed by the Effect on Performance of Primary Tasks Paired with a Secondary Identification Task

	Primary Task			Secondary Task		
Type	Stable	Degraded	Enhanced	Stable	Degraded	Enhanced
Detection		Price (1975)			Price (1975)	
Driving		Hicks and Wierwille (1979)			Wierwille, Gutmann, Hicks, and Muto (1977)	
Identification	Allport, Antonis, and Reynolds (1972)			Allport et al. (1972)		
Memory		Mitsuda (1968)				
Monitoring					Savage, Wierwille and Cordes (1978)	
Spatial transformation		Fournier and Stager (1976)			Fournier and Stager (1976)	
Tracking		Gabay and Merhav (1977)		Gabay and Merhav (1977)		

Source: From Lysaght, R.J., Hill, S.G., Dick, A.O., Plamondon, B.D., Linton, P.M., Wierwille, W.W., Zaklad, A.L., Bittner, A.C., and Wherry, R.J. *Operator workload: Comprehensive review and evaluation of operator workload methodologies (Technical Report 851).* Alexandria, VA: Army Research Institute for the Behavioral and Social Sciences; June 1989, p. 246.

SOURCES

Allport, D.A., Antonis, B., and Reynolds, P. On the division of attention: A disproof of the single channel hypothesis. *Quarterly Journal of Experimental Psychology.* 24, 225–235, 1972.

Fournier, B.A. and Stager, P. Concurrent validation of a dual-task selection test. *Journal of Applied Psychology.* 5, 589–595, 1976.

Gabay, E. and Merhav, S.J. Identification of a parametric model of the human operator in closed-loop control tasks. *IEEE Transactions on Systems, Man, and Cybernetics.* SMC-7, 284–292, 1977.

Hicks, T.G. and Wierwille, W.W. Comparison of five mental workload assessment procedures in a moving-base driving simulator. *Human Factors.* 21, 129–142, 1979.

Lysaght, R.J., Hill, S.G., Dick, A.O., Plamondon, B.D., Linton, P.M., Wierwille, W.W., Zaklad, A.L., Bittner, A.C., and Wherry, R.J. *Operator workload: Comprehensive review and evaluation of operator workload methodologies (Technical Report 851).* Alexandria, VA: Army Research Institute for the Behavioral and Social Sciences; June 1989.

Mitsuda, M. Effects of a subsidiary task on backward recall. *Journal of Verbal Learning and Verbal Behavior.* 7, 722–725, 1968.

Price, D.L. The effects of certain gimbal orders on target acquisition and workload. *Human Factors.* 20, 649–654, 1975.

Savage, R.E., Wierwille, W.W., and Cordes, R.E. Evaluating the sensitivity of various measures of operator workload using random digits as a secondary task. *Human Factors.* 20, 649–654, 1978.

Wierwille, W.W. and Connor, S.A. Evaluation of 20 workload measures using a psychomotor task in a moving-base aircraft simulator. *Human Factors.* 25(1), 1–16, 1983.

Wierwille, W.W., Gutmann, J.C., Hicks, T.G., and Muto, W.H. Secondary task measurement of workload as a function of simulated vehicle dynamics and driving conditions. *Human Factors.* 19, 557–565, 1977.

3.2.9 Lexical Decision Secondary Task

General description. "Typically, the subject is briefly presented with a sequence of letters and must judge whether this letter sequence forms a word or a nonword" (Lysaght et al., 1989, p. 233).

Strengths and limitations. "This task is thought to impose heavy demands on semantic memory processes" (Lysaght et al., 1989, p. 233).

Data requirements. Mean RT for correct responses is used as data for this task.

Thresholds. Not stated.

SOURCE

Lysaght, R.J., Hill, S.G., Dick, A.O., Plamondon, B.D., Linton, P.M., Wierwille, W.W., Zaklad, A.L., Bittner, A.C., and Wherry, R.J. *Operator workload: Comprehensive review and evaluation of operator workload methodologies (Technical Report 851).* Alexandria, VA: Army Research Institute for the Behavioral and Social Sciences; June 1989.

3.2.10 Memory-Scanning Secondary Task

General description. These secondary tasks require a subject to memorize a list of letters, numbers, and/or shapes and then indicate whether a probe stimulus is a member of that set. Typically, there is a linear relation between the number of items in the memorized list and RT to the probe stimulus. A variation of this task involves recalling one of the letters based on its position in the presented set.

Strengths and limitations. The slope of the linear function may reflect the memory-scanning rate. Fisk and Hodge (1992) reported no significant differences in RT in a single-task performance of a memory-scanning task after 32 d without practice. But Carter, Krause, and Harbeson (1986) warned that the slope may be less reliable than the RTs used to calculate the slope.

These tasks may not produce good estimates of memory load, because of the following: (1) addition of memory load does not affect the slope and (2) the slope is affected by stimulus-response compatibility effects. Further, Wierwille and Connor (1983), however, reported that a memory-scanning task was not sensitive to workload. Their primary task was to control a moving-base aircraft simulator. Workload was varied by manipulating pitch-stability level, wind-gust disturbance level, and crosswind direction and velocity. However, Park and Lee (1992) reported that memory tasks significantly predicted flight performance of pilot trainees.

Based on 25 studies using a memory secondary task, Lysaght et al. (1989) reported that performance remained stable on tracking, mental math, monitoring, problem solving, and driving primary tasks; degraded on tracking, choice RT, memory, monitoring, problem-solving, detection, identification, classification, and distraction primary tasks; and improved on a tracking primary task. Performance of the memory secondary task remained stable with a tracking primary task and degraded when paired with tracking, choice RT, memory, mental math, monitoring, detection, identification, classification, and driving primary tasks (see table 7).

Data requirements. RT to items from the memorized list must be at an asymptote to ensure that no additional learning will take place during the experiment. The presentation of the probe stimulus is intrusive and thus may be difficult to use in nonlaboratory settings.

Thresholds. 40 ms for RT.

TABLE 7

References Listed by the Effect on Performance of Primary Tasks Paired with a Secondary Memory Task

	Primary task			Secondary task		
Type	**Stable**	**Degraded**	**Enhanced**	**Stable**	**Degraded**	**Enhanced**
Choice RT		Broadbent and Gregory (1965) Keele and Boies (1973)			Broadbent and Gregory (1965)	
Classification		Wickens et al. (1981)			Wickens et al. (1981)	
Detection		Wickens et al. (1981)			Wickens et al. (1981)	
Distraction		Broadbent and Heron (1962)				
Driving	Brown (1962, 1965, 1966) Brown and Poulton (1961) Kantowitz (1995) Wetherell (1981)	Richard, Wright, Ee, Prime, Shimizu, and Vavrik (2002; task was to search visual driving scene)		Brown (1965)	Brown (1962, 1965, 1966) Brown and Poulton (1961) Wetherell (1981)	
Identification		Klein (1976)			Allpor, Antonis, and Reynolds (1972)	
Memory		Broadbent and Heron (1962) Chow and Murdock (1975)		Shulman and Greenberg (1971)		
Mental math	Mandler and Worden (1973)			Mandler and Worden (1973)		

Continued

TABLE 7 (Continued)

References Listed by the Effect on Performance of Primary Tasks Paired with a Secondary Memory Task

Type	Primary task			Secondary task		
	Stable	**Degraded**	**Enhanced**	**Stable**	**Degraded**	**Enhanced**
Monitoring	Chechile, Butler, Gutowski, and Palmer (1979) Moskowitz and McGlothlin (1974)	Chiles and Alluisi (1979)		Chechile et al. (1979) Chiles and Alluisi (1979) Mandler and Worden (1973) Moskowitz and McGlothlin (1974)		
Problem solving	Daniel, Florek, Kosinar, and Strizenec (1969)	Stager and Zufelt (1972)				
Tracking	Finkelman and Glass (1970) Zeitlin and Finkelman (1975)	Heimstra (1970) Huddleston and Wilson (1971) Noble, Trumbo, and Fowler (1967) Trumbo and Milone (1971) Wickens and Kessel (1980) Wickens, Mountford, and Schreiner (1981)	Tsang and Wickens (1984)	Noble et al. (1967) Trumbo and Milone (1971)	Finkelman and Glass (1970) Heimstra (1970) Huddleston and Wilson (1971) Tsang and Wickens (1984) Wickens and Kessel (1980) Wickens et al. (1981)	

Source: From Lysaght, R.J., Hill, S.G., Dick, A.O., Plamondon, B.D., Linton, P.M., Wierwille, W.W., Zaklad, A.L., Bittner, A.C., and Wherry, R.J. *Operator workload: Comprehensive review and evaluation of operator workload methodologies (Technical Report 851).* Alexandria, VA: Army Research Institute for the Behavioral and Social Sciences; June 1989, p. 246.

SOURCES

Allport, D.A., Antonis, B., and Reynolds, P. On the division of attention: A disproof of the single channel hypothesis. *Quarterly Journal of Experimental Psychology.* 24, 225–235, 1972.

Broadbent, D.E. and Gregory, M. On the interaction of S-R compatibility with other variables affecting reaction time. *British Journal of Psychology.* 56, 61–67, 1965.

Broadbent, D.E. and Heron, A. Effects of a subsidiary task on performance involving immediate memory by younger and older men. *British Journal of Psychology.* 53, 189–198, 1962.

Brown, I.D. Measuring the "spare mental capacity" of car drivers by a subsidiary auditory task. *Ergonomics.* 5, 247–250, 1962.

Brown, I.D. A comparison of two subsidiary tasks used to measure fatigue in car drivers. *Ergonomics.* 8, 467–473, 1965.

Brown, I.D. Subjective and objective comparisons of successful and unsuccessful trainee drivers. *Ergonomics.* 9, 49–56, 1966.

Brown, I.D. and Poulton, E.C. Measuring the spare "mental capacity" of car drivers by a subsidiary task. *Ergonomics.* 4, 35–40, 1961.

Carter, R.C., Krause, M., and Harbeson, M.M. Beware the reliability of slope scores for individuals. *Human Factors.* 28(6), 673–683, 1986.

Chechile, R.A., Butler, K., Gutowski, W., and Palmer, E.A. Division of attention as a function of the number of steps, visual shifts and memory load. *Proceedings of the 15th Annual Conference on Manual Control* (pp. 71–81). Dayton, OH: Wright State University, 1979.

Chiles, W.D. and Alluisi, E.A. On the specification of operator or occupational workload performance-measurement methods. *Human Factors.* 21, 515–528, 1979.

Chow, S.L. and Murdock, B.B. The effect of a subsidiary task on iconic memory. *Memory and Cognition.* 3, 678–688, 1975.

Daniel, J., Florek, H., Kosinar, V., and Strizenec, M. Investigation of an operator's characteristics by means of factorial analysis. *Studia Psychologica.* 11, 10–22, 1969.

Finkelman, J.M. and Glass, D.C. Reappraisal of the relationship between noise and human performance by means of a subsidiary task measure. *Journal of Applied Psychology.* 54, 211–213, 1970.

Fisk, A.D. and Hodge, K.A. Retention of trained performance in consistent mapping search after extended delay. *Human Factors.* 34(2), 147–164, 1992.

Heimstra, N.W. The effects of "stress fatigue" on performance in a simulated driving situation. *Ergonomics.* 13, 209–218, 1970.

Huddleston, J.H.F. and Wilson, R.V. An evaluation of the usefulness of four secondary tasks in assessing the effect of a lag in simulated aircraft dynamics. *Ergonomics.* 14, 371–380, 1971.

Kantowitz, B.H. Simulator evaluation of heavy-vehicle driver workload. *Proceedings of the Human Factors and Ergonomics Society 39th Annual Meeting.* 2, 1107–1111, 1995.

Keele, S.W. and Boies, S.J. Processing demands of sequential information. *Memory and Cognition.* 1, 85–90, 1973.

Klein, G.A. Effect of attentional demands on context utilization. *Journal of Educational Psychology.* 68, 25–31, 1976.

Lysaght, R.J., Hill, S.G., Dick, A.O., Plamondon, B.D., Linton, P.M., Wierwille, W.W., Zaklad, A.L., Bittner, A.C., and Wherry, R.J. *Operator workload: Comprehensive review and evaluation of operator workload methodologies (Technical Report 851).* Alexandria, VA: Army Research Institute for the Behavioral and Social Sciences; June 1989.

Mandler, G. and Worden, P.E. Semantic processing without permanent storage. *Journal of Experimental Psychology.* 100, 277–283, 1973.

Moskowitz, H. and McGlothlin, W. Effects of marijuana on auditory signal detection. *Psychopharmacologia*. 40, 137–145, 1974.
Noble, M., Trumbo, D., and Fowler, F. Further evidence on secondary task interference in tracking. *Journal of Experimental Psychology*. 73, 146–149, 1967.
Park, K.S. and Lee, S.W. A computer-aided aptitude test for predicting flight performance of trainees. *Human Factors*. 34(2), 189–204, 1992.
Richard, C.M., Wright, R.D., Ee, C., Prime, S.L., Shimizu, Y., and Vavrik, J. Effect of a concurrent auditory task on visual search performance in a driving-related image-flicker task. *Human Factors*. 44(1), 108–119, 2002.
Shulman, H.G. and Greenberg, S.N. Perceptual deficit due to division of attention between memory and perception. *Journal of Experimental Psychology*. 88, 171–176, 1971.
Stager, P. and Zufelt, K. Dual-task method in determining load differences. *Journal of Experimental Psychology*. 94, 113–115, 1972.
Trumbo, D. and Milone, F. Primary task performance as a function of encoding, retention, and recall in a secondary task. *Journal of Experimental Psychology*. 91, 273–279, 1971.
Tsang, P.S. and Wickens, C.D. The effects of task structures on time-sharing efficiency and resource allocation optimality. *Proceedings of the 20th Annual Conference on Manual Control* (pp. 305–317). Moffett Field, CA: Ames Research Center, 1984.
Wetherell, A. The efficacy of some auditory-vocal subsidiary tasks as measures of the mental load on male and female drivers. *Ergonomics*. 24, 197–214, 1981.
Wickens, C.D. and Kessel, C. Processing resource demands of failure detection in dynamic systems. *Journal of Experimental Psychology: Human Perception and Performance*. 6, 564–577, 1980.
Wickens, C.D., Mountford, S.J., and Schreiner, W. Multiple resources, task-hemispheric integrity, and individual differences in time sharing. *Human Factors*. 23, 211–229, 1981.
Wierwille, W.W. and Connor, S.A. Evaluation of 20 workload measures using a psychomotor task in a moving-base aircraft simulator. *Human Factors*. 25(1), 1–16, 1983.
Zeitlin, L.R. and Finkelman, J.M. Research note: Subsidiary task techniques of digit generation and digit recall indirect measures of operator loading. *Human Factors*. 17, 218–220, 1975.

3.2.11 Mental Mathematics Secondary Task

General description. Subjects are asked to perform arithmetic operations (i.e., addition, subtraction, multiplication, and division) on sets of visually or aurally presented digits.

Strengths and limitations. The major strength of this workload measure is its ability to discriminate between good and poor operators and high and low workload. For example, Ramacci and Rota (1975) required pilot applicants to perform progressive subtraction during their initial flight training. They reported that the number of subtractions performed increased although the percentage of errors decreased with flight experience. Further, successful applicants performed more subtractions and had a lower percentage of errors than those applicants who were not accepted.

Green and Flux (1977) required pilots to add the digit three to aurally presented digits during a simulated flight. They reported increased performance time of the secondary task as the workload associated with the primary task increased. Huddleston and Wilson (1971) asked pilots to determine if digits were odd or even, their sum was odd or even, two consecutive digits were the same or different, or every other digit was the same or different. Again, secondary task performance discriminated between high and low workload on the primary task. The major disadvantage of secondary tasks is their intrusion into the primary task. Harms (1986) reported similar results for a driving task. However, Andre, Heers, and Cashion (1995) reported greater rmse in roll, pitch, and yaw in a primary simulated flight task when paired with a mental mathematics secondary task (i.e., fuel range).

Mental mathematics tasks have also been used in the laboratory. For example, Kramer, Wickens, and Donchin (1984) reported a significant increase in tracking error on the primary task when the secondary task was counting flashes. In addition, Damos (1985) required subjects to calculate the absolute difference between the digit currently presented visually and the digit that had preceded it. In the single-task condition, percent correct scores were significantly related to trial. Correct RT scores were related to trial and trial by pacing condition. In the dual-task condition, percent correct scores were not significantly related to trial, behavior pattern, or pacing condition. Correct RT in the dual-task condition, however, was related to trial, trial by pacing, and trial by behavior pattern.

On the basis of 15 studies that used a mental math secondary task, Lysaght et al. (1989) reported that performance remained the same for tracking, driving, and tapping primary tasks and degraded for tracking, choice RT, memory, monitoring, simple RT, and detection primary tasks. Performance of the mental math secondary task remained stable with a tracking primary task; degraded with tracking, choice RT, monitoring, detection, driving, and tapping primary tasks; and improved with tracking primary task (see table 8).

Data requirements. The following data are calculated: number of correct responses, mean RT for correct responses, and number of incorrect responses (Lysaght et al., 1989, p. 235). The researcher should compare primary task performance with and without a secondary task to ensure that subjects are not sacrificing primary task performance to enhance secondary task performance.

Thresholds. Not stated.

TABLE 8

References Listed by the Effect on Performance of Primary Tasks Paired with a Secondary Mental Math Task

Type	Primary Task			Secondary Task		
	Stable	**Degraded**	**Enhanced**	**Stable**	**Degraded**	**Enhanced**
Choice RT		Chiles and Jennings (1970) Fisher (1975) Keele (1967)			Fisher (1975) Keele (1967) Schouten, Kalsbeek, and Leopold (1962)	
Detection		Jaschinski (1982)			Jaschinski (1982)	
Driving	Brown and Poulton (1961) Wetherell (1981)			Verwey (2000)*	Brown and Poulton (1961) Wetherell (1981)	
Memory		Roediger, Knight, and Kantowitz (1977) Silverstein and Glanzer (1971)				
Monitoring		Chiles and Jennings (1970) Kahneman, Beatty, and Pollack (1967)			Chiles and Jennings (1970) Kahneman et al. (1967)	
Simple RT		Chiles and Jennings (1970) Green and Flux (1977)* Wierwille and Connor (1983)*			Green and Flux (1977)*	

Simulated flight task	Green and Flux (1977)* Wierwille and Connor (1983)*	Andre, Heers, and Cashion (1995)*			Green and Flux (1977)*	
Tapping	Kantowitz and Knight (1974)				Kantowitz and Knight (1974, 1976)	
Tracking	Huddleston and Wilson (1971)	Bahrick, Noble, and Fitts (1954) Chiles and Jennings (1970) Heimstra (1970) McLeod (1973) Wickens, Mountford, and Schreiner (1981)	Kramer, Wickens, and Donchin (1984)	Bahrick et al. (1954) Heimstra (1970)	Huddleston and Wilson (1971) McLeod (1973) Wickens et al. (1981)	Chiles and Jennings (1970)

* Not included in Lysaght, R.J., Hill, S.G., Dick, A.O., Plamondon, B.D., Linton, P.M., Wierwille, W.W., Zaklad, A.L., Bittner, A.C., and Wherry, R.J. *Operator workload: Comprehensive review and evaluation of operator workload methodologies (Technical Report 851).* Alexandria, VA: Army Research Institute for the Behavioral and Social Sciences; June 1989.

Source: From Lysaght, R.J., Hill, S.G., Dick, A.O., Plamondon, B.D., Linton, P.M., Wierwille, W.W., Zaklad, A.L., Bittner, A.C., and Wherry, R.J. *Operator workload: Comprehensive review and evaluation of operator workload methodologies (Technical Report 851).* Alexandria, VA: Army Research Institute for the Behavioral and Social Sciences; June 1989, p. 247.

SOURCES

Andre, A.D., Heers, S.T., and Cashion, P.A. Effects of workload preview on task scheduling during simulated instrument flight. *International Journal of Aviation Psychology*. 5(1), 5–23, 1995.

Bahrick, H.P., Noble, M., and Fitts, P.M. Extra-task performance as a measure of learning task. *Journal of Experimental Psychology*. 4, 299–302, 1954.

Brown, I.D. and Poulton, E.C. Measuring the spare "mental capacity" of car drivers by a subsidiary task. *Ergonomics*. 4, 35–40, 1961.

Chiles, W.D. and Jennings, A.E. Effects of alcohol on complex performance. *Human Factors*. 12, 605–612, 1970.

Damos, D. The relation between the type A behavior pattern, pacing, and subjective workload under single- and dual-task conditions. *Human Factors*. 27(6), 675–680, 1985.

Fisher, S. The microstructure of dual task interaction. 1. The patterning of main task response within secondary-task intervals. *Perception*. 4, 267–290, 1975.

Green, R. and Flux, R. Auditory communication and workload. *Proceedings of NATO Advisory Group for Aerospace Research and Development Conference on Methods to Assess Workload, AGARD-CPP-216*, A4-1-A4-8, 1977.

Harms, L. Drivers' attentional response to environmental variations: A dual-task real traffic study. In Gale, A.G., Freeman, M.H., Haslegrave, C.M., Smith, P., and Taylor, S.P. (Eds.) *Vision in vehicles* (pp. 131–138). Amsterdam: North Holland, 1986.

Heimstra, N.W. The effects of "stress fatigue" on performance in a simulated driving situation. *Ergonomics*. 13, 209–218, 1970.

Huddleston, J.H.F. and Wilson, R.V. An evaluation of the usefulness of four secondary tasks in assessing the effect of a log in simulated aircraft dynamics. *Ergonomics*. 14, 371–380, 1971.

Jaschinski, W. Conditions of emergency lighting. *Ergonomics*. 25, 363–372, 1982.

Kahneman, D., Beatty, J., and Pollack, I. Perceptual deficit during a mental task. *Science*. 157, 218–219, 1967.

Kantowitz, B.H. and Knight, J.L. Testing tapping time-sharing. *Journal of Experimental Psychology*. 103, 331–336, 1974.

Kantowitz, B.H. and Knight, J.L. Testing tapping time sharing: II. Auditory secondary task. *Acta Psychologica*. 40, 343–362, 1976.

Keele, S.W. Compatibility and time-sharing in serial reaction time. *Journal of Experimental Psychology*. 75, 529–539, 1967.

Kramer, A.F., Wickens, C.D., and Donchin, E. *Performance and enhancements under dual-task conditions*. Annual Conference on Manual Control, June 1984, 21–35.

Lysaght, R.J., Hill, S.G., Dick, A.O., Plamondon, B.D., Linton, P.M., Wierwille, W.W., Zaklad, A.L., Bittner, A.C., and Wherry, R.J. *Operator workload: Comprehensive review and evaluation of operator workload methodologies (Technical Report 851)*. Alexandria, VA: Army Research Institute for the Behavioral and Social Sciences; June 1989.

McLeod, P.D. Interference of "attend to and learn" tasks with tracking. *Journal of Experimental Psychology*. 99, 330–333, 1973.

Ramacci, C.A. and Rota, P. Flight fitness and psycho-physiological behavior of applicant pilots in the first flight missions. *Proceedings of NATO Advisory Group for Aerospace Research and Development*. 153, B8, 1975.

Roediger, H.L., Knight, J.L., and Kantowitz, B.H. Inferring delay in short-term memory: The issue of capacity. *Memory and Cognition*. 5, 167–176, 1977.

Schouten, J.F., Kalsbeek, J.W.H., and Leopold, F.F. On the evaluation of perceptual and mental load. *Ergonomics*. 5, 251–260, 1962.

Silverstein, C. and Glanzer, M. Concurrent task in free recall: Differential effects of LTS and STS. *Psychonomic Science*. 22, 367–368, 1971.

Verwey, W.B. On-line driver workload estimation: Effects of road situation and age on secondary task measures. *Ergonomics*. 43(2): 187–209, 2000.
Wetherell, A. The efficacy of some auditory-vocal subsidiary tasks as measures of the mental load on male and female drivers. *Ergonomics*. 24, 197–214, 1981.
Wickens, C.D., Mountford, S.J., and Schreiner, W. Multiple resources, task-hemispheric integrity, and individual differences in time-sharing. *Human Factors*. 23, 211–229, 1981.
Wierwille, W.W. and Connor, S. Evaluation of 20 workload measures using a psychomotor task in a moving base aircraft simulator. *Human Factors*. 25, 1–16, 1983.

3.2.12 Michon Interval Production Secondary Task

General description. "The Michon paradigm of interval production requires the subject to generate a series of regular time intervals by executing a motor response (i.e., a single finger tap [every] 2 sec.). No sensory input is required." (Lysaght et al., 1989, p. 233). On the basis of a review of six studies in which the Michon Interval Production task was the secondary task, Lysaght et al. (1989) stated that, in dual-task pairings, performance of flight simulation and driving primary tasks remained stable; performance of monitoring, problem-solving, detection, psychomotor, Sternberg, tracking, choice RT, memory, and mental math primary tasks degraded; and performance of simple RT primary task improved. In these same pairings, performance of the Michon Interval Production task remained stable with monitoring, Sternberg, flight simulation, and memory primary tasks; and degraded with problem-solving, simple RT, detection, psychomotor, flight simulation, driving, tracking, choice RT, and mental math primary tasks (see table 9).

Strengths and limitations. "This task is thought to impose heavy demand on motor output/response resources. It has been demonstrated with high demand primary tasks that subjects exhibit irregular or variable tapping rates." (Lysaght et al., 1989, p. 233). Crabtree, Bateman, and Acton (1984) reported that scores on this secondary task discriminated the workload of three switch-setting tasks. Johannsen, Pfendler, and Stein (1976) reported similar results for autopilot evaluation in a fixed-based flight simulator. Wierwille, Rahimi, and Casali (1985), however, reported that tapping regularity was not affected by variations in the difficulty of a mathematical problem to be solved during simulated flight. Further, Drury (1972) recommended a correlation to provide a nondimensional measure.

Data requirements. Michon (1966) stated that a functional description of behavior is needed for the technique. The following data are calculated: mean interval per trial, standard deviation of interval per trial, and sum of differences between successive intervals per minute of total time (Lysaght et al., 1989, p. 235).

Thresholds. Not stated.

TABLE 9

References Listed by the Effect on Performance of Primary Tasks Paired with a Secondary Michon Interval Production Task

	Primary Task			Secondary Task		
Type	**Stable**	**Degraded**	**Enhanced**	**Stable**	**Degraded**	**Enhanced**
Choice RT		Michon (1964)			Michon (1964)	
Detection		Michon (1964)			Michon (1964)	
Driving	Brown (1967)	Brown, Simmonds, and Tickner (1967)*			Brown (1967); Brown et al. (1967)*	
Flight simulation	Wierwille, Casali, Connor, and Rahimi (1985)			Wierwille, Casali, Connor, and Rahimi (1985)	Wierwille et al. (1985)	
Memory		Roediger, Knight, and Kantowitz (1977)		Roediger, Knight, and Kantowitz (1977)		
Mental math		Michon (1964)			Michon (1964)	
Monitoring		Shingledecker, Acton, and Crabtree (1983)		Shingledecker et al. (1983)		
Problem solving		Michon (1964)			Michon (1964)	
Psychomotor		Michon (1964)			Michon (1964)	
Simple RT			Vroon (1973)		Vroon (1973)	
Sternberg		Shingledecker et al. (1983)		Shingledecker et al. (1983)		
Tracking		Shingledecker et al. (1983)			Shingledecker et al. (1983)	

* Not included in Lysaght, R.J., Hill, S.G., Dick, A.O., Plamondon, B.D., Linton, P.M., Wierwille, W.W., Zaklad, A.L., Bittner, A.C., and Wherry, R.J. *Operator workload: Comprehensive review and evaluation of operator workload methodologies (Technical Report 851)*. Alexandria, VA: Army Research Institute for the Behavioral and Social Sciences; June 1989.

Source: From Lysaght, R.J., Hill, S.G., Dick, A.O., Plamondon, B.D., Linton, P.M., Wierwille, W.W., Zaklad, A.L., Bittner, A.C., and Wherry, R.J. *Operator workload: Comprehensive review and evaluation of operator workload methodologies (Technical Report 851)*. Alexandria, VA: Army Research Institute for the Behavioral and Social Sciences; June 1989, p. 245.

SOURCES

Brown, I.D. Measurement of control skills, vigilance, and performance on a subsidiary task during twelve hours of car driving. *Ergonomics*. 10, 665–673, 1967.

Brown, I.D., Simmonds, D.C.V., and Tickner, A.H. Measurement of control skills, vigilance, and performance on a subsidiary task during 12 hours of car driving. *Ergonomics*, 10, 655–673, 1967.

Crabtree, M.S., Bateman, R.P., and Acton, W. Benefits of using objective and subjective workload measures. *Proceedings of the 28th Annual Meeting of the Human Factors Society* (pp. 950–953). Santa Monica, CA; Human Factors Society, 1984.

Drury, C.G. Note on Michon's measure of tapping irregularity. *Ergonomics*. 15(2): 195–197, 1972.

Johannsen, G., Pfendler, C., and Stein, W. Human performance and workload in simulated landing approaches with autopilot failures. In Moray, N. (Ed.) *Mental workload, its theory and measurement* (pp. 101–104). New York: Plenum Press, 1976.

Lysaght, R.J., Hill, S.G., Dick, A.O., Plamondon, B.D., Linton, P.M., Wierwille, W.W., Zaklad, A.L., Bittner, A.C., and Wherry, R.J. *Operator workload: Comprehensive review and evaluation of operator workload methodologies (Technical Report 851).* Alexandria, VA: Army Research Institute for the Behavioral and Social Sciences; June 1989.

Michon, J.A. A note on the measurement of perceptual motor load. *Ergonomics*. 7, 461–463, 1964.

Michon, J.A. Tapping regularity as a measure of perceptual motor load. *Ergonomics*. 9(5): 401–412, 1966.

Roediger, H.L., Knight, J.L., and Kantowitz, B.H. Inferring decay in short-term memory: The issue of capacity. *Memory and Cognition*. 5, 167–176, 1977.

Shingledecker, C.A., Acton, W., and Crabtree, M.S. *Development and application of a criterion task set for workload metric evaluation (SAE Technical Paper No. 831419).* Warrendale, PA: Society of Automotive Engineers, 1983.

Vroon, P.A. Tapping rate as a measure of expectancy in terms of response and attention limitation. *Journal of Experimental Psychology*. 101, 183–185, 1973.

Wierwille, W.W., Casali, J.G., Connor, S.A., and Rahimi, M. Evaluation of the sensitivity and intrusion of mental workload estimation techniques. In Roner, W. (Ed.) *Advances in man-machine systems research, Volume 2* (pp. 51–127). Greenwich, CT: J.A.I. Press, 1985.

Wierwille, W.W., Rahimi, M., and Casali, J.G. Evaluation of 16 measures of mental workload using a simulated flight task emphasizing mediational activity. *Human Factors*. 27(5), 489–502, 1985.

3.2.13 Monitoring Secondary Task

General description. Subjects are asked to respond either manually or verbally to the onset of visual or auditory stimuli. Both the time to respond and the accuracy of the response have been used as workload measures.

Strengths and limitations. The major advantage of the monitoring-task technique is its relevance to system safety. It is also able to discriminate between levels of automation and workload. For example, Anderson and Toivanen (1970) used a force-paced digit-naming task as a secondary task to investigate the effects of varying levels of automation in a helicopter simulator. Bortolussi, Kantowitz, and Hart (1986) reported significant differences in two- and four-choice visual RT in easy and difficult flight scenarios. Bortolussi, Hart, and Shively (1987) reported significantly longer RTs in a four-choice RT task during a high-difficulty scenario than during a low-difficulty one. Brown (1969) studied it in relation to flicker.

On the basis of the results of 36 studies that included a secondary monitoring task, Lysaght et al. (1989) reported that performance remained stable on tracking, choice RT, memory, mental math, problem-solving, identification, and driving primary tasks; degraded on tracking, choice RT, memory, monitoring, detection, and driving primary tasks; and improved on a monitoring primary task. Performance of the monitoring secondary task remained stable when paired with tracking, memory, monitoring, flight simulation, and driving primary tasks; degraded when paired with tracking, choice RT, mental math, monitoring, problem-solving, detection, identification and driving primary tasks; and improved when paired with tracking and driving primary tasks (see table 10).

Data requirements. The experimenter should calculate the following: number of correct detections, number of incorrect detections, number of errors of omission, mean RT for correct detections, and mean RT for incorrect detections (Lysaght et al., 1989, p. 235).

The Knowles (1963) guidelines are appropriate in selecting a vigilance task. In addition, the modality of the task must not interfere with performance of the primary task, for example, requiring a verbal response while a pilot is communicating with Air Traffic Control (ATC) or other crew members.

Thresholds. Not stated.

TABLE 10
References Listed by the Effect on Performance of Primary Tasks Paired with a Secondary Monitoring Task

	Primary Task			Secondary Task		
Type	**Stable**	**Degraded**	**Enhanced**	**Stable**	**Degraded**	**Enhanced**
Choice RT	Boggs and Simon (1968)	Hilgendorf (1967)			Hilgendorf (1967)	
Detection		Dewar, Ells, and Mundy (1976) Tyler and Halcomb (1974)			Tyler and Halcomb (1974)	
Driving	Brown (1962, 1967) Hoffman and Jorbert (1966) Wetherell (1981)	Brown (1965)		Hoffman and Jorbert (1966)	Brown (1962, 1965)	Brown (1967)
Flight simulation				Soliday and Schohan (1965)		
Identification	Dornic (1980)				Dornic (1980) Chiles, Jennings, and Alluisi (1979)	
Memory	Tyler and Halcomb (1974)	Chow and Murdock (1975) Lindsay and Norman (1969) Mitsuda (1968)		Lindsay and Norman (1968)		
Mental math	Dornic (1980)				Chiles, Jennings, and Alluisi (1979) Dornic (1980)	
Monitoring		Chechile, Butler, Gutowski, and Palmer (1979) Fleishman (1965)	McGrath (1965)	Stager and Muter (1971)	Chechile et al. (1979) Hohmuth (1970) Long (1976)	

Continued

TABLE 10 (Continued)

References Listed by the Effect on Performance of Primary Tasks Paired with a Secondary Monitoring Task

	Primary Task			Secondary Task		
Type	**Stable**	**Degraded**	**Enhanced**	**Stable**	**Degraded**	**Enhanced**
		Goldstein and Dorfman (1978) Hohmuth (1970) Long (1976) Stager and Muter (1971)				
Problem solving	Wright, Holloway, and Aldrich (1974)				Chiles et al. (1979) Wright et al. (1974)	
Tracking	Bell (1978) Figarola and Billings (1966) Gabriel and Burrows (1968) Huddleston and Wilson (1971) Kelley and Wargo (1967) Kyriakides and Leventhal (1967) Schori and Jones (1975)	Bergeron (1968) Heimstra (1970) Herman (1965) Kramer, Wickens, and Donchin (1984) Malmstrom, Reed, and Randle (1983) Monty and Ruby (1965) Putz and Rothe (1974)		Figarola and Billings (1966) Kramer et al. (1984) Malmstrom et al. (1983)	Bell (1978) Bergeron (1968) Gabriel and Burrows (1968) Herman (1965) Huddleston and Wilson (1971) Kelley and Wargo (1967) Kyriakides and Leventhal (1967) Monty and Ruby (1965) Putz and Rothe (1974) Schori and Jones (1975)	Heimstra (1970)

* Not included in Lysaght, R.J., Hill, S.G., Dick, A.O., Plamondon, B.D., Linton, P.M., Wierwille, W.W., Zaklad, A.L., Bittner, A.C., and Wherry, R.J. *Operator workload: Comprehensive review and evaluation of operator workload methodologies (Technical Report 851)*. Alexandria, VA: Army Research Institute for the Behavioral and Social Sciences; June 1989.

Source: From Lysaght, R.J., Hill, S.G., Dick, A.O., Plamondon, B.D., Linton, P.M., Wierwille, W.W., Zaklad, A.L., Bittner, A.C., and Wherry, R.J. *Operator workload: Comprehensive review and evaluation of operator workload methodologies (Technical Report 851)*. Alexandria, VA: Army Research Institute for the Behavioral and Social Sciences; June 1989, p. 246.

SOURCES

Anderson, P.A. and Toivanen, M.L. *Effects of varying levels of autopilot assistance and workload on pilot performance in the helicopter formation flight mode (Technical Report JANAIR 680610).* Washington, D.C.: Office of Naval Research, March 1970.

Bell, P.A. Effects of noise and heat stress on primary and subsidiary task performance. *Human Factors.* 20, 749–752, 1978.

Bergeron, H.P. Pilot response in combined control tasks. *Human Factors.* 10, 277–282, 1968.

Boggs, D.H. and Simon, J.R. Differential effect of noise on tasks of varying complexity. *Journal Applied Psychology.* 52, 148–153, 1968.

Bortolussi, M.R., Hart, S.G., and Shively, R.J. Measuring moment-to-moment pilot workload using synchronous presentations of secondary tasks in a motion-base trainer. *Proceedings of the 4th Symposium on Aviation Psychology.* 651–657, 1987.

Bortolussi, M.R., Kantowitz, B.H., and Hart, S.G. Measuring pilot workload in a motion base trainer: A comparison of four techniques. *Applied Ergonomics.* 17, 278–283, 1986.

Brown, I.D. Measuring the "spare mental capacity" of car drivers by a subsidiary auditory task. *Ergonomics.* 5, 247–250, 1962.

Brown, I.D. A comparison of two subsidiary tasks used to measure fatigue in car drivers. *Ergonomics.* 8, 467–473, 1965.

Brown, I.D. Measurement of control skills, vigilance, and performance on a subsidiary task during twelve hours of car driving. *Ergonomics.* 10, 665–673, 1967.

Brown, J.L. Flicker and intermittent stimulation. In Graham, C.H. (Ed.) *Vision and visual perception.* New York: Wiley, 1969.

Chechile, R.A., Butler, K., Gutowski, W., and Palmer, E.A. Division of attention as a function of the number of steps, visual shifts, and memory load. *Proceedings of the 15th Annual Conference on Manual Control* (pp. 71–81). Dayton, OH: Wright State University, 1979.

Chiles, W.D., Jennings, A.E., and Alluisi, E.C. Measurement and scaling of workload in complex performance. *Aviation, Space, and Environmental Medicine.* 50, 376–381, 1979.

Chow, S.L and Murdock, B.B. The effect of a subsidiary task on iconic memory. *Memory and Cognition.* 3, 678–688, 1975.

Dewar, R.E., Ells, J.E., and Mundy, G. Reaction time as an index of traffic sign perception. *Human Factors.* 18, 381–392, 1976.

Dornic, S. Language dominance, spare capacity and perceived effort in bilinguals. *Ergonomics.* 23, 369–377, 1980.

Figarola, T.R. and Billings, C.E. Effects of meprobamate and hypoxia on psychomotor performance. *Aerospace Medicine.* 37, 951–954, 1966.

Fleishman, E.A. The prediction of total task performance from prior practice on task components. *Human Factors.* 7, 18–27, 1965.

Gabriel, R.F. and Burrows, A.A. Improving time-sharing performance of pilots through training. *Human Factors.* 10, 33–40, 1968.

Goldstein, I.L. and Dorfman, P.W. Speed and load stress as determinants of performance in a time sharing task. *Human Factors.* 20, 603–609, 1978.

Heimstra, N.W. The effects of "stress fatigue" on performance in a simulated driving situation. *Ergonomics.* 13, 209–218, 1970.

Herman, L.M. Study of the single channel hypothesis and input regulation within a continuous, simultaneous task situation. *Quarterly Journal of Experimental Psychology.* 17, 37–46, 1965.

Hilgendorf, E.L. Information processing practice and spare capacity. *Australian Journal of Psychology.* 19, 241–251, 1967.

Hoffman, E.R. and Jorbert, P.N. The effect of changes in some vehicle handling variables on driver steering performance. *Human Factors.* 8, 245–263, 1966.

Hohmuth, A.V. Vigilance performance in a bimodal task. *Journal of Applied Psychology*. 54, 520–525, 1970.

Huddleston, J.H.F. and Wilson, R.V. An evaluation of the usefulness of four secondary tasks in assessing the effect of a lag in simulated aircraft dynamics. *Ergonomics*. 14, 371–380, 1971.

Kelley, C.R. and Wargo, M.J. Cross-adaptive operator loading tasks. *Human Factors*. 9, 395–404, 1967.

Knowles, W.B. Operator loading tasks. *Human Factors*. 5, 151–161, 1963.

Kramer, A.F., Wickens, C.D., and Donchin, E. Performance enhancements under dual-task conditions. *Proceedings of the 20th Annual Conference on Manual Control* (pp. 21–35). Moffett Field, CA: Ames Research Center, 1984.

Kyriakides, K. and Leventhal, H.G. Some effects of intrasound on task performance. *Journal of Sound and Vibration*. 50, 369–388, 1977.

Lindsay, P.H. and Norman, D.A. Short-term retention during a simultaneous detection task. *Perception and Psychophysics*. 5, 201–205, 1969.

Long, J. Effect on task difficulty on the division of attention between nonverbal signals: Independence or interaction? *Quarterly Journal of Experimental Psychology*. 28, 179–193, 1976.

Lysaght, R.J., Hill, S.G., Dick, A.O., Plamondon, B.D., Linton, P.M., Wierwille, W.W., Zaklad, A.L., Bittner, A.C., and Wherry, R.J. *Operator workload: Comprehensive review and evaluation of operator workload methodologies (Technical Report 851)*. Alexandria, VA: Army Research Institute for the Behavioral and Social Sciences; June 1989.

Malmstrom, F.V., Reed, L.E., and Randle, R.J. Restriction of pursuit eye movement range during a concurrent auditory task. *Journal of Applied Psychology*. 68, 565–571, 1983.

McGrath, J.J. Performance sharing in an audio-visual vigilance task. *Human Factors*. 7, 141–153, 1965.

Mitsuda, M. Effects of a subsidiary task on backward recall. *Journal of Verbal Learning and Verbal Behavior*. 7, 722–725, 1968.

Monty, R.A. and Ruby, W.J. Effects of added workload on compensatory tracking for maximum terrain following. *Human Factors*. 7, 207–214, 1965.

Putz, V.R. and Rothe, R. Peripheral signal detection and concurrent compensatory tracking. *Journal of Motor Behavior*. 6, 155–163, 1974.

Schori, T.R. and Jones, B.W. Smoking and workload. *Journal of Motor Behavior*. 7, 113–120, 1975.

Soliday, S.M. and Schohan, B. Task loading of pilots in simulated low-altitude high-speed flight. *Human Factors*. 7, 45–53, 1965.

Stager, P. and Muter, P. Instructions and information processing in a complex task. *Journal of Experimental Psychology*. 87, 291–294, 1971.

Tyler, D.M. and Halcomb, C.G. Monitoring performance with a time-shared encoding task. *Perceptual and Motor Skills*. 38, 383–386, 1974.

Wetherell, A. The efficacy of some auditory vocal subsidiary tasks as measures of the mental load on male and female drivers. *Ergonomics*. 24, 197–214, 1981.

Wright, P., Holloway, C.M. and Aldrich, A.R. Attending to visual or auditory verbal information while performing other concurrent tasks. *Quarterly Journal of Experimental Psychology*. 26, 454–463, 1974.

3.2.14 Multiple Task Performance Battery of Secondary Tasks

General description. The Multiple Task Performance Battery (MTPB) requires subjects to time-share three or more of the following tasks: (1) light and dial monitoring, (2) mental math, (3) pattern discrimination, (4) target identification, (5) group problem solving, and (6) two-dimensional compensatory tracking. The monitoring tasks are used as secondary tasks and performance associated with these tasks as measures of workload.

Strengths and limitations. Increasing the number of tasks being time-shared does increase the detection time associated with the monitoring task. The MTPB may be difficult to implement in nonlaboratory settings. Lysaght et al. (1989) reported the results of one study (Alluisi, 1971) in which the MTPB was paired with itself. In dual-task performance, performance of both the primary and the secondary MTPB tasks degraded.

Data requirements. The MTPB requires individual programming and analysis of six tasks as well as coordination among them during the experiment.

Thresholds. Not stated.

SOURCES

Alluisi, E.A. and Morgan, B.B. Effects on sustained performance of time-sharing a three-phase code transformation task (3P-Cotran). *Perceptual and Motor Skills.* 33, 639–651, 1971.

Lysaght, R.J., Hill, S.G., Dick, A.O., Plamondon, B.D., Linton, P.M., Wierwille, W.W., Zaklad, A.L., Bittner, A.C., and Wherry, R.J. *Operator workload: Comprehensive review and evaluation of operator workload methodologies (Technical Report 851).* Alexandria, VA: Army Research Institute for the Behavioral and Social Sciences; June 1989.

3.2.15 Occlusion Secondary Task

General description. "The subject's view of a visual display is obstructed (usually by a visor). These obstructions are either initiated by the subject or imposed by the experimenter in order to determine the viewing time needed to perform a task adequately" (Lysaght et al., 1989, p. 234).

Strengths and limitations. This task can be extremely disruptive of primary-task performance.

On the basis of the results of four studies in which a secondary occlusion task was used, Lysaght et al. (1989) reported that performance remained stable on monitoring and driving primary tasks and degraded on driving primary tasks. Performance of the secondary occlusion task degraded when paired with primary driving tasks (see table 11).

Data requirements. The following data are used to assess performance of this task: mean voluntary occlusion time and percent looking time/total time (Lysaght et al., 1989, p. 236).

Thresholds. Not stated.

TABLE 11
References Listed by the Effect on Performance of Primary Tasks Paired with a Secondary Occlusion Task

	Primary Task			Secondary Task		
Type	**Stable**	**Degraded**	**Enhanced**	**Stable**	**Degraded**	**Enhanced**
Driving	Farber and Gallagher (1972)	Hicks and Wierwille (1979) Senders, Kristofferson, Levison, Dietrich, and Ward (1967)			Farber and Gallagher (1972) Senders et al. (1967)	
Monitoring	Gould and Schaffer (1967)					

Source: From Lysaght, R.J., Hill, S.G., Dick, A.O., Plamondon, B.D., Linton, P.M., Wierwille, W.W., Zaklad, A.L., Bittner, A.C., and Wherry, R.J. *Operator workload: Comprehensive review and evaluation of operator workload methodologies (Technical Report 851).* Alexandria, VA: Army Research Institute for the Behavioral and Social Sciences; June 1989, p. 250.

SOURCES

Farber, E. and Gallagher, V. Attentional demand as a measure of the influence of visibility conditions on driving task difficulty. *Highway Research Record.* 414, 1–5, 1972.

Gould, J.D. and Schaffer, A. The effects of divided attention on visual monitoring of multichannel displays. *Human Factors.* 9, 191–202, 1967.

Hicks, T.G. and Wierwille, W.W. Comparison of five mental workload assessment procedures in a moving-base driving simulator. *Human Factors.* 21, 129–143, 1979.

Lysaght, R.J., Hill, S.G., Dick, A.O., Plamondon, B.D., Linton, P.M., Wierwille, W.W., Zaklad, A.L., Bittner, A.C., and Wherry, R.J. *Operator workload: Comprehensive review and evaluation of operator workload methodologies (Technical Report 851).* Alexandria, VA: Army Research Institute for the Behavioral and Social Sciences; June 1989.

Senders, J.W., Kristofferson, A.B., Levison, W.H., Dietrich, C.W., and Ward, J.L. The attentional demand of automobile driving. *Highway Research Record.* 195, 15–33, 1967.

3.2.16 Problem-Solving Secondary Task

General description. "The subject engages in a task which requires verbal or spatial reasoning. For example, the subject might attempt to solve anagram or logic problems" (Lysaght et al., 1989, p. 233).

Strengths and limitations. "This class of tasks is thought to impose heavy demands on central processing resources" (Lysaght et al., 1989, p. 233). On the basis of eight studies in which a problem-solving secondary task was used, Lysaght et al. (1989) reported that performance remained stable on a primary monitoring

TABLE 12

References Listed by the Effect on Performance of Primary Tasks Paired with a Secondary Problem-Solving Task

	Primary Task			Secondary Task		
Type	**Stable**	**Degraded**	**Enhanced**	**Stable**	**Degraded**	**Enhanced**
Choice RT					Schouten, Kalsbeek, and Leopold (1962)	
Driving		Wetherell (1981)			Wetherell (1981)	
Memory		Trumbo, Noble, and Swink (1967)			Trumbo et al. (1967)	
Monitoring	Gould and Schaffer (1967) Smith, Lucaccini, Groth, and Lyman (1966)					Smith et al. (1966)
Problem solving					Chiles and Alluisi (1979)	
Tracking		Trumbo et al. (1967)			Trumbo et al. (1967)	

Source: From Lysaght, R.J., Hill, S.G., Dick, A.O., Plamondon, B.D., Linton, P.M., Wierwille, W.W., Zaklad, A.L., Bittner, A.C., and Wherry, R.J. *Operator workload: Comprehensive review and evaluation of operator workload methodologies (Technical Report 851).* Alexandria, VA: Army Research Institute for the Behavioral and Social Sciences; June 1989, p. 250.

task and degraded on driving, tracking, and memory primary tasks. Performance of the secondary problem-solving task remained stable when paired with a primary tracking task; degraded when paired with problem-solving, driving, choice RT, and memory primary tasks; and improved when paired with a primary monitoring task (see table 12).

Data requirements. The following data are used for these tasks: number of correct responses, number of incorrect responses, and mean RT for correct responses (Lysaght et al., 1989, p. 233).

Thresholds. Not stated.

SOURCES

Chiles, W.D. and Alluisi, E.A. On the specification of operator or occupational workload with performance-measurement methods. *Human Factors.* 21, 515–528, 1979.

Gould, J.D. and Schaffer, A. The effects of divided attention on visual monitoring of multichannel displays. *Human Factors.* 9, 191–202, 1967.

Lysaght, R.J., Hill, S.G., Dick, A.O., Plamondon, B.D., Linton, P.M., Wierwille, W.W., Zaklad, A.L., Bittner, A.C., and Wherry, R.J. *Operator workload: Comprehensive review and evaluation of operator workload methodologies (Technical Report 851).* Alexandria, VA: Army Research Institute for the Behavioral and Social Sciences; June 1989.

Schouten, J.F., Kalsbeek, J.W.H., and Leopold, F.F. On the evaluation of perceptual and mental load. *Ergonomics.* 5, 251–260, 1962.

Smith, R.L., Lucaccini, L.F., Groth, H., and Lyman, J. Effects of anticipatory alerting signals and a compatible secondary task on vigilance performance. *Journal of Applied Psychology.* 50, 240–246, 1966.

Trumbo, D., Noble, M., and Swink, J. Secondary task interference in the performance of tracking tasks. *Journal of Experimental Psychology.* 73, 232–240, 1967.

Wetherell, A. The efficacy of some auditory-vocal subsidiary tasks as measures of mental load on male and female drivers. *Ergonomics.* 24, 197–214, 1981.

3.2.17 Production/Handwriting Secondary Task

General description. "The subject is required to produce spontaneous handwritten passages of prose" (Lysaght et al., 1989, p. 234).

Strengths and limitation. "With primary tasks that impose a high workload, subject's handwriting is thought to deteriorate (i.e., semantic and grammatical errors) under such conditions" (Lysaght et al., 1989, p. 234). Lysaght et al. (1989) cite a study reported by Schouten, Kalsbeek, and Leopold (1962) in which a spontaneous writing secondary task was paired with a choice RT primary task. Performance on the secondary task degraded.

Data requirements. The number of semantic and grammatical errors is used as data for this task (Lysaght et al., 1989, p. 236).

Thresholds. Not stated.

SOURCES

Lysaght, R.J., Hill, S.G., Dick, A.O., Plamondon, B.D., Linton, P.M., Wierwille, W.W., Zaklad, A.L., Bittner, A.C., and Wherry, R.J. *Operator workload: Comprehensive review and evaluation of operator workload methodologies (Technical Report 851).* Alexandria, VA: Army Research Institute for the Behavioral and Social Sciences; June 1989.

Schouten, J.F., Kalsbeek, J.W.H., and Leopold, F.F. On the evaluation of perceptual and mental load. *Ergonomics.* 15, 251–260, 1962.

3.2.18 Psychomotor Secondary Task

General description. "The subject must perform a psychomotor task such as sorting different types of metal screws by size" (Lysaght et al., 1989, p. 233).

Strengths and limitations. "Tasks of this nature are thought to reflect psychomotor skills" (Lysaght et al., 1989, p. 233). On the basis of three studies in which a psychomotor secondary task was used, Lysaght et al. (1989) reported that performance of a tracking primary task degraded. Performance of the secondary psychomotor task degraded when paired with either a tracking or choice RT primary task (see table 13).

Data requirements. The number of completed items is used to assess performance of this task.

Thresholds. Not stated.

TABLE 13
References Listed by the Effect on Performance of Primary Tasks Paired with a Secondary Psychomotor Task

	Primary Task			Secondary Task		
Type	Stable	Degraded	Enhanced	Stable	Degraded	Enhanced
Choice RT					Schouten, Kalsbeek, and Leopold (1962)	
Tracking		Bergeron (1968) Wickens (1976)			Bergeron (1968)	

Source: From Lysaght, R.J., Hill, S.G., Dick, A.O., Plamondon, B.D., Linton, P.M., Wierwille, W.W., Zaklad, A.L., Bittner, A.C., and Wherry, R.J. *Operator workload: Comprehensive review and evaluation of operator workload methodologies (Technical Report 851).* Alexandria, VA: Army Research Institute for the Behavioral and Social Sciences; June 1989, p. 251.

SOURCES

Bergeron, H.P. Pilot response in combined control tasks. *Human Factors.* 10, 277–282, 1968.

Lysaght, R.J., Hill, S.G., Dick, A.O., Plamondon, B.D., Linton, P.M., Wierwille, W.W., Zaklad, A.L., Bittner, A.C., and Wherry, R.J. *Operator workload: Comprehensive review and evaluation of operator workload methodologies (Technical Report 851).* Alexandria, VA: Army Research Institute for the Behavioral and Social Sciences; June 1989.

Schouten, J.F., Kalsbeek, J.W.H., and Leopold, F.F. On the evaluation of perceptual and mental load. *Ergonomics.* 5, 251–260, 1962.

Wickens, C.D. The effects of divided attention on information processing in manual tracking. *Journal of Experimental Psychology: Human Perception and Performance.* 2, 1–12, 1976.

3.2.19 Randomization Secondary Task

General description. "The subject must generate a random sequence of numbers, for example. It is postulated that with increased workload levels subjects will generate repetitive responses (i.e., lack randomness in responses)" (Lysaght et al., 1989, p. 232).

Strengths and limitations. The task is extremely intrusive, and calculating "randomness" difficult and time consuming. On the basis of five studies that used a randomization secondary task, Lysaght et al. (1989) reported that performance remained stable on tracking, card-sorting, and driving primary tasks; and degraded on tracking and memory primary tasks. Performance of the secondary randomization task remained stable when paired with a tracking primary task and degraded when paired with tracking and card-sorting primary tasks (see table 14).

Data requirements. The experimenter must calculate a percent redundancy score in bits of information.

Thresholds. Not stated.

TABLE 14
References Listed by the Effect on Performance of Primary Tasks Paired with a Secondary Randomization Task

	Primary Task			Secondary Task		
Type	**Stable**	**Degraded**	**Enhanced**	**Stable**	**Degraded**	**Enhanced**
Card sorting	Baddeley (1966)				Baddeley (1966)	
Driving	Wetherell (1981)					
Memory		Trumbo and Noble (1970)				
Tracking	Zeitlin and Finkelman (1975)	Truijens, Trumbo, and Wagenaar (1976)		Zeitlin and Finkelman (1975)	Truijens et al. (1976)	

Source: From Lysaght, R.J., Hill, S.G., Dick, A.O., Plamondon, B.D., Linton, P.M., Wierwille, W.W., Zaklad, A.L., Bittner, A.C., and Wherry, R.J. *Operator workload: Comprehensive review and evaluation of operator workload methodologies (Technical Report 851).* Alexandria, VA: Army Research Institute for the Behavioral and Social Sciences; June 1989, p. 250.

SOURCES

Baddeley, A.D. The capacity for generating information by randomization. *Quarterly Journal of Experimental Psychology.* 18, 119–130, 1966.

Lysaght, R.J., Hill, S.G., Dick, A.O., Plamondon, B.D., Linton, P.M., Wierwille, W.W., Zaklad, A.L., Bittner, A.C., and Wherry, R.J. *Operator workload: Comprehensive review and evaluation of operator workload methodologies (Technical Report 851).* Alexandria, VA: Army Research Institute for the Behavioral and Social Sciences; June 1989.

Truijens, C.L., Trumbo, D.A., and Wagenaar, W.A. Amphetamine and barbiturate effects on two tasks performed singly and in combination. *Acta Psychologica.* 40, 233–244, 1976.

Trumbo, D. and Noble, M. Secondary task effects on serial verbal learning. *Journal of Experimental Psychology.* 85, 418–424, 1970.

Wetherell, A. The efficacy of some auditory-vocal subsidiary tasks as measures of the mental load on male and female drivers. *Ergonomics.* 24, 197–214, 1981.

Zeitlin, L.R. and Finkelman, J.M. Research note: Subsidiary task techniques of digit generation and digit recall indirect measures of operator loading. *Human Factors.* 17, 218–220, 1975.

3.2.20 Reading Secondary Task

General description. Subjects are asked to read digits or words aloud from a visual display. Measures can include number of digits or words read, longest interval between spoken responses, longest string of consecutive digits or words, and the number of times three consecutive digits or words occur (Savage, Wierwille, and Cordes, 1978).

Strengths and limitations. This task has been sensitive to the difficulty of a monitoring task. There were significant differences in the number of random digits spoken, the longest consecutive string of spoken digits, and the number of times three consecutive digits were spoken (Savage, Wierwille, and Cordes, 1978). The longest interval between spoken responses was not significantly different between various levels of primary task difficulty (i.e., monitoring of 2, 3, or 4 m).

Wierwille, Gutmann, Hicks, and Muto (1977) asked subjects to read random digits aloud while driving in a simulator while steering ratio and wind disturbance level were manipulated. They reported that this secondary task was significantly affected by both independent variables. They concluded that this secondary task was simple to implement but may not be able to detect small changes in disturbance. In a follow-on study in 1978, Wierwille and Gutmann reported that this secondary task degraded primary-task performance, but only at low levels of workload.

Data requirements. Spoken responses must be recorded, timed, and tabulated.

Thresholds. Not stated.

SOURCES

Savage, R.E., Wierwille, W.W., and Cordes, R.E. Evaluating the sensitivity of various measures of operator workload using random digits as a secondary task. *Human Factors.* 20(6), 649–654, 1978.

Wierwille, W.W. and Gutmann, J.C. Comparison of primary and secondary task measures as a function of simulated vehicle dynamics and driving conditions. *Human Factors.* 20(2), 233–244, 1978.

Wierwille, W.W., Gutmann, J.C., Hicks, T.G., and Muto, W.H. Secondary task measurement of workload as a function of simulated vehicle dynamics and driving conditions. *Human Factors.* 19(6), 557–565, 1977.

3.2.21 Simple Reaction-Time Secondary Task

General description. "The subject is presented with one discrete stimulus (either visual or auditory) and generates one response to this stimulus" (Lysaght et al., 1989, p. 232).

Strengths and limitations. This task minimizes central processing and response selection demands on the subject (Lysaght et al., 1989, p. 232).

On the basis of 10 studies in which a simple RT secondary task was used, Lysaght et al. (1989) reported that performance remained stable on choice RT and classification primary tasks; degraded on tracking, classification, and lexical decision tasks; and improved on detection and driving primary tasks. Performance of the secondary simple RT task degraded when paired with tracking, choice RT, memory, detection, classification, driving, and lexical decision primary tasks; and improved when paired with a tracking primary task (see table 15).

Lisper, Laurell, and van Loon (1986) reported that subjects who had longer RTs on driving with simple auditory RT task on a closed course were more likely to fall asleep during an on-road drive.

Andre, Heers, and Cashion (1995) reported significant increase in pitch, roll, and yaw error of a simulated flight task while performing a secondary simple RT task.

Data requirements. The experimenter must be able to calculate mean RT for correct responses and the number of correct responses.

Thresholds. Not stated.

TABLE 15

References Listed by the Effect on Performance of Primary Tasks Paired with a Secondary Simple RT Task

	Primary Task			Secondary Task		
Type	**Stable**	**Degraded**	**Enhanced**	**Stable**	**Degraded**	**Enhanced**
Choice RT	Becker (1976)				Becker (1976)	
Classification	Comstock (1973)	Miller (1975)			Comstock (1973) Miller (1975)	
Detection			Laurell and Lisper (1978)		Laurell and Lisper (1978)	
Driving			Laurell and Lisper (1978)		Laurell and Lisper (1978) Lisper, Laurell, and Stening (1973)	
Lexical decision		Becker (1976)			Becker (1976)	
Memory		Dodds, Clark-Carter, and Howarth (1986)*			Martin and Kelly (1974)	

Simulated flight task		Andre, Heers, and Cashion (1995)*				
Tracking	Martin, Long, and Broome (1984) for verbal response on secondary task	Heimstra (1970) Kelly and Klapp (1985) Klapp, Kelly, Battiste, and Dunbar (1984) Martin, Long, and Broome (1984) for pointing response on secondary task Wickens and Gopher (1977)		Martin, Long, and Broome (1984)	Wickens and Gopher (1977)	Heimstra (1970)

* Not included in Lysaght, R.J., Hill, S.G., Dick, A.O., Plamondon, B.D., Linton, P.M., Wierwille, W.W., Zaklad, A.L., Bittner, A.C., and Wherry, R.J. *Operator workload: Comprehensive review and evaluation of operator workload methodologies (Technical Report 851).* Alexandria, VA: Army Research Institute for the Behavioral and Social Sciences; June 1989.

Source: From Lysaght, R.J., Hill, S.G., Dick, A.O., Plamondon, B.D., Linton, P.M., Wierwille, W.W., Zaklad, A.L., Bittner, A.C., and Wherry, R.J. *Operator workload: Comprehensive review and evaluation of operator workload methodologies (Technical Report 851).* Alexandria, VA: Army Research Institute for the Behavioral and Social Sciences; June 1989, p. 251.

SOURCES

Andre, A.D., Heers, S.T., and Cashion, P.A. Effects of workload preview on task scheduling during simulated instrument flight. *International Journal of Aviation Psychology*. 5(1), 5–23, 1995.

Becker, C.A. Allocation of attention during visual word recognition. *Journal of Experimental Psychology: Human Perception and Performance*. 2, 556–566, 1976.

Comstock, E.M. Processing capacity in a letter-matching task. *Journal of Experimental Psychology*. 100, 63–72, 1973.

Dodds, A.G., Clark-Carter, D., and Howarth, C.I. The effects of precueing on vibrotactile reaction times: implications for a guidance device for blind people. *Ergonomics*. 29(9): 1063–1071, 1986.

Heimstra, N.W. The effects of "stress fatigue" on performance in a simulated driving situation. *Ergonomics*. 13, 209–213, 1970.

Kelly, P.A. and Klapp, S.T. Hesitation in tracking induced by a concurrent manual task. *Proceedings of the 21st Annual Conference on Manual Control* (pp. 19.1–19.3). Columbus, OH: Ohio State University, 1985.

Klapp, S.T., Kelly, P.A., Battiste, V., and Dunbar, S. Types of tracking errors induced by concurrent secondary manual task. *Proceedings of the 20th Annual Conference on Manual Control* (pp. 299–304). Moffett Field, CA: Ames Research Center, 1984.

Laurell, H. and Lisper, H.L. A validation of subsidiary reaction time against detection of roadside obstacles during prolonged driving. *Ergonomics*. 21, 81–88, 1978.

Lisper, H.L., Laurell, H., and Stening, G. Effects of experience of the driver on heart-rate, respiration-rate, and subsidiary reaction time in a three-hour continuous driving task. *Ergonomics*. 16, 501–506, 1973.

Lisper, H.O., Laurell, H., and van Loon, J. Relation between time to falling asleep behind the wheel on a closed course and changes in subsidiary reaction time during prolonged driving on a motorway. *Ergonomics*. 29(3), 445–453, 1986.

Lysaght, R.J., Hill, S.G., Dick, A.O., Plamondon, B.D., Linton, P.M., Wierwille, W.W., Zaklad, A.L., Bittner, A.C., and Wherry, R.J. *Operator workload: Comprehensive review and evaluation of operator workload methodologies (Technical Report 851)*. Alexandria, VA: Army Research Institute for the Behavioral and Social Sciences; June 1989.

Martin, D.W. and Kelly, R.T. Secondary task performance during directed forgetting. *Journal of Experimental Psychology*. 103, 1074–1079, 1974.

Martin, J., Long, J., and Broome, D. The division of attention between a primary tracing task and secondary tasks of pointing with a stylus or speaking in a simulated ship's-gunfire-control task. *Ergonomics*. 27(4), 397–408, 1984.

Miller, K. Processing capacity requirements of stimulus encoding. *Acta Psychologica*. 39, 393–410, 1975.

Wickens, C.D. and Gopher, D. Control theory measures of tracking as indices of attention allocation strategies. *Human Factors*. 19, 349–365, 1977.

3.2.22 Simulated Flight Secondary Task

General description. "Depending on the purpose of the particular study, the subject is required to perform various maneuvers (e.g., landing approaches) under different types of conditions such as instrument flight rules or simulated crosswind conditions" (Lysaght et al., 1989, p. 234).

Strengths and limitations. This task requires extensive subject training.

Data requirements. The experimenter should record the following data: mean error from required altitude, root-mean-square localizer error, root-mean-square glide-slope error, and number of control movements, and attitude high-pass mean square (Lysaght et al., 1989, p. 236).

Thresholds. Not stated.

SOURCE

Lysaght, R.J., Hill, S.G., Dick, A.O., Plamondon, B.D., Linton, P.M., Wierwille, W.W., Zaklad, A.L., Bittner, A.C., and Wherry, R.J. *Operator workload: Comprehensive review and evaluation of operator workload methodologies (Technical Report 851).* Alexandria, VA: Army Research Institute for the Behavioral and Social Sciences; June 1989.

3.2.23 Spatial-Transformation Secondary Task

General description. "The subject must judge whether information (data)—provided by an instrument panel or radar screen—matches information which is spatially depicted by pictures or drawings of aircraft" (Lysaght et al., 1989, p. 233). Lysaght et al. (1989) cite work by Vidulich and Tsang (1985) in which performance of a spatial transformation secondary task was degraded when paired with a primary tracking task.

Kramer, Wickens, and Donchin (1984) reported a significant decrease in tracking error on a primary task when subjects performed translational changes of the cursor as a secondary task.

Strengths and limitations. "This task involves perceptual and cognitive processes" (Lysaght et al., 1989, p. 233). Damos (1986) paired a visual matrix rotation task with a subtraction task. She reported no significant differences between speech responses or manual responses on the first experiment. There was a significant decrease in RTs for correct speech responses to the subtraction task when the responses were adjusted for the delay in the speech recognition system being used.

Data requirements. The following data are used to assess performance of this task: mean RT for correct responses, number of correct responses, and number of incorrect responses (Lysaght et al., 1989, p. 236).

Thresholds. Not stated.

SOURCES

Damos, D. The effect of using voice generation and recognition systems on the performance of dual task. *Ergonomics.* 29(11), 1359–1370, 1986.

Kramer, A.F., Wickens, C.D., and Donchin, E. Performance enhancements under dual-task conditions. *Annual Conference on Manual Control.* 21–35, 1984.

Lysaght, R.J., Hill, S.G., Dick, A.O., Plamondon, B.D., Linton, P.M., Wierwille, W.W., Zaklad, A.L., Bittner, A.C., and Wherry, R.J. *Operator workload: Comprehensive review and evaluation of operator workload methodologies (Technical Report 851).* Alexandria, VA: Army Research Institute for the Behavioral and Social Sciences; June 1989.

Vidulich, M.A. and Tsang, P.S. Evaluation of two cognitive abilities tests in a dual-task environment. *Proceedings of the 21st Annual Conference on Manual Control.* 12.1–12.10, 1985.

3.2.24 Speed-Maintenance Secondary Task

General description. "The subject must operate a control knob to maintain a designated constant speed. This task is a psychomotor type task" (Lysaght et al., 1989, p. 234).

Strengths and limitations. This task provides a constant estimate of reserve response capacity but may be extremely intrusive on primary-task performance.

Data requirements. Response is used as data for this task.

Thresholds. Not stated.

SOURCE

Lysaght, R.J., Hill, S.G., Dick, A.O., Plamondon, B.D., Linton, P.M., Wierwille, W.W., Zaklad, A.L., Bittner, A.C., and Wherry, R.J. *Operator workload: Comprehensive review and evaluation of operator workload methodologies (Technical Report 851).* Alexandria, VA: Army Research Institute for the Behavioral and Social Sciences; June 1989.

3.2.25 Sternberg Memory Secondary Task

General description. The Sternberg (1966) recognition task presents a subject with a series of single letters. After each letter, the subject indicates whether that letter was or was not part of a previously memorized set of letters. RT is typically measured at two memory set sizes, two and four, and is plotted against set size (see figure 6). Differences in the slope (*b* in figure 6) across various design configurations indicate differences in central information processing demands. Changes in the intercept (*a*1 and *a*2 in figure 6) suggest differences in either sensory or response demands. Additional data for this task include number of correct responses and RTs for correct responses.

Strengths and limitations. The Sternberg task is sensitive to workload imposed by wind conditions, handling qualities, and display configurations. For example, Wolf (1978) reported that the longest RTs to the Sternberg task occurred in the high gust, largest memory set (4), and poorest handling qualities condition. Schiflett (1980) reported both increased Sternberg RT and errors with degraded handling qualities during approach and landing tasks. Similarly, Schiflett, Linton, and Spicuzza (1982) reported increases in four Sternberg measures (RT for correct responses, intercept, slope, and percent errors) as handling qualities were degraded. Poston and Dunn (1986) used the Sternberg task to assess a kinesthetic tactile display. They recorded response speed and accuracy.

A second advantage of the task is that it is minimally intrusive on performance of a primary flight task (Schiflett, Linton, and Spicuzza, 1980; Dellinger and Taylor, 1985). Used alone, the Sternberg provided reliable measure of mercury exposure (Smith and Langolf, 1981).

However, there are several problems associated with the use of the task. Based on data collected in a flight simulator (Taylor, Dellinger, Richardson, Weller, Porges, Wickens, LeGrand, and Davis, 1985), RTs increase with workload but more so for negative than for positive responses. Further, work by Micalizzi (1981) suggests that performance of the Sternberg task is poorest when it is time-shared with a failure detection task. Gawron, Schiflett, Miller, Ball, Slater, Parker, Lloyd, Travale, and

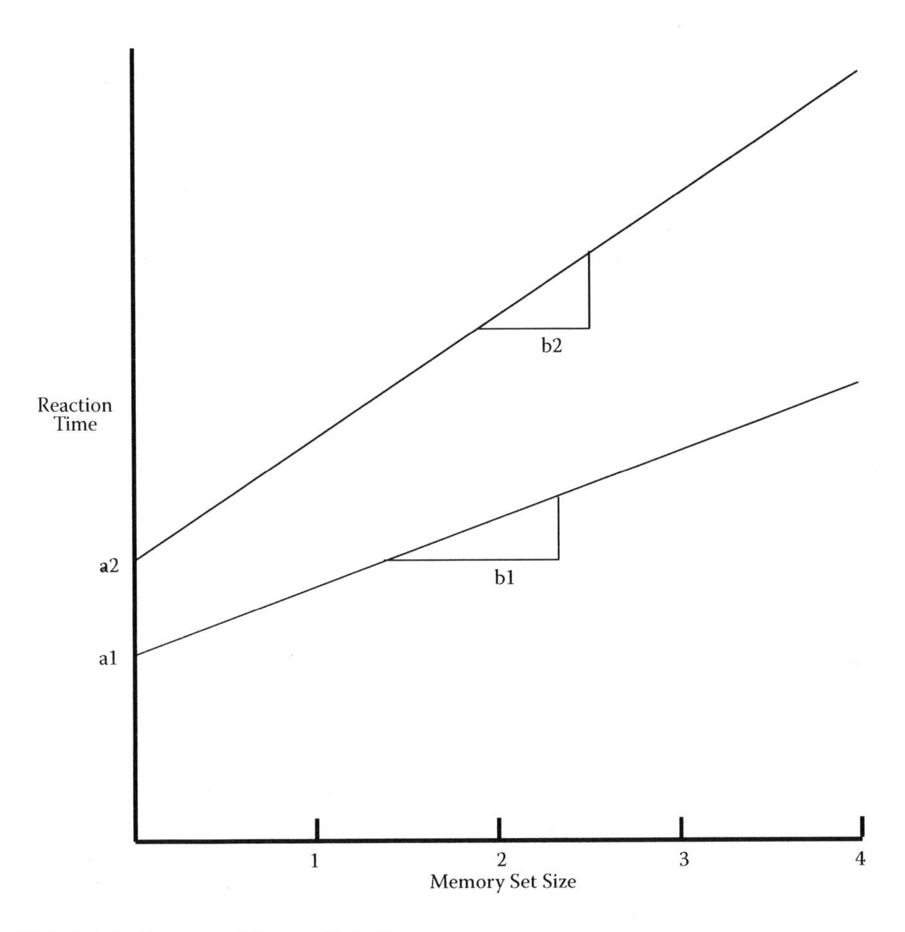

FIGURE 6 Sternberg Memory Task Data.

Spicuzza (1988) used the Sternberg memory search task to access the in-flight effects of pyridostigmine bromide. There were no significant effects of drug or crew position (pilot or copilot). This finding may be due to a true lack of a drug or crew position effect or to insensitivity of the measure. In an earlier study, Knotts and Gawron (1983) reported that the presence of the Peripheral Vision Display (PVD) reduced Sternberg RTs for one subject but not for another. They also reported that performance of the Sternberg task improved throughout the flight program and suggested extensive training before using the task in flight.

Micalizzi and Wickens (1980) compared Sternberg RT under single- and dual-task conditions. There were three manipulations of the Sternberg task: no mask, single mask, or double mask. They reported a significant increase in Sternberg RT under dual-task condition for the two masked conditions as compared to the no-mask condition.

On the basis of the results of four studies in which a secondary Sternberg task was used, Lysaght et al. (1989) reported performance degraded on tracking, choice RT, and driving primary tasks; and improved on a tracking primary task. Performance of the secondary Sternberg task degraded when paired with a primary tracking task (see table 16) or a simulated flight task.

TABLE 16

References Listed by the Effect on Performance of Primary Tasks Paired with a Secondary Sternberg Task

	Primary Task			Secondary Task		
Type	**Stable**	**Degraded**	**Enhanced**	**Stable**	**Degraded**	**Enhanced**
Choice RT		Hart, Shively, Vidulich, and Miller (1985)				
Driving		Wetherell (1981)				
Mental math		Payne, Peters, Birkmire, Bonto, Anatasi, and Wenger (1994)*				
Simple RT		Payne, Peters, Birkmire, Bonto, Anatasi, and Wenger (1994)*				
Simulated flight	Crawford, Pearson, and Hoffman (1978)* Wierwille and Connor (1983)	O'Donnell (1976)*			Schiflett, Linton, and Spicuzza (1982)* Hyman, Collins, Taylor, Domino, and Nagel (1988)	
Tracking	Tsang, Velaquez, and Vidulich (1996)	Wickens and Yeh (1985) Vidulich and Wickens (1986)	Briggs, Peters, and Fisher (1972)		Briggs et al. (1972) Wickens and Yeh (1985)	

* Not included in Lysaght, R.J., Hill, S.G., Dick, A.O., Plamondon, B.D., Linton, P.M., Wierwille, W.W., Zaklad, A.L., Bittner, A.C., and Wherry, R.J. *Operator workload: Comprehensive review and evaluation of operator workload methodologies (Technical Report 851)*. Alexandria, VA: Army Research Institute for the Behavioral and Social Sciences; June 1989.

Source: From Lysaght, R.J., Hill, S.G., Dick, A.O., Plamondon, B.D., Linton, P.M., Wierwille, W.W., Zaklad, A.L., Bittner, A.C., and Wherry, R.J. *Operator workload: Comprehensive review and evaluation of operator workload methodologies (Technical Report 851)*. Alexandria, VA: Army Research Institute for the Behavioral and Social Sciences; June 1989, p. 252.

Vidulich and Wickens (1986) reported that the "addition of the tracking task tends to overwhelm distinctions among the Sternberg task configurations" (p. 293). Further, Manzey, Lorenz, Schiewe, Finell, and Thiele (1995) reported significant differences in dual-task performance of the Sternberg, and tracking prior and during space flight. However, there was no difference in single-task performance of the Sternberg.

The Sternberg has also been used as a primary task to evaluate the effects of altitude on short-term memory. Kennedy, Dunlap, Banderet, Smith, and Houston (1989) reported significant increases in time per response and decreases in number correct. Rolnick and Bles (1989) also used the Sternberg but evaluated the effect of tilting and closed cabin conditions simulating shipboard conditions. RT increased in the closed cabin condition compared to the no-motion condition but showed no significant difference with the artificial horizon or windows conditions.

Van de Linde (1988) used a Sternberg-like test to assess the effect of wearing a respirator on the ability to recall numbers and letters from memory, and to concentrate. Memory sets were 1, 2, 3, or 4 with 24 targets mixed with 120 nontargets. Letter targets took longer to find than number targets. There was no significant difference between trials 1 and 3. However, there was a 12% decrease in time while wearing the respirator.

Tsang and Velazquez (1996) used the Sternberg combined with tracking to evaluate a workload profile measure consisting of ratings of perceptual/central, response, spatial, verbal, visual, and manual. They concluded that the variance accounted for by the memory set size variations of the Sternberg was comparable to the subjective workload ratings.

Data requirements. The slopes and intercepts of the RT are used as data for this task. Wickens, Hyman, Dellinger, Taylor, and Meador (1986) have recommended that the Sternberg letters in the memorized set be changed after 50 trials. Schiflett (1983) recommended the use of an adaptive interstimulus interval (ISI). Adaptation would be based on response accuracy. Knotts and Gawron (1983) recommended an extensive training on the Sternberg task before data collection.

Thresholds. Not stated.

SOURCES

Briggs, G.E., Peters, G.L., and Fisher, R.P. On the locus of the divided-attention effects. *Perception and Psychophysics*. 11, 315–320, 1972.

Crawford, B.M., Pearson, W.H., and Hoffman, M. *Multipurpose digital switching and flight control workload (AMRL-TR-78-43).* Wright-Patterson Air Force Base, OH: Air Force Aerospace Medical Research Laboratory, 1978.

Dellinger, J.A. and Taylor, H.L. The effects of atropine sulfate on flight simulator performance on respiratory period interactions. *Aviation, Space, and Environmental Medicine*, 1985.

Gawron, V.J., Schiflett, S., Miller, J., Ball, J., Slater, T., Parker, F., Lloyd, M., Travale, D., and Spicuzza, R.J. *The effect of pyridostigmine bromide on in-flight aircrew performance (USAFSAM-TR-87-24).* Brooks AFB, TX: School of Aerospace Medicine; January 1988.

Hart, S.G., Shively, R.J., Vidulich, M.A., and Miller, R.C. The effects of stimulus modality and task integrity: Predicting dual-task performance and workload from single-task levels. *Proceedings of the 21st Annual Conference on Manual Control* (pp. 5.1–5.18). Columbus, OH: NASA Ames Research Center and Ohio State University, 1985.

Hyman, F.C., Collins, W.E., Taylor, H. L., Domino, E.F., and Nagel, R.J. Instrument flight performance under the influence of certain combinations of antiemetic drugs. *Aviation, Space, and Environmental Medicine.* 59(6), 533–539, 1988.

Kennedy, R.S., Dunlap, W.P., Banderet, L.E., Smith, M.G., and Houston, C.S. Cognitive performance deficits in a simulated climb of Mount Everest: Operation Everest II. *Aviation, Space, and Environmental Medicine.* 60(2), 99–104, 1989.

Knotts, L.H. and Gawron, V.J. *A preliminary flight evaluation of the peripheral vision display using the NT-33A aircraft (Report 6645-F-13).* Buffalo, NY: Calspan; December 1983.

Lysaght, R.J., Hill, S.G., Dick, A.O., Plamondon, B.D., Linton, P.M., Wierwille, W.W., Zaklad, A.L., Bittner, A.C., and Wherry, R.J. *Operator workload: Comprehensive review and evaluation of operator workload methodologies (Technical Report 851).* Alexandria, VA: Army Research Institute for the Behavioral and Social Sciences; June 1989.

Manzey, D., Lorenz, B., Schiewe, A., Finell, G., and Thiele, G. Dual-task performance in space: Results from a single-case study during a short-term space mission. *Human Factors.* 37(4), 667–681, 1995.

Micalizzi, J. *The structure of processing resource demands in monitoring automatic systems (Technical Report 81-2T).* Wright-Patterson AFB, OH: Air Force Institute of Technology, 1981.

Micalizzi, J. and Wickens, C.D. *The application of additive factors methodology to workload assessment in a dynamic system monitoring task.* Champaign, IL: University of Illinois, Engineering-Psychology Research Laboratory, TREPL-80-2/ONR- 80-2, December 1980.

O'Donnell, R.D. Secondary task assessment of cognitive workload in alternative cockpit configurations. In Hartman, B.O. (Ed.) *Higher mental functioning in operational environments* (pp. C10/1–C10/5). AGARD Conference Proceedings Number 181. Neuilly sur Seine, France: Advisory Group for Aerospace Research and Development, 1976.

Payne, D.G., Peters, L.J., Birkmire, D.P., Bonto, M.A., Anatasi, J.S., and Wenger, M.J. Effects of speech intelligibility level on concurrent visual task performance. *Human Factors.* 36(3), 441–475, 1994.

Poston, A.M. and Dunn, R.S. *Helicopter flight evaluation of kinesthetic tactual displays: An interim report (HEL-TN-3-86).* Aberdeen Proving Ground, MD: Human Engineering Laboratory; March 1986.

Rolnick, A. and Bles, W. Performance and well-being under titling conditions: The effects of visual reference and artificial horizon. *Aviation, Space, and Environmental Medicine.* 60(2), 779–785, 1989.

Schiflett, S.G. *Evaluation of a pilot workload assessment device to test alternate display formats and control handling qualities.* Patuxent River, MD: Naval Air Test Center, SY-33R-80, July 1980.

Schiflett, S.G. Theoretical development of an adaptive secondary task to measure pilot workload for flight evaluations. *Proceedings of the 27th Annual Meeting of the Human Factors Society.* 602–607, 1983.

Schiflett, S., Linton, P.M, and Spicuzza, R.J *Evaluation of a pilot workload assessment device to test alternate display formats and control handling qualities.* Proceedings of North Atlantic Treaty Organization (NATO) Advisory Group for Aerospace Research and Development (AGARD) (Paper Number 312). Neuilly-sur-Seine, France: AGARD, 1980.

Schiflett, S.G., Linton, P.M., and Spicuzza, R.J. Evaluation of a pilot workload assessment device to test alternative display formats and control handling qualities. *Proceedings of the AIAA Workshops on flight testing to identify pilot workload and pilot dynamics* (pp. 222–233). Edwards Air Force Base, CA: Air Force Flight Test Center, 1982.

Smith, P.J. and Langolf, G.D. The use of Sternberg's memory-scanning paradigm in assessing effects of chemical exposure. *Human Factors*. 23(6), 701–708, 1981.
Sternberg, S. High speed scanning in human memory. *Science*. 153, 852–654, 1966.
Taylor, H.L., Dellinger, J.A., Richardson, B.C., Weller, M.H., Porges, S.W., Wickens, C.D., LeGrand, J.E., and Davis, J.M. *The effect of atropine sulfate on aviator performance (Technical Report APL-TR-85-1)*. Champaign, IL: University of Illinois, Aviation Research Laboratory; March 1985.
Tsang, P.S. and Velazquez, V.L. Diagnosticity and multidimensional subjective workload ratings. *Ergonomics*. 39(3), 358–381, 1996.
Tsang, P.S., Velaquez, V.L., and Vidulich, M.A. Viability of resource theories in explaining time-sharing performance. *Acta Psychologica*. 91(2), 175–206, 1996.
Van de Linde, F.J.G. Loss of performance while wearing a respirator does not increase during a 22.5-hour wearing period. *Aviation, Space, and Environmental Medicine*. 59(3), 273–277, 1988.
Vidulich, M.A. and Wickens, C.D. Causes of dissociation between subjective workload measures and performance. *Applied Ergonomics*. 17(4), 291–296, 1986.
Wetherell, A. The efficacy of some auditory-vocal subsidiary tasks as measures of the mental load on male and female drivers. *Ergonomics*. 24, 197–214, 1981.
Wickens, C.D., Hyman, F., Dellinger, J., Taylor, H., and Meador, M. The Sternberg memory search task as an index of pilot workload. *Ergonomics*. 29, 1371–1383, 1986.
Wickens, C.D. and Yeh, Y. POCs and performance decrements: A reply to Kantowitz and Weldon. *Human Factors*. 27, 549–554, 1985.
Wierwille, W.W. and Connor, S. Evaluation of 20 workload measures using a psychomotor task in a moving base aircraft simulator. *Human Factors*. 25, 1–16, 1983.
Wolf, J.D. *Crew workload assessment: Development of a measure of operator workload (AFFDL-TR-78-165)*. Wright-Patterson Air Force Base, OH: Air Force Flight Dynamics Laboratory; December 1978.

3.2.26 Three-Phase Code Transformation Secondary Task

General description. "The subject operates the 3P-Cotran which is a workstation consisting of three indicator lights, a response board for subject responses and a memory unit that the subject uses to save his/her responses. The subject must engage in a 3-phase problem solving task by utilizing information provided by the indicator lights and recording solutions onto the memory unit" (Lysaght et al., 1989, p. 234).

Strengths and limitations. "It is a synthetic work battery used to study work behavior and sustained attention" (Lysaght et al., 1989, p. 234).

Data requirements. The following data are used to evaluate performance of this task: mean RT for different phases of response required and number of errors (resets) for different phases of response required (Lysaght et al., 1989, p. 236).

Thresholds. Not stated.

SOURCE

Lysaght, R.J., Hill, S.G., Dick, A.O., Plamondon, B.D., Linton, P.M., Wierwille, W.W., Zaklad, A.L., Bittner, A.C., and Wherry, R.J. *Operator workload: Comprehensive review and evaluation of operator workload methodologies (Technical Report 851)*. Alexandria, VA: Army Research Institute for the Behavioral and Social Sciences; June 1989.

3.2.27 Time-Estimation Secondary Task

General description. Subjects are asked to produce a given time interval, usually 10 s, from the start of a stimulus, usually a tone. Measures for this task include the number of incomplete estimates and/or the length of the estimates.

Strengths and limitations. The technique is sensitive to workload (Hart, 1978; Wierwille, Casali, Connor, and Rahimi, 1985). For example, Bortolussi, Kantowitz, and Hart (1986) found a significant increase in 10 s time production intervals between easy and difficult flight scenarios. Bortolussi, Hart, and Shively (1987) reported similar results for a 5 s time production task.

In addition, the length of the time interval produced decreased from the beginning to the end of each flight. Gunning (1978) asked pilots to indicate when 10 s had passed after an auditory tone. He reported that both the number of incomplete estimates and the length of the estimates increased as workload increased. Similarly, Madero, Sexton, Gunning, and Moss (1979) reported that the number of incomplete time estimates increased over time in an aerial delivery mission. These researchers also calculated a time estimation ratio (the length of the time estimate in flight divided by the baseline estimate). This measure was also sensitive to workload, with significant increases occurring between cruise and Initial Point (IP), and cruise and Computed Air Release Point (CARP).

Connor and Wierwille (1983) recorded time estimation mean, standard deviation, absolute error, and rmse of completed estimates during three levels of gust and aircraft stability (load). There was only one significant load effect: the standard deviation of time estimates decreased from the low to the medium load then increased from the medium to high load. This same measure was sensitive to communication load, danger, and navigation load. Specifically, Casali and Wierwille (1983) reported significant increases in time estimation standard deviation as communication load increased. Casali and Wierwille (1984) found significant increases between low- and high-danger conditions. Wierwille, Rahimi, and Casali (1985) reported the same results for navigation load, and Hartzell (1979) for difficulty of precision hover maneuvers in a helicopter simulator.

Hauser, Childress, and Hart (1983) reported less variability in time estimates made using a counting technique than those made without a counting technique. Bobko, Bobko, and Davis (1986) reported, however, that verbal time estimates of a fixed time interval decrease as screen size (0.13, 0.28, and 0.58 diagonal meters) increased. In addition, men gave significantly shorter time estimates than women.

On the basis of the results of four studies in which a time estimation secondary task was used, Lysaght et al. (1989) reported that performance on a primary flight simulation task remained stable but degraded on a primary monitoring task. Performance of the secondary time estimation task degraded when paired with either a monitoring or a flight simulation primary task (see table 17).

Data requirements. Although some researchers have reported significant differences in several time estimation measures, the consistency of the findings using time estimation standard deviation suggests that this may be the best time estimation measure.

Thresholds. Not stated.

TABLE 17

References Listed by the Effect on Performance of Primary Tasks Paired with a Secondary Time Estimation Task

	Primary Task			Secondary Task		
Type	**Stable**	**Degraded**	**Enhanced**	**Stable**	**Degraded**	**Enhanced**
Flight simulation	Bortolussi, Hart, and Shively (1987) Bortolussi, Kantowitz, and Hart (1986) Casali and Wierwille (1983)* Kantowitz, Bortolussi, and Hart (1987)* Connor and Wierwille (1983)* Wierwille, Rahimi, and Casali (1985)				Bortolussi et al. (1987, 1986) Gunning (1978)* Wierwille et al. (1985)	
Monitoring		Liu and Wickens (1987)			Liu and Wickens (1987)	

* Not included in Lysaght, R.J., Hill, S.G., Dick, A.O., Plamondon, B.D., Linton, P.M., Wierwille, W.W., Zaklad, A.L., Bittner, A.C., and Wherry, R.J. *Operator workload: Comprehensive review and evaluation of operator workload methodologies (Technical Report 851).* Alexandria, VA: Army Research Institute for the Behavioral and Social Sciences; June 1989.

Source: From Lysaght, R.J., Hill, S.G., Dick, A.O., Plamondon, B.D., Linton, P.M., Wierwille, W.W., Zaklad, A.L., Bittner, A.C., and Wherry, R.J. *Operator workload: Comprehensive review and evaluation of operator workload methodologies (Technical Report 851).* Alexandria, VA: Army Research Institute for the Behavioral and Social Sciences; June 1989, p. 252.

SOURCES

Bobko, D.J., Bobko, P., and Davis, M.A. Effect of visual display scale on duration estimates. *Human Factors*. 28(2), 153–158, 1986.

Bortolussi, M.R., Hart, S.G., and Shively, R.J. Measuring moment-to-moment pilot workload using synchronous presentations of secondary tasks in a motion-base trainer. *Proceedings of the 4th Symposium on Aviation Psychology*. Columbus, OH: Ohio State University, 1987.

Bortolussi, M.R., Kantowitz, B.H., and Hart, S.G. Measuring pilot workload in a motion base trainer: A comparison of four techniques. *Applied Ergonomics*. 17, 278–283, 1986.

Casali, J.G. and Wierwille, W.W. A comparison of rating scale, secondary task, physiological, and primary task workload estimation techniques in a simulated flight emphasizing communications load. *Human Factors*. 25, 623–641, 1983.

Casali, J.G. and Wierwille, W.W. On the measurement of pilot perceptual workload: A comparison of assessment techniques addressing sensitivity and intrusion issues. *Ergonomics*. 27, 1033–1050, 1984.

Connor, S.A. and Wierwille, W.W. *Comparative evaluation of twenty pilot workload assessment measures using a psychomotor task in a moving base aircraft simulator (Report 166457)*. Moffett Field, CA: NASA Ames Research Center; January 1983.

Gunning, D. Time estimation as a technique to measure workload. *Proceedings of the Human Factors Society 22nd Annual Meeting*. 41–45, 1978.

Hart, S.G. Subjective time estimation as an index of workload. *Proceedings of the symposium on man-system interface: Advances in workload study*. 115–131, 1978.

Hartzell, E.J. Helicopter pilot performance and workload as a function of night vision symbologies. *Proceedings of the 18th IEEE Conference on Decision and Control Volumes*. 995–996, 1979.

Hauser, J.R., Childress, M.E., and Hart, S.G. Rating consistency and component salience in subjective workload estimation. *Proceedings of the Annual Conference on Manual Control*. 127–149, 1983.

Kantowitz, B.H., Bortolussi, M.R., and Hart, S.G. Measuring workload in a motion base simulation III. Synchronous secondary task. *Proceedings of the Human Factors Society 31st Annual Meeting*. Santa Monica, CA: Human Factors Society, 1987.

Liu, Y.Y. and Wickens, C.D. *Mental workload and cognitive task automation: An evaluation of subjective and time estimation metrics (NASA 87-2)*. Campaign, IL: University of Illinois Aviation Research Laboratory, 1987.

Lysaght, R.J., Hill, S.G., Dick, A.O., Plamondon, B.D., Linton, P.M., Wierwille, W.W., Zaklad, A.L., Bittner, A.C., and Wherry, R.J. *Operator workload: Comprehensive review and evaluation of operator workload methodologies (Technical Report 851)*. Alexandria, VA: Army Research Institute for the Behavioral and Social Sciences; June 1989.

Madero, R.P., Sexton, G.A., Gunning, D., and Moss, R. *Total aircrew workload study for the AMST (AFFDL-TR-79-3080, Volume 1)*. Wright-Patterson Air Force Base, OH: Air Force Flight Dynamics Laboratory, February 1979.

Wierwille, W.W., Casali, J.G., Connor, S.A., and Rahimi, M. Evaluation of the sensitivity and intrusion of mental workload estimation technique. In Roner, W. (Ed.), *Advances in man-machine systems research (Vol. 2*, pp. 51–127). Greenwich, CT: JAI Press, 1985.

Wierwille, W.W. and Connor, S.A. Evaluation of 20 workload measures using a psychomotor task in a moving base aircraft simulator. *Human Factors*. 25, 1–16, 1983.

Wierwille, W.W., Rahimi, M., and Casali, J.G. Evaluation of 16 measures of mental workload using a simulated flight task emphasizing mediational activity. *Human Factors*. 27, 489–502, 1985.

3.2.28 Tracking Secondary Task

General description. "The subject must follow or track a visual stimulus (target) which is either stationary or moving by means of positioning an error cursor on the stimulus using a continuous manual response device" (Lysaght et al., 1989, p. 232). Tracking tasks require nullifying an error between a desired and an actual location. The Critical Tracking Task (CTT) has been widely used.

Strengths and limitations. Tracking tasks provide a continuous measure of workload. This is especially important in subjects with large variations in workload. Several investigators (Corkindale, Cumming, and Hammerton-Fraser, 1969; Spicuzza, Pinkus, and O'Donnell, 1974) have used a secondary tracking task to successfully evaluate workload in an aircraft simulator. Spicuzza et al. (1974) concluded that a secondary tracking task was a sensitive measure of workload. Clement (1976) used a cross-coupled tracking task in a STOL (Short Take-Off and Landing) simulator to evaluate horizontal situation displays. Park and Lee (1992) reported that tracking task performance distinguished between passing and failing groups of flight students. Manzey, Lorenz, Schiewe, Finell, and Thiele (1995) reported significant decreases in performance of tracking and tracking with the Sternberg tasks between pre- and post-space flight.

The technique may be useful in ground-based simulators but inappropriate for use in flight. For example, Ramacci and Rota (1975) required flight students to perform a secondary tracking task during their initial flights. The researchers were unable to quantitatively evaluate the scores on this task because of artifacts of air turbulence and subject fatigue. Further, Williges and Wierwille (1979) state that hardware constraints make in-flight use of a secondary tracking task unfeasible and potentially unsafe.

In addition, Damos, Bittner, Kennedy, and Harbeson (1981) reported that dual performance of a critical tracking task improved over 15 testing sessions, suggesting a long training time to asymptote. Robinson and Eberts (1987) reported degraded performance on tracking when paired with a speech warning rather than a pictorial warning.

Andre, Heers, and Cashion (1995) reported increased pitch, roll, and yaw rmse in a primary simulated flight task when time-shared with a secondary target acquisition task.

Korteling (1991, 1993) reported age differences in dual-task performance of two one-dimensional compensatory tracking tasks.

On the basis of the results of 12 studies in which a secondary tracking task was used, Lysaght et al. (1989) reported that performance remained stable on tracking, monitoring, and problem-solving primary tasks; degraded on tracking, choice RT, memory, simple RT, detection, and classification primary tasks; and improved on a tracking primary task. Performance of the secondary tracking task degraded when paired with tracking, choice RT, memory, monitoring, problem-solving, simple RT, detection, and classification primary tasks and improved when paired with a tracking primary task (see table 18).

Data requirements. The experimenter should calculate the following: integrated errors in mils (rmse), total time on target, total time of target, number of times of target, and number of target hits (Lysaght et al., 1989, p. 235). A secondary tracking task is most appropriate in systems when a continuous measure of workload

TABLE 18

References Listed by the Effect on Performance of Primary Tasks Paired with a Secondary Tracking Task

	Primary Task			Secondary Task		
Type	**Stable**	**Degraded**	**Enhanced**	**Stable**	**Degraded**	**Enhanced**
Choice RT		Looper (1976) Whitaker (1979)			Hansen (1982) Whitaker (1979)	
Classification		Wickens, Mountford, and Schreiner (1981)			Wickens et al. (1981)	
Detection		Wickens et al. (1981)			Wickens et al. (1981) Robinson and Eberts (1987)*	
Memory		Johnston, Greenberg, Fisher, and Martin (1970)			Johnston et al. (1970)	
Monitoring	Griffiths and Boyce (1971)				Griffiths and Boyce (1971)	

Problem solving	Wright, Holloway, and Aldrich (1974)			Wright et al. (1974)	
Simple RT		Schmidt, Kleinbeck, and Brockman (1984)		Schmidt et al. (1984)	
Simulated flight task		Andre, Heers, and Cashion (1995)*			
Tracking	Mirchandani (1972)	Gawron (1982)* Hess and Teichgraber (1974) Wickens and Kessel (1980) Wickens, Mountford, and Schreiner (1981)	Tsang and Wickens (1984)	Gawron (1982)* Tsang and Wickens (1984) Wickens et al. (1981)	Mirchandani (1972)

* Not included in Lysaght, R.J., Hill, S.G., Dick, A.O., Plamondon, B.D., Linton, P.M., Wierwille, W.W., Zaklad, A.L., Bittner, A.C., and Wherry, R.J. *Operator workload: Comprehensive review and evaluation of operator workload methodologies (Technical Report 851).* Alexandria, VA: Army Research Institute for the Behavioral and Social Sciences; June 1989.

Source: From Lysaght, R.J., Hill, S.G., Dick, A.O., Plamondon, B.D., Linton, P.M., Wierwille, W.W., Zaklad, A.L., Bittner, A.C., and Wherry, R.J. *Operator workload: Comprehensive review and evaluation of operator workload methodologies (Technical Report 851).* Alexandria, VA: Army Research Institute for the Behavioral and Social Sciences; June 1989, p. 251.

is required. It is recommended that known forcing functions be used rather than unknown, quasi-random disturbances.

Thresholds. Not stated.

SOURCES

Andre, A.D., Heers, S.T., and Cashion, P.A. Effects of workload preview on task scheduling during simulated instrument flight. *International Journal of Aviation Psychology*. 5(1), 5–23, 1995.

Clement, W.F. Investigating the use of a moving map display and a horizontal situation indicator in a simulated powered-lift short-haul operation. *Proceedings of the 12th Annual NASA-University Conference on Manual Control*. 201–224, 1976.

Corkindale, K.G.G., Cumming, F.G., and Hammerton-Fraser, A.M. Physiological assessment of a pilot's stress during landing. *Proceedings of the NATO Advisory Group for Aerospace Research and Development*. 56, 1969.

Damos, D., Bittner, A.C., Kennedy, R.S., and Harbeson, M.M. Effects of extended practice on dual-task tracking performance, *Human Factors*. 23(5), 625–631, 1981.

Gawron, V.J. Performance effects of noise intensity, psychological set, and task type and complexity. *Human Factors*. 24(2), 225–243, 1982.

Griffiths, I.D. and Boyce, P.R. Performance and thermal comfort. *Ergonomics*. 14, 457–468, 1971.

Hansen, M.D. Keyboard design variables in dual-task. *Proceedings of the 18th Annual Conference on Manual Control* (pp. 320–326). Dayton, OH: Flight Dynamics Laboratory, 1982.

Hess R.A. and Teichgraber, W.M. Error quantization effects in compensatory tracking tasks. *IEEE Transactions on Systems, Man, and Cybernetics*. SMC-4, 343–349, 1974.

Johnston, W.A., Greenberg, S.N., Fisher, R.P., and Martin, D.W. Divided attention: A vehicle for monitoring memory processes. *Journal of Experimental Psychology*. 83, 164–171, 1970.

Korteling, J.E. Effects of skill integration and perceptual competition on age-related differences in dual-task performance. *Human Factors*. 33(1), 35–44, 1991.

Korteling, J.E. Effects of age and task similarity on dual-task performance. *Human Factors*. 35(1), 99–113, 1993.

Looper, M. The effect of attention loading on the inhibition of choice reaction time to visual motion by concurrent rotary motion. *Perception and Psychophysics*. 20, 80–84, 1976.

Lysaght, R.J., Hill, S.G., Dick, A.O., Plamondon, B.D., Linton, P.M., Wierwille, W.W., Zaklad, A.L., Bittner, A.C., and Wherry, R.J. *Operator workload: Comprehensive review and evaluation of operator workload methodologies (Technical Report 851)*. Alexandria, VA: Army Research Institute for the Behavioral and Social Sciences; June 1989.

Manzey, D., Lorenz, B., Schiewe, A., Finell, G., and Thiele, G. Dual-task performance in space: Results from a single-case study during a short-term space mission. *Human Factors*. 37(4), 667–681, 1995.

Mirchandani, P.B. An auditory display in a dual axis tracking task. *IEEE Transactions on Systems, Man, and Cybernetics*. 2, 375–380, 1972.

Park, K.S. and Lee, S.W. A computer-aided aptitude test for predicting flight performance of trainees. *Human Factors*. 34(2), 189–204, 1992.

Ramacci, C.A. and Rota, P. Flight fitness and psycho-physiological behavior of applicant pilots in the first flight missions. *Proceedings of NATO Advisory Group for Aerospace Research and Development*. 153, B8, 1975.

Robinson, C.P. and Eberts, R.E. Comparison of speech and pictorial displays in a cockpit environment. *Human Factors*. 29(1), 31–44, 1987.

Schmidt, K.H., Kleinbeck, U., and Brockman, W. Motivational control of motor performance by goal setting in a dual-task situation. *Psychological Research.* 46, 129–141, 1984.

Spicuzza, R.J., Pinkus, A.R., and O'Donnell, R.D. *Development of performance assessment methodology for the digital avionics information system.* Dayton, OH: Systems Research Laboratories; August 1974.

Tsang, P.S. and Wickens, C.D. The effects of task structures on time-sharing efficiency and resource allocation optimality. *Proceedings of the 20th Annual Conference on Manual Control* (pp. 305–317). Moffett Field, CA: Ames Research Center, 1984.

Whitaker, L.A. Dual task interference as a function of cognitive processing load. *Acta Psychologica.* 43, 71–84, 1979.

Wickens, C.D. and Kessel, C. Processing resource demands of failure detection in dynamic systems. *Journal of Experimental Psychology: Human Perception and Performance.* 6, 564–577, 1980.

Wickens, C.D., Mountford, S.J., and Schreiner, W. Multiple resources, task-hemispheric integrity, and individual differences in time-sharing. *Human Factors.* 23, 211–229, 1981.

Williges, R.C. and Wierwille, W.W. Behavioral measures of aircrew mental workload. *Human Factors.* 21, 549–574, 1979.

Wright, P., Holloway, C.M., and Aldrich, A.R. Attending to visual or auditory verbal information while performing other concurrent tasks. *Quarterly Journal of Experimental Psychology.* 26, 454–463, 1974.

3.2.29 Workload Scale Secondary Task

General description. A workload scale was developed by tallying the number of persons who performed better in each task combination of the Multiple Task Performance Battery (MTPB) and converting these proportions to *z* scores (Chiles and Alluisi, 1979). The resulting *z* scores are multiplied by −1, so that the most negative score is associated with the lowest workload.

Strengths and limitations. Workload scales are easy to calculate but must satisfy the following conditions: (1) linear additivity and (2) no interaction between tasks. Some task combinations may violate these assumptions. Further, the intrusiveness of secondary tasks may preclude their use in nonlaboratory settings.

Data requirements. Performance of multiple combinations of tasks is required.

Thresholds. Dependent on task and task combinations being used.

SOURCE

Chiles, W.D. and Alluisi, E.A. On the specification of operator or occupational workload with performance-measurement methods. *Human Factors.* 21 (5), 515–528, 1979.

3.3 SUBJECTIVE MEASURES OF WORKLOAD

There are five types of subjective measures of workload. The first is comparison measures, in which the subject is asked which of two tasks has the higher workload. This type of measure is described in section 3.3.1. The second type is a decision tree, in which the subject is stepped through a series of discrete questions to reach a single workload rating (see section 3.3.2). The third type of subjective workload measure is a set of subscales, each of which is designed to measure different aspects of workload (see section 3.3.3). The fourth type is a single number, which as the name implies requires the subject to give only one number to rate the workload (section 3.3.4). The final type of subjective measure of workload is task-analysis based. These measures break the tasks into subtasks and subtask requirements for workload evaluation (see section 3.3.5). A summary is provided in table 19.

Casali and Wierwille (1983) identified several advantages of subjective measures: "inexpensive, unobtrusive, easily administered, and readily transferable to full-scale aircraft and to a wide range of tasks" (p. 640). Gopher (1983) concluded that subjective measures "are well worth the bother" (p. 19). Wickens (1984) states that subjective measures have high face validity. Muckler and Seven (1992) state that subjective measures may be essential.

O'Donnell and Eggemeier (1986), however, identified six limitations of subjective measures of workload: (1) potential confounding of mental and physical workload, (2) difficulty in distinguishing external demand or task difficulty from actual workload, (3) unconscious processing of information that the operator cannot rate subjectively, (4) dissociation of subjective ratings and task performance, (5) requirement of well-defined question, and (6) dependence on short-term memory. Eggemeier (1981) described two additional issues: (1) developing a generalized measure of subjective mental workload and (2) identifying factors related to the subjective experience of workload.

In addition, Meshkati, Hancock, and Rahimi (1990) warn that raters may interpret the words in a rating scale differently, thus leading to inconsistent results.

TABLE 19
Comparison of Subjective Measures of Workload

Section	Measure	Reliability	Task Time	Ease of Scoring
3.3.1.1	Analytical Hierarchy Process	High	Requires rating pairs of tasks	Computer-scored
3.3.5.1	Arbeitswissenshaftliches Erhebungsverfahren zur Tatigkeitsanalyze	High	Requires rating 216 items	Requires multivariate statistics
3.3.2.1	Bedford Workload Scale	High	Requires two decisions	No scoring needed
3.3.5.2	Computerized Rapid Analysis of Workload	Unknown	None	Requires detailed mission timeline
3.3.4.1	Continuous Subjective Assessment of Workload	High	Requires programming computer prompts	Computer-scored
3.3.2.2	Cooper–Harper Rating Scale	High	Requires three decisions	No scoring needed
3.3.3.1	Crew Status Survey	High	Requires one decision	No scoring needed
3.3.4.2	Dynamic Workload Scale	High	Requires ratings by pilot and observer whenever workload changes	No scoring needed
3.3.4.3	Equal-Appearing Intervals	Unknown	Requires ratings in several categories	No scoring needed
3.3.3.2	Finegold Workload Rating Scale	High	Requires five ratings	Requires calculating an average
3.3.3.3	Flight Workload Questionnaire	May evoke response bias	Requires four ratings	No scoring needed
3.3.4.4	Hart and Bortolussi Rating Scale	Unknown	Requires one rating	No scoring needed
3.3.3.4	Hart and Hauser Rating Scale	Unknown	Requires six ratings	Requires interpolating quantity from mark on scale
3.3.2.3	Honeywell Cooper–Harper Rating Scale	Unknown	Requires three decisions	No scoring needed
3.3.4.5	Instantaneous Self Assessment (ISA)	High	Requires rating of 1 to 5	No scoring needed
3.3.1.2	Magnitude Estimation	Moderate	Requires comparison to a standard	No scoring needed
3.3.5.3	McCracken–Aldrich Technique	Unknown	May require months of preparation	Requires computer programmer

Continued

TABLE 19 (Continued)
Comparison of Subjective Measures of Workload

Section	Measure	Reliability	Task Time	Ease of Scoring
3.3.4.6	McDonnell Rating Scale	Unknown	Requires three or four decisions	No scoring needed
3.3.2.4	Mission Operability Assessment Technique	Unknown	Requires two ratings	Requires conjoint measurement techniques
3.3.2.5	Modified Cooper–Harper Rating Scale	High	Requires three decisions	No scoring needed
3.3.3.5	Multi-Descriptor Scale	Low	Requires six ratings	Requires calculating an average
3.3.3.6	Multidimensional Rating Scale	High	Requires eight ratings	Requires measurement of line length
3.3.3.7	Multiple Resources Questionnaire	Moderate	Requires 16 ratings	No scoring needed
3.3.3.8	NASA Bipolar Rating Scale	High	Requires ten ratings	Requires weighting procedure
3.3.3.9	NASA Task Load Index	High	Requires six ratings	Requires weighting procedure
3.3.4.7	Overall Workload Scale	Moderate	Requires one rating	No scoring needed
3.3.4.8	Pilot Objective/ Subjective Workload Assessment Technique	High	Requires one rating	No scoring needed
3.3.1.3	Pilot Subjective Evaluation	Unknown	Requires rating systems on four scales and completion of questionnaire	Requires extensive interpretation
3.3.3.10	Profile of Mood States	High	Requires about 10 min to complete	Requires manual or computer scoring
3.3.2.6	Sequential Judgment Scale	High	Requires rating each task	Requires measurement and conversion to percent
3.3.3.11	Subjective Workload Assessment Technique	High	Requires prior card sort and three ratings	Requires computer scoring
3.3.1.4	Subjective Workload Dominance Technique	High	Requires N(N-1)/2 paired comparisons	Requires calculating geometric means
3.3.5.4	Task Analysis Workload	Unknown	May require months of preparation	Requires detailed task analysis
3.3.4.9	Utilization	High	None	Requires regression

TABLE 19 (Continued)
Comparison of Subjective Measures of Workload

Section	Measure	Reliability	Task Time	Ease of Scoring
3.3.3.12	Workload/ Compensation/ Interference/Technical Effectiveness	Unknown	Requires ranking 16 matrix cells	Requires complex mathematical processing
3.3.5.5	Zachary/Zaklad Cognitive Analysis	Unknown	May require months of preparation	Requires detailed task analysis

SOURCES

Casali, J.G. and Wierwille, W.W. A comparison of rating scale, secondary task, physiological, and primary-task workload estimation techniques in a simulated flight task emphasizing communications load. *Human Factors*. 25, 623–642, 1983.

Eggemeier, F.T. Current issues in subjective assessment of workload. *Proceedings of the Human Factors Society 25th Annual Meeting*. 513–517, 1981.

Gopher, D. *The workload book: Assessment of operator workload to engineering systems (NASA-CR-166596)*. Moffett Field, CA: NASA Ames Research Center; November 1983.

Meshkati, N., Hancock, P.A., and Rahimi, M. Techniques in mental workload assessment. In Wilson, J.R. and Corlett, E.N. (Eds.). *Evaluation of human work: A practical ergonomics methodology*. New York: Taylor and Francis, 1990.

Muckler, F.A. and Seven, S.A. Selecting performance measures "objective" versus "subjective" measurement. *Human Factors*. 34, 441–455, 1992.

O'Donnell, R.D. and Eggemeier, F.T. Workload assessment methodology. In Boff, K.R., Kaufman, L. and Thomas, J.P. (Eds.) *Handbook of perception and human performance*. New York: Wiley and Sons, 1986.

Wickens, C.D. *Engineering psychology and human performance*. Columbus, OH: Charles E. Merrill, 1984.

3.3.1 Comparison Subjective Workload Measures

Comparison measures require the subject to identify which of two tasks has the higher workload. Examples include the Analytical Hierarchy Process (section 3.3.1.1), Magnitude Estimation (section 3.3.1.2), Pilot Subjective Evaluation (section 3.3.1.3), and Subjective Workload Dominance (section 3.3.1.4).

3.3.1.1 Analytical Hierarchy Process

General description. The analytical hierarchy process (AHP) uses the method of paired comparisons to measure workload. Specifically, subjects rate which of a pair of conditions has the higher workload. All combinations of conditions must be compared. Therefore, if there are n conditions, the number of comparisons is $0.5n(n - 1)$.

Strengths and limitations. Lidderdale (1987) found a high consensus in the ratings of both pilots and navigators for a low-level tactical mission. Vidulich and Tsang (1987) concluded that AHP ratings were more valid and reliable than either an overall workload rating or NASA-TLX. Vidulich and Bortolussi (1988) reported that AHP ratings were more sensitive to attention than secondary RTs. Vidulich and Tsang (1988) reported high test-retest reliability. Bortolussi and Vidulich (1991) reported significantly higher workload using speech controls than manual controls in a combat helicopter simulated mission. AHP accounted for 64.2% of the variance in mission phase (Vidulich and Bortolussi, 1988). AHP was also sensitive to the degree of automation in a combat helicopter simulation (Bortolussi and Vidulich, 1989). It has been used in China to determine worker salaries based on "job intensity" derived from physical and mental loading, environmental conditions, and danger (Shen, Meng, and Yan, 1990).

Metta (1993) used the AHP to develop a rank ordering of computer interfaces. She identified the following advantages of AHP: (1) easy to quantify consistency in human judgments, (2) yields useful results in spite of small sample sizes and low probability of statistical significant results, and (3) requires no statistical assumptions.

However, complex mathematical procedures must be employed (Lidderdale, 1987; Lidderdale and King, 1985; Saaty, 1990).

Data requirements. Four steps are required to use the AHP. First, a set of instructions must be written. A verbal review of the instructions should be conducted after the subjects have read the instructions to ensure their understanding of the task. Second, a set of evaluation sheets must be designed to collect the subjects' data. An example is presented in figure 7. Each sheet has the two conditions to be compared in separate columns, one on the right side of the page, the other on the left. A 17-point rating scale is placed between the two sets of conditions. The scale uses five descriptors in a predefined order and allows a single point between each for mixed ratings. Vidulich (1988) defined the scale descriptors (see table 20). Budescu, Zwick, and Rapoport (1986) provide critical value tables for detecting inconsistent judgments and subjects.

Third, the data must be scored. The scores range from +8 (absolute dominance of the left-side condition over the right-side condition) to –8 (absolute dominance of the right-side condition over the left-side condition). Finally, the scores are input, in matrix form, into a computer program. The output of this program is a scale weight for each condition and three measures of goodness of fit.

Thresholds. Not stated.

FIGURE 7 Example AHP Rating Scale.

TABLE 20
Definitions of AHP Scale Descriptors

EQUAL	The two task combinations are absolutely equal in the amount of workload generated by the simultaneous tasks.
WEAK	Experience and judgment slightly suggest that one of the combinations of tasks has more workload than the other.
STRONG	Experience and judgment strongly suggest that one of the combinations has higher workload.
VERY STRONG	One task combination is strongly dominant in the amount of workload, and this dominance is clearly demonstrated in practice.
ABSOLUTE	The evidence supporting the workload dominance of one task combination is the highest possible order of affirmation (adapted from Vidulich, 1988, p. 5).

SOURCES

Bortolussi, M.R. and Vidulich, M.A. The effects of speech controls on performance in advanced helicopters in a double stimulation paradigm. *Proceedings of the International Symposium on Aviation Psychology.* 216–221, 1991.

Bortolussi, M.R. and Vidulich, M.A. The benefits and costs of automation in advanced helicopters: An empirical study. *Proceedings of the Fifth International Symposium on Aviation Psychology.* 594–559, 1989.

Budescu, D.V., Zwick, R., and Rapoport, A. A comparison of the Eigen value method and the geometric mean procedure for ratio scaling. *Applied Psychological Measurement.* 10, 68–78, 1986.

Lidderdale, I.G. Measurement of aircrew workload during low-level flight, practical assessment of pilot workload (AGARD-AG-282). *Proceedings of NATO Advisory Group for Aerospace Research and Development (AGARD).* Neuilly-sur-Seine, France: AGARD, 1987.

Lidderdale, I.G. and King, A.H. *Analysis of subjective ratings using the analytical hierarchy process: A microcomputer program.* High Wycombe, England: OR Branch NFR, HQ ST C, RAF, 1985.

Metta, D.A. An application of the Analytic Hierarchy Process: A rank-ordering of computer interfaces. *Human Factors.* 35(1), 141–157, 1993.

Saaty, T.L. *The analytical hierarchy process.* New York: McGraw-Hill, 1980.

Shen, R., Meng, X., and Yan, Y. Analytic hierarchy process applied to synthetically evaluate the labor intensity of jobs. *Ergonomics.* 33(7), 867–874, 1990.

Vidulich, M.A. *Notes on the AHP procedure; 1988.* (Available from Dr. Michael A. Vidulich, Human Engineering Division, AAMRL/HEG, Wright-Patterson Air Force Base, OH 45433-6573).

Vidulich, M.A. and Bortolussi, M.R. A dissociation of objective and subjective workload measures in assessing the impact of speech controls in advanced helicopters. *Proceedings of the Human Factors Society 32nd Annual Meeting.* 1471–1475, 1988.

Vidulich, M.A. and Tsang, P.S. Absolute magnitude estimation and relative judgment approaches to subjective workload assessment. *Proceedings of the Human Factors Society 31st Annual Meeting.* 1057–1061, 1987.

Vidulich, M.A. and Tsang, P.S. Evaluating immediacy and redundancy in subjective workload techniques. *Proceedings of the Twenty-Third Annual Conference on Manual Control*, 1988.

3.3.1.2 Magnitude Estimation

General description. Subjects are required to estimate workload numerically in relation to a standard.

Strengths and limitations. Borg (1978) successfully used this method to evaluate workload. Helm and Heimstra (1981) reported a high correlation between workload estimates and task difficulty. Masline (1986) reported sensitivity comparable to estimates from the equal-appearing intervals method and SWAT. Gopher and Braune (1984), however, found a low correlation between workload estimates and reaction-time performance. In contrast, Kramer, Sirevaag, and Braune (1987) reported good correspondence to performance in a fixed-based flight simulator. Hart and Staveland (1988) suggest that the presence of a standard enhances interrater reliability. O'Donnell and Eggemeier (1986), however, warned that subjects may be unable to retain an accurate memory of the standard over the course of an experiment.

Data requirements. A standard must be well-defined.

Thresholds. Not stated.

SOURCES

Borg, C.G. Subjective aspects of physical and mental load. *Ergonomics*. 21, 215–220, 1978.

Gopher, D. and Braune, R. On the psychophysics of workload: Why bother with subjective measures? *Human Factors*. 26, 519–532, 1984.

Hart, S.G. and Staveland, L.E. Development of NASA-TLX (Task Load Index): Results of empirical and theoretical research. In Hancock, P.A. and Meshkati, N. (Eds.) *Human mental workload*. Amsterdam: Elsevier, 1988.

Helm, W. and Heimstra, N.W. *The relative efficiency of psychometric measures of task difficulty and task performance in predictive task performance (Report No. HFL-81-5)*. Vermillion, S.D.: University of South Dakota, Psychology Department, Human Factors Laboratory, 1981.

Kramer, A.F., Sirevaag, E.J., and Braune, R. A psychophysical assessment of operator workload during simulated flight missions. *Human Factors*. 29, 145–160, 1987.

Masline, P.J. A comparison of the sensitivity of interval scale psychometric techniques in the assessment of subjective workload. Unpublished master's thesis. Dayton, OH: University of Dayton, 1986.

O'Donnell, R.D. and Eggemeier, F.T. Workload assessment methodology. In Boff, K.R., Kaufman, L., and Thomas, J. (Eds.) *Handbook of perception and human performance. Vol. 2, Cognitive processes and performance*. New York: Wiley, 1986.

3.3.1.3 Pilot Subjective Evaluation

General description. The Pilot Subjective Evaluation (PSE) workload scale (see figure 8) was developed by Boeing for use in the certification of the Boeing 767 aircraft. The scale is accompanied by a questionnaire. Both the scale and the questionnaire are completed with reference to an existing aircraft selected by the pilot.

Strengths and limitations. Fadden (1982) and Ruggerio and Fadden (1987) stated that the ratings of workload greater than the reference aircraft were useful in identifying aircraft design deficiencies.

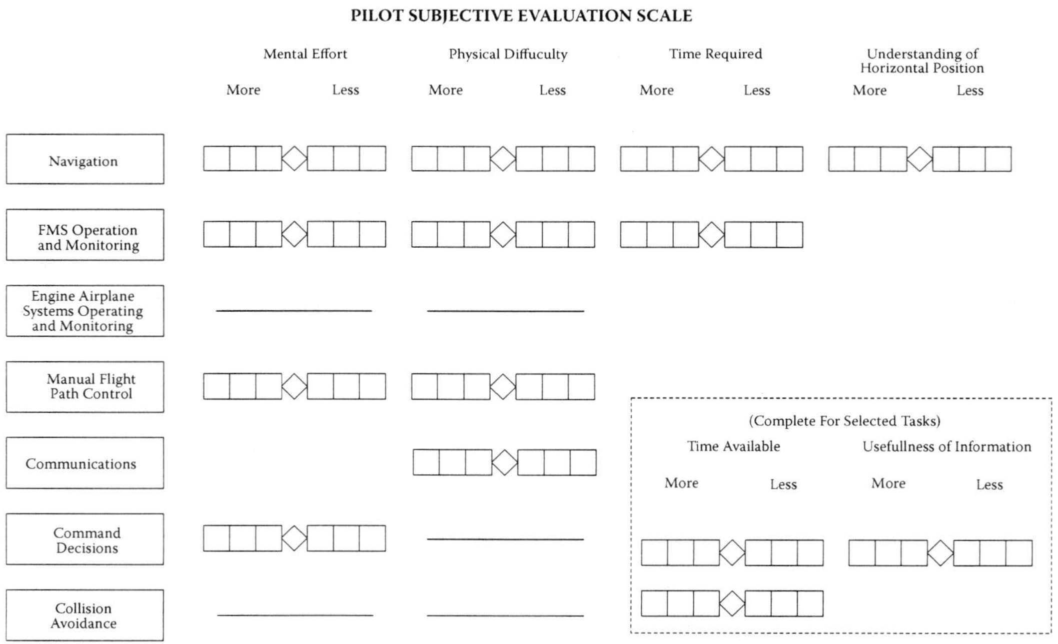

PILOT SUBJECTIVE EVALUATION SCALE

Mental Effort
More Less
Physical Diffuculty
More Less
Time Required
More Less
Understanding of Horizontal Position
More Less

Navigation

FMS Operation and Monitoring

Engine Airplane Systems Operating and Monitoring

Manual Flight Path Control

Communications

Command Decisions

Collision Avoidance

(Complete For Selected Tasks)
Time Available
More Less
Usefullness of Information
More Less

FIGURE 8 Pilot Subjective Evaluation Scale (from Lysaght, R.J., Hill, S.G., Dick, A.O., Plamondon, B.D., Linton, P.M., Wierwille, W.W., Zaklad, A.L., Bittner, A.C., and Wherry, R.J. *Operator workload: Comprehensive review and evaluation of operator workload methodologies (Technical Report 851)*. Alexandria, VA: Army Research Institute for the Behavioral and Social Sciences; June 1989, p. 107).

Data requirements. Each subject must complete both the PSE scale and the questionnaire.

Thresholds. 1, minimum workload; 7, maximum workload.

SOURCES

Fadden, D. *Boeing Model 767 flight deck workload assessment methodology.* Paper presented at the SAE Guidance and Control System Meeting, Williamsburg, VA, 1982.

Lysaght, R.J., Hill, S.G., Dick, A.O., Plamondon, B.D., Linton, P.M., Wierwille, W.W., Zaklad, A.L., Bittner, A.C., and Wherry, R.J. *Operator workload: Comprehensive review and evaluation of operator workload methodologies (Technical Report 851).* Alexandria, VA: Army Research Institute for the Behavioral and Social Sciences; June 1989.

Ruggerio, F. and Fadden, D. Pilot subjective evaluation of workload during a flight test certification programme. In Roscoe, A.H. (Ed.) *The practical assessment of pilot workload.* AGARD-ograph 282 (pp. 32–36). Neuilly-sur-Seine, France: AGARD, 1987.

3.3.1.4 Subjective Workload Dominance

General description. The Subjective Workload Dominance (SWORD) technique uses judgment matrices to assess workload.

Strengths and limitations. SWORD is a sensitive and reliable workload measure (Vidulich, 1989). It has also been useful in projecting workload associated with various HUD formats (Vidulich, Ward, and Schueren, 1991). In addition, Tsang and Vidulich (1994) reported significant differences in SWORD ratings as a function of tracking task condition. The test-retest reliability was +0.937.

Data requirements. There are three required steps: (1) a rating scale listing all possible pairwise comparisons of the tasks performed must be completed, (2) a judgment matrix comparing each task to every other task must be filled in with each subject's evaluation of the tasks, and (3) ratings must be calculated using a geometric means approach.

Thresholds. Not stated.

SOURCES

Tsang, P.S. and Vidulich, M.A. The roles of immediacy and redundancy in relative subjective workload assessment. *Human Factors.* 36(3), 503–513, 1994.

Vidulich, M.A. The use of judgment matrices in subjective workload assessment: The subjective workload dominance (SWORD) technique. *Proceedings of the Human Factors Society 33rd Annual Meeting.* 1406–1410, 1989.

Vidulich, M.A., Ward, G.F., and Schueren, J. Using the Subjective Workload Dominance (SWORD) technique for projective workload assessment. *Human Factors.* 33(6), 677–691, 1991.

3.3.2 Decision Tree Subjective Workload Measures

Decision tree subjective measures of workload require the subject to step through a series of discrete questions to reach a single workload rating. Examples include the following: the Bedford Workload Scale (section 3.3.2.1), Cooper–Harper Rating Scale (section 3.3.2.2), Honeywell Cooper–Harper Rating Scale (section 3.3.2.3), Mission Operability Assessment Technique (section 3.3.2.4), Modified Cooper–Harper Rating Scale (section 3.3.2.5), and Sequential Judgment Scale (section 3.3.2.6).

3.3.2.1 Bedford Workload Scale

General description. Roscoe (1984) described a modification of the Cooper–Harper Rating Scale created by trial and error with the help of test pilots at the Royal Aircraft Establishment at Bedford, England. The Bedford Workload Scale (see figure 9) retained the binary decision tree and the four- and ten-rank ordinal structures of the Cooper–Harper Rating Scale. The three-rank ordinal structure asked pilots to assess whether (1) it was possible to complete the task, (2) the workload was tolerable, and (3) the workload was satisfactory without reduction. The rating-scale end points were: *workload insignificant* to *task abandoned*. In addition to the structure, the Cooper–Harper (1969) definition of pilot workload was used: "… the integrated mental and physical effort required to satisfy the perceived demands of a specified flight task" (Roscoe, 1984, p. 12–8). The concept of spare capacity was used to help define levels of workload.

Strengths and limitations. The Bedford Workload Scale was reported to be welcomed by pilots. Roscoe (1984) reported that pilots found the scale "easy to use without the need to always refer to the decision tree." He also noted that it was necessary to accept ratings of 3.5 from the pilots. These statements suggest that the pilots emphasized the ten- rather than the four-rank ordinal structure of the Bedford Workload Scale. Roscoe (1984) found that pilot workload ratings and heart rates varied in similar ways during close-coupled in-flight maneuvers in a BAE 125 twinjet aircraft. He felt that the heart-rate information complemented and increased the value of subjective workload ratings. He also noted the lack of absolute workload information provided by the Bedford Workload Scale and by heart-rate data. Wainwright (1987) used the scale during certification of the BAE 146 aircraft. Tsang and Johnson (1987) concluded that the Bedford Workload Scale provided a good measure of spare capacity. More recently, Oman, Kendra, Hayashi, Stearns, and Burki-Cohen (2001) reported no increase in Bedford Workload Scale ratings between different vertical display formats for fixed-wing aircraft. There were reductions in root-mean-square altitude scores, however.

Roscoe (1987) reported that the scale was well accepted by aircrews. Corwin, Sandry-Garza, Biferno, Boucek, Logan, Jonsson, and Metalis (1989) concluded that the Bedford Workload Scale is a reliable and valid measure of workload based on flight simulator data. Vidulich and Bortolussi (1988) reported significant differences in Bedford Workload Scale ratings across four flight segments. However, the workload during hover was rated less than that during hover with a simultaneous communication task. Further, the scale was insensitive to differences in both control configurations and combat countermeasure conditions. Lidderdale (1987) reported that postflight ratings were very difficult for aircrews to make. In addition, Vidulich (1991) questions whether the scale measures space capacity.

Svensson, Angelborg-Thanderz, Sjoberg, and Olsson (1997) reported the reliability of the Bedford Workload Scale to be +0.82 among 18 pilots flying simulated

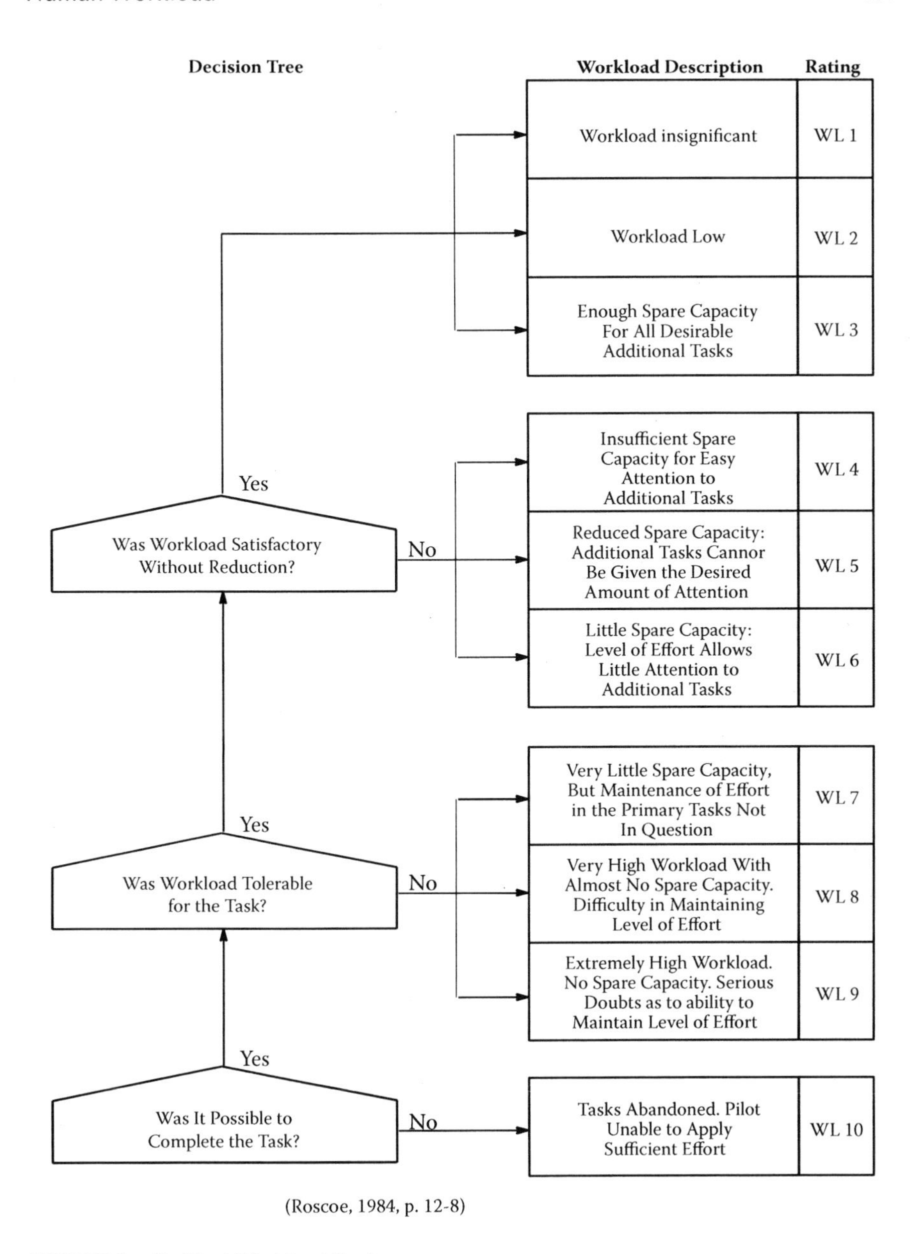

FIGURE 9 Bedford Workload Scale.

low-level, high-speed missions. The correlation with the NASA TLX was +0.826 and with the SWAT was +0.687.

In a more recent study, Ed George (personal communication, 2002) analyzed the responses to a survey examining the terminology used in the Bedford Workload Scale. Of the 20 (United States) U.S. Air Force pilots, 4 reported confusion between WL1 and WL2, 2 between WL2 and WL3, 10 between WL4 and WL5, 6 between WL5 and WL6, 2 between WL6 and WL7, and 3 between WL8 and WL9.

Data requirements. Roscoe (1984) suggested the use of short, well-defined flight tasks to enhance the reliability of subjective workload ratings. Harris, Hill, Lysaght, and Christ (1992) state that "some practice" is necessary to become familiar with the scale. They also suggest the use of nonparametric analysis technique because the Bedford Workload Scale is not an interval scale.

Thresholds. The minimum value is 1, and the maximum is 10.

SOURCES

Corwin, W.H., Sandry-Garza, D.L., Biferno, M.H., Boucek, G.P., Logan, A.L., Jonsson, J.E., and Metalis, S.A. *Assessment of crew workload measurement methods, techniques and procedures.* Vol. I—*Process, methods, and results (WRDC-TR-89-7006).* Wright-Patterson Air Force Base, OH, 1989.

Harris, R.M., Hill, S.G., Lysaght, R.J., and Christ, R.E. *Handbook for operating the OWL & NEST technology (ARI Research Note 92-49).* Alexandria, VA: United States Army Research Institute for the Behavioral and Social Sciences, 1992.

Lidderdale, I.G. Measurement of aircrew workload during low-level flight, practical assessment of pilot workload (AGARD-AG-282). *Proceedings of NATO Advisory Group for Aerospace Research and Development (AGARD).* Neuilly-sur-Seine, France: AGARD, 1987.

Oman, C.M., Kendra, A.J., Hayashi, M., Stearns, M.J., and Burki-Cohen, J. Vertical navigation displays: Pilot performance and workload during simulated constant angle of descent GPS approaches. *International Journal of Aviation Psychology.* 11 (1), 15–31, 2001.

Roscoe, A.H. Assessing pilot workload in flight: Flight test techniques. *Proceedings of NATO Advisory Group for Aerospace Research and Development (AGARD) (AGARD-CP-373).* Neuilly-sur-Seine, France: AGARD, 1984.

Roscoe, A.H. In-flight assessment of workload using pilot ratings and heart rate. In A.H. Roscoe (Ed.) *The practical assessment of pilot workload. AGARDograph No. 282* (pp. 78–82). Neuilly-sur-Seine, France: AGARD, 1987.

Svensson, E., Angelborg-Thanderz, M., Sjoberg, L., and Olsson, S. Information complexity–mental workload and performance in combat aircraft. *Ergonomics.* 40(3), 362–380, 1997.

Tsang, P.S. and Johnson, W. Automation: Changes in cognitive demands and mental workload. *Proceedings of the 4th Symposium on Aviation Psychology.* Columbus, OH: Ohio State University, 1987.

Vidulich, M.A. The Bedford Scale: Does it measure spare capacity? *Proceedings of the 6th International Symposium on Aviation Psychology.* 2, 1136–1141, 1991.

Vidulich, M.A. and Bortolussi, M.R. Control configuration study. *Proceedings of the American Helicopter Society National Specialist's Meeting.* Automation Applications for Rotorcraft, 1988.

Wainwright, W. Flight test evaluation of crew workload. In Roscoe, A.H. (Ed.) *The practical assessment of pilot workload. AGARDograph No. 282* (pp. 60–68). Neuilly-sur-Seine, France: AGARD, 1987.

3.3.2.2 Cooper–Harper Rating Scale

General description. The Cooper–Harper Rating Scale is a decision tree that uses adequacy for the task, aircraft characteristics, and demands on the pilot to rate handling qualities of an aircraft (see figure 10).

Strengths and limitations. The Cooper–Harper Rating Scale is the current standard for evaluating aircraft-handling qualities. It reflects differences in both perfor-

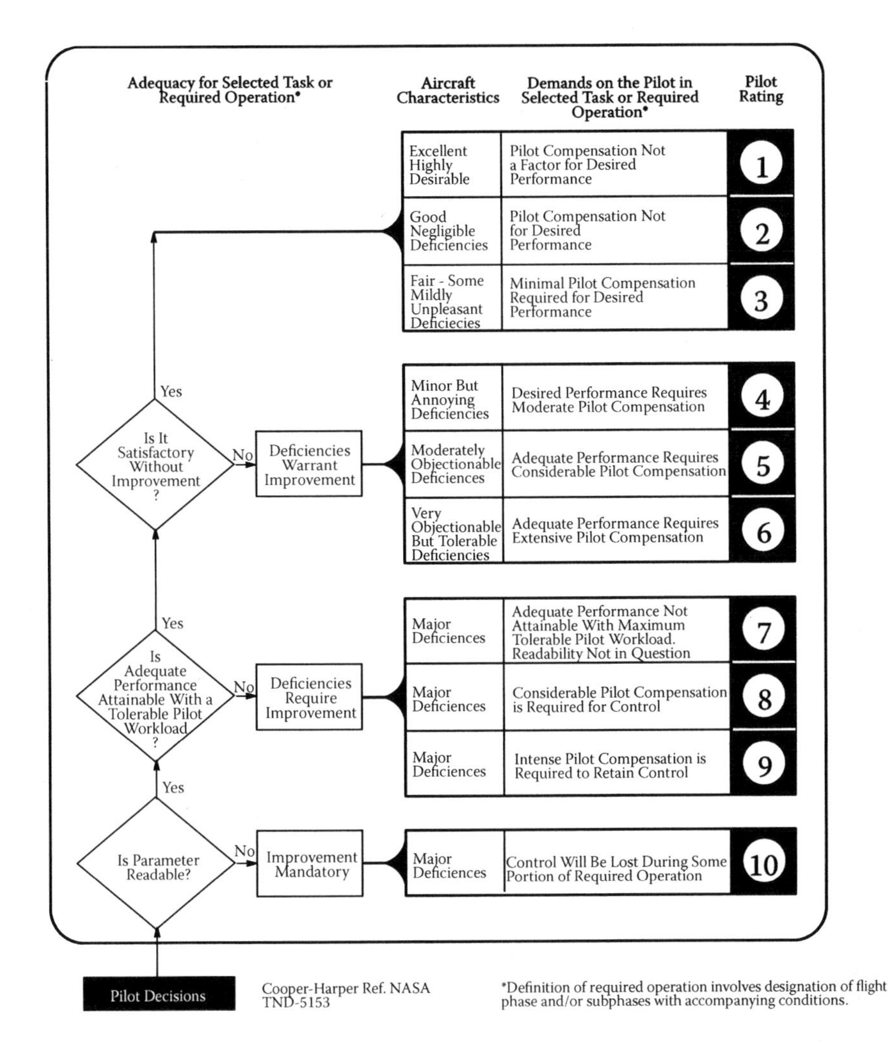

FIGURE 10 Cooper–Harper Rating Scale.

mance and workload and is behaviorally anchored. It requires minimum training, and a briefing guide has been developed (see Cooper and Harper, 1969, pp. 34–39). Cooper–Harper ratings have been sensitive to variations in controls, displays, and aircraft stability (Crabtree, 1975; Krebs and Wingert, 1976; Labacqz and Aiken, 1975; Schultz, Newell, and Whitbeck, 1970; Wierwille and Connor, 1983). Harper and Cooper (1984) describe a series of evaluations of the rating scale.

Connor and Wierwille (1983) reported significant increases in Cooper–Harper Rating Scale ratings as the levels of wind gust increased and/or as the aircraft pitch stability decreased. Ntuen, Park, Strickland, and Watson (1996) reported increases in Cooper–Harper Rating Scale ratings as instability in a compensatory tracking task increased. The highest ratings were for acceleration control, and the lowest for position control, and rate control was in the middle.

Data requirement. The scale provides ordinal data that must be analyzed accordingly. The Cooper–Harper Rating Scale should be used for workload assessment only if handling difficulty is the major determinant of workload. The task must be fully defined for a common reference.

Thresholds. Ratings vary from 1 (excellent, highly desirable) to 10 (major deficiencies). Noninteger ratings are not allowed.

SOURCES

Connor, S.A. and Wierwille, W.W. *Comparative evaluation of twenty pilot workload assessment measures using a psychomotor task in a moving base aircraft simulator (Report 166457).* Moffett Field, CA: NASA Ames Research Center; January 1983.

Cooper, G.E. and Harper, R.P. *The use of pilot rating in the evaluation of aircraft handling qualities (AGARD Report 567).* London: Technical Editing and Reproduction Ltd.; April 1969.

Crabtree, M.S. *Human factors evaluation of several control system configurations, including workload sharing with force wheel steering during approach and flare (AFFDL-TR-75-43).* Wright-Patterson Air Force Base, OH: Flight Dynamics Laboratory; April 1975.

Harper, R.P. and Cooper, G.E. *Handling qualities and pilot evaluation. AIAA, AHS, ASEE, Aircraft Design Systems and Operations Meeting,* AIAA Paper 84–2442, 1984.

Krebs, M.J. and Wingert, J.W. *Use of the oculometer in pilot workload measurement (NASA CR-144951).* Washington, DC: National Aeronautics and Space Administration; February 1976.

Lebacqz, J.V. and Aiken, E.W. *A flight investigation of control, display, and guidance requirements for decelerating descending VTOL instrument transitions using the X-22A variable stability aircraft (AK-5336-F-1).* Buffalo, NY: Calspan Corporation; September 1975.

Ntuen, C.A., Park, E., Strickland, D. and Watson, A.R. *A frizzy model for workload assessment in complex task situations.* IEEE 0-8186-7493, August 1996, pp. 101–107.

Schultz, W.C., Newell, F.D., and Whitbeck, R.F. A study of relationships between aircraft system performance and pilot ratings. *Proceedings of the Sixth Annual NASA University Conference on Manual Control, Wright-Patterson Air Force Base*, OH. 339–340; April 1970.

Wierwille, W.W. and Connor, S.A. Evaluation of 20 workload measures using a psychomotor task in a moving-base aircraft simulator. *Human Factors.* 25(1), 1–16, 1983.

3.3.2.3 Honeywell Cooper–Harper Rating Scale

General description. This rating scale (see figure 11) uses a decision-tree structure for assessing overall task workload.

Strengths and limitations. The Honeywell Cooper–Harper Rating Scale was developed by Wolf (1978) to assess overall task workload. North, Stackhouse, and Graffunder (1979) used the scale to assess workload associated with various Vertical Take-Off and Landing (VTOL) aircraft displays. For the small subset of conditions analyzed, the scale ratings correlated well with performance.

Data requirements. Subjects must answer three questions related to task performance. The ratings are ordinal and must be treated as such in subsequent analyses.

Thresholds. The minimum value is 1 and the maximum is 9.

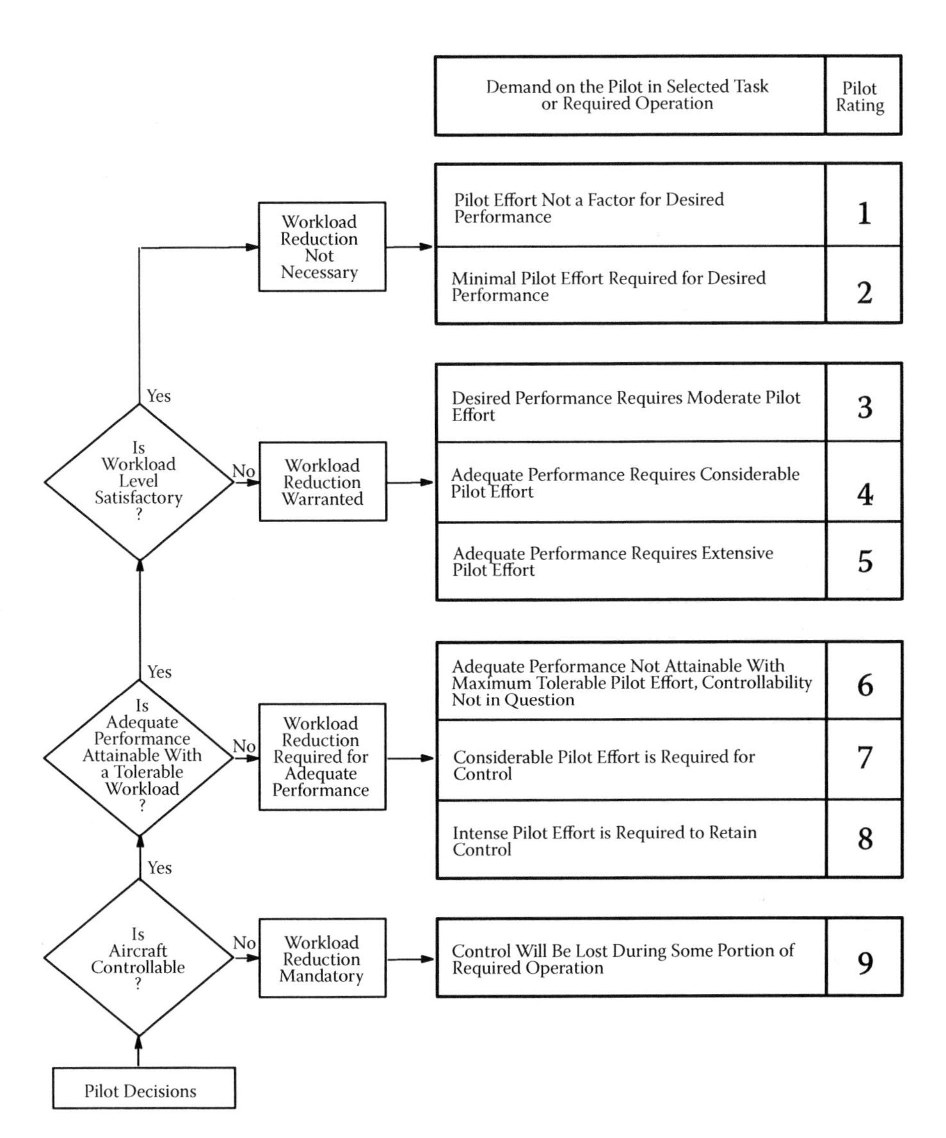

FIGURE 11 Honeywell Cooper–Harper Rating Scale.

SOURCES

Lysaght, R.J., Hill, S.G., Dick, A.O., Plamondon, B.D., Linton, P.M., Wierwille, W.W., Zaklad, A.L., Bittner, A.C., and Wherry, R.J. *Operator workload: Comprehensive review and evaluation of operator workload methodologies (Technical Report 851).* Alexandria, VA: Army Research Institute for the Behavioral and Social Sciences; June 1989.

North, R.A., Stackhouse, S.P., and Graffunder, K. *Performance, physiological and oculometer evaluations of VTOL landing displays (NASA Contractor Report 3171).* Hampton, VA: NASA Langley Research Center, 1979.

Wolf, J.D. *Crew workload assessment: Development of a measure of operator workload (AFFDL-TR-78-165)*. Wright-Patterson AFB, OH: Air Force Flight Dynamics Laboratory, 1978.

3.3.2.4 Mission Operability Assessment Technique

General description. The Mission Operability Assessment Technique includes two four-point ordinal rating scales, one for pilot workload and the other for technical effectiveness (see table 21). Subjects rate both pilot workload and technical effectiveness for each subsystem identified in the task analysis of the aircraft.

Strengths and limitations. Interrater reliabilities are high for most but not all tasks (Donnell, 1979; Donnell, Adelman, and Patterson, 1981; Donnell and O'Connor, 1978).

Data requirements. Conjoint measurement techniques are applied to the individual pilot workload and subsystem technical effectiveness ratings to develop an overall interval scale of systems capability.

Thresholds. Not stated.

TABLE 21
Mission Operability Assessment Technique Pilot Workload and Subsystem Technical Effectiveness Rating Scales

Pilot Workload

1. The pilot workload (PW)/compensation (C)/interference (I) required to perform the designated task is *extreme*. This is a *poor* rating on the PW/C/I dimension.
2. The pilot workload /compensation/interference required to perform the designated task is *high*. This is a *fair* rating on the PW/C/I dimension.
3. The pilot workload /compensation/interference required to perform the designated task is *moderate*. This is a *good* rating on the PW/C/I dimension.
4. The pilot workload /compensation/interference required to perform the designated task is *low*. This is an *excellent* rating on the PW/C/I dimension.

Subsystem Technical Effectiveness

1. The technical effectiveness of the required subsystem is *inadequate* for performing the designated task. Considerable redesign is necessary to attain task requirements. This is a *poor* rating on the subsystem technical effectiveness scale.
2. The technical effectiveness of the required subsystem is *adequate* for performing the designated task. Some redesign is necessary to attain task requirements. This is a *fair* rating on the subsystem technical effectiveness scale.
3. The technical effectiveness of the required subsystem *enhances individual task performance*. No redesign is necessary to attain task requirements. This is a *good* rating on the subsystem technical effectiveness scale.
4. The technical effectiveness of the required subsystem *allows for the integration of multiple tasks*. No redesign is necessary to attain task requirements. This is an *excellent* rating on the subsystem effectiveness scale. (O'Donnell and Eggemeier, 1986, pp. 42–16)

SOURCES

Donnell, M.L. *An application of decision-analytic techniques to the test and evaluation of a major air system Phase III (TR-PR-79-6-91).* McLean, VA: Decisions and Designs; May 1979.

Donnell, M.L., Adelman, L, and Patterson, J.F. *A systems operability measurement algorithm (SOMA): Application, validation, and extensions (TR-81-11-156).* McLean, VA: Decisions and Designs; April 1981.

Donnell, M.L. and O'Connor, M.F. *The application of decision analytic techniques to the test and evaluation phase of the acquisition of a major air system Phase II (TR-78-3-25).* McLean, VA: Decisions and Designs; April 1978.

O'Donnell, R.D. and Eggemeier, F.T. Workload assessment methodology. In Boff, K.R., Kaufman, L., and Thomas, J.P. (Eds.) *Handbook of perception and human performance.* New York: John Wiley, 1986.

3.3.2.5 Modified Cooper–Harper Rating Scale

General description. Wierwille and Casali (1983) noted that the Cooper–Harper Rating Scale represented a combined handling-qualities/workload rating scale. They found that it was sensitive to psychomotor demands on an operator, especially for aircraft-handling qualities. They wanted to develop an equally useful scale for the estimation of workload associated with cognitive functions, such as "perception, monitoring, evaluation, communications, and problem-solving." The Cooper–Harper Rating Scale terminology was not suited to this purpose. A Modified Cooper–Harper Rating Scale (see figure 12) was developed to "increase the range of applicability to situations commonly found in modern systems." Modifications included are the following: (1) changing the rating-scale end points to *very easy* and *impossible,* (2) asking the pilot to rate mental workload level rather than controllability, and (3) emphasizing difficulty rather than deficiencies. In addition, Wierwille and Casali (1983) defined mental effort as "minimal" in rating 1, whereas mental effort is not defined as minimal until rating 3 in the original Cooper–Harper Rating Scale. Further, adequate performance begins at rating 3 in the Modified Cooper–Harper Rating Scale but at rating 5 in the original scale.

Strengths and limitations. Investigations were conducted to assess the modified Cooper–Harper Rating Scale. They focused on perception (e.g., aircraft engine instruments out of limits during simulated flight), cognition (e.g., arithmetic problem solving during simulated flight), and communications (e.g., detection of, comprehension of, and response to own aircraft call sign during simulated flight).

The Modified Cooper–Harper Rating Scale is sensitive to various types of workloads. For example, Casali and Wierwille (1983a, 1983b) reported that Modified Cooper–Harper Rating Scale ratings increased as the communication load increased. Wierwille, Rahimi, and Casali (1985) reported significant increase in workload as navigation load increased. Casali and Wierwille (1984) reported significant increases in ratings as the number of danger conditions increased. Skipper, Rieger, and Wierwille (1986) reported significant increases in ratings in both high communication and high navigation loads. Wolf (1978) reported the highest workload ratings in the highest workload flight condition (i.e., high wind gust and poor handling qualities).

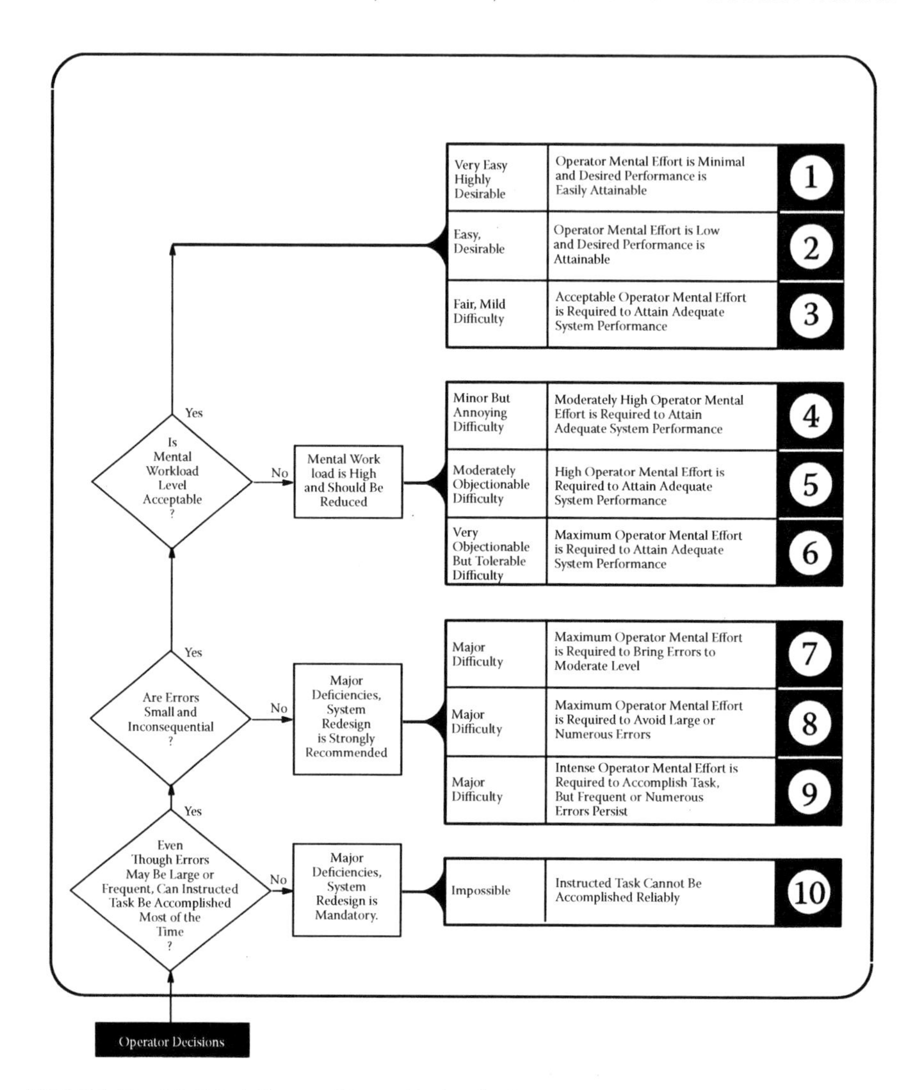

FIGURE 12 Modified Cooper–Harper Rating Scale.

Bittner, Byers, Hill, Zaklad, and Christ (1989) reported reliable differences between mission segments in a mobile air defense system. Byers, Bittner, Hill, Zaklad, and Christ (1988) reported reliable differences between crew positions in a remotely piloted vehicle system. These results suggested that the Modified Cooper–Harper Rating Scale is a valid, statistically reliable indicator of overall mental workload. However, it carries with it the underlying assumptions that high workload is the only determinant of the need for changing the control/display configuration. In spite of that assumption it has been widely used in cockpit evaluations and comparisons. For example, Itoh, Hayashi, Tsuki, and Saito (1990) used the Modified Cooper–Harper Rating Scale to compare workload in the Boeing 747-400 electromechanical displays and the Boeing 767 integrated CRT displays. There was no significant difference in workload ratings.

Wierwille, Casali, Connor, and Rahimi (1985) concluded that the Modified Cooper–Harper Rating Scale provided consistent and sensitive ratings of workload across a range of tasks. Wierwille, Skipper, and Rieger (1985) reported the best consistency and sensitivity with the Modified Cooper–Harper Rating Scale from five alternative tests. Warr, Colle, and Reid (1986) reported that the Modified Cooper–Harper Rating Scale ratings were as sensitive to task difficulty as SWAT ratings. Kilmer, Knapp, Burdsal, Borresen, Bateman, and Malzahn (1988), however, reported that the Modified Cooper–Harper Rating Scale was less sensitive than SWAT ratings to changes in tracking task difficulties. Hill, Iavecchia, Byers, Bittner, Zaklad, and Christ (1992) reported that the Modified Cooper–Harper Rating Scale was not as sensitive or as operator accepted as the NASA TLX or the Overall Workload Scale.

Papa and Stoliker (1988) tailored the Modified Cooper–Harper Rating Scale to evaluate the Low Altitude Navigation and Targeting Infrared System for Night (LANTIRN) on an F-16 aircraft. In another aircraft application, Jennings, Craig, Cheung, Rupert, and Schultz (2004) reported no difference in Modified Cooper–Harper Workload Scale ratings of 11 pilots flying a helicopter using a Tactical Situational Awareness System.

Data requirements. Wierwille and Casali (1983) recommend the use of the Modified Cooper–Harper Rating Scale in experiments where overall mental workload is to be assessed. They emphasize the importance of proper instructions to the subjects. Because the scale was designed for use in experimental situations, it may not be appropriate in situations requiring an absolute diagnosis of a subsystem. Harris, Hill, Lysaght, and Christ (1992) recommend the use of nonparametric analysis techniques because the Modified Cooper–Harper Rating Scale is not an interval scale.

Thresholds. Not stated.

SOURCES

Bittner, A.C., Byers, J.C., Hill, S.G., Zaklad, A.L., and Christ, R.E. Generic workload ratings of a mobile air defense system (LOS-F-H). *Proceedings of the 33rd Annual Meeting of the Human Factors Society* (pp. 1476–1480). Santa Monica, CA: Human Factors Society, 1989.

Byers, J.C., Bittner, A.C., Hill, S.G., Zaklad, A.L., and Christ, R.E. Workload assessment of a remotely piloted vehicle (RPV) system. *Proceedings of the 32nd Annual Meeting of the Human Factors Society* (pp. 1145–1149). Santa Monica, CA: Human Factors Society, 1988.

Casali, J.G. and Wierwille, W.W. A comparison of rating scale, secondary task, physiological, and primary-task workload estimation techniques in a simulated flight task emphasizing communications load. *Human Factors.* 25, 623–642, 1983a.

Casali, J.G. and Wierwille, W.W. Communications-imposed pilot workload: A comparison of sixteen estimation techniques. *Proceedings of the Second Symposium on Aviation Psychology.* 223–235, 1983b.

Casali, J.G. and Wierwille, W.W. On the comparison of pilot perceptual workload: A comparison of assessment techniques addressing sensitivity and intrusion issues. *Ergonomics.* 27, 1033–1050, 1984.

Harris, R.M., Hill, S.G., Lysaght, R.J., and Christ, R.E. *Handbook for operating the OWL-KNEST technology (ARI Research Note 92-49).* Alexandria, VA: United States Army Research Institute for the Behavioral and Social Sciences, 1992.

Hill, S.G., Iavecchia, H.P., Byers, J.C., Bittner, A.C., Zaklad, A.L., and Christ, R.E. Comparison of four subjective workload rating scales. *Human Factors.* 34, 429–439, 1992.
Itoh, Y., Hayashi, Y., Tsukui, I., and Saito, S. The ergonomics evaluation of eye movement and mental workload in aircraft pilots. *Ergonomics*, 1990, 33(6), 719–733.
Jennings, S., Craig, G., Cheung, B., Rupert, A., and Schultz, K. Flight-test of a tactile situational awareness system in a land-based deck landing task. *Proceedings of the Human Factors and Ergonomics Society 48th Annual Meeting.* 142–146, 2004.
Kilmer, K.J., Knapp, R., Burdsal, C., Borresen, R., Bateman, R., and Malzahn, D. Techniques of subjective assessment: A comparison of the SWAT and modified Cooper-Harper scale. *Proceedings of the Human Factors Society 32nd Annual Meeting.* 155–159, 1988.
Papa, R.M. and Stoliker, J.R. Pilot workload assessment: A flight test approach. Washington, DC: *American Institute of Aeronautics and Astronautics.* 88–2105, 1988.
Skipper, J.H., Rieger, C.A., and Wierwille, W.W. Evaluation of decision-tree rating scales for mental workload estimation. *Ergonomics.* 29, 585–599, 1986.
Warr, D., Colle, H. and Reid, G. *A comparative evaluation of two subjective workload measures: The subjective workload assessment technique and the Modified Cooper–Harper scale.* Paper presented at the Symposium on Psychology in Department of Defense. Colorado Springs, CO: US Air Force Academy, 1986.
Wierwille, W.W. and Casali, J.G. A validated rating scale for global mental workload measurement applications. *Proceedings of the 27th Annual Meeting of the Human Factors Society.* 129–133. Santa Monica, CA: Human Factors Society, 1983.
Wierwille, W.W., Casali, J.G., Connor, S.A., and Rahimi, M. Evaluation of the sensitivity and intrusion of mental workload estimation techniques. In Romer, W. (Ed.) *Advances in man-machine systems research. Vol. 2* (pp. 51-127). Greenwich, CT: J.A.I. Press, 1985.
Wierwille, W.W., Rahimi, M., and Casali, J.G. Evaluation of 16 measures of mental workload using a simulated flight task emphasizing mediational activity. *Human Factors.* 27(5), 489–502, 1985.
Wierwille, W.W., Skipper, J. and Reiger, C. *Decision tree rating scales for workload estimation: theme and variations (N85-11544)*, Blacksburg, VA: Vehicle Simulation Laboratory, 1985.
Wolf, J.D. *Crew workload assessment: Development of a measure of operator workload (AFFDL-TR-78-165).* Wright-Patterson AFB, OH: Air Force Flight Dynamics Laboratory; December 1978.

3.3.2.6 Sequential Judgment Scale

General description. Pitrella and Kappler (1988) developed the Sequential Judgment Scale to measure the difficulty of driver vehicle handling. It was designed to meet the following rating scale guidelines: "1) use continuous instead of category scale formats, 2) use both verbal descriptors and numbers at scale points, 3) use descriptors at all major scale markings, 4) use horizontal rather than vertical scale formats, 5) either use extreme or no descriptors at end points, 6) use short, precise, and value-unloaded descriptors, 8) select and use equidistant descriptors, 9) use psychologically-scaled descriptors, 10) use positive numbers only, 11) have desirable qualities increase to the right, 12) use descriptors free of evaluation demands and biases, 13) use 11 or more scale points as available descriptors permit, and 14) minimize rater workload with suitable aids" (Pfendler, Pitrella, and Wiegand, 1994, p. 28). The scale has interval scale properties. It exists in both 11- and 15-point versions in German, Dutch, and English. The 15-point English version is presented in figure 13.

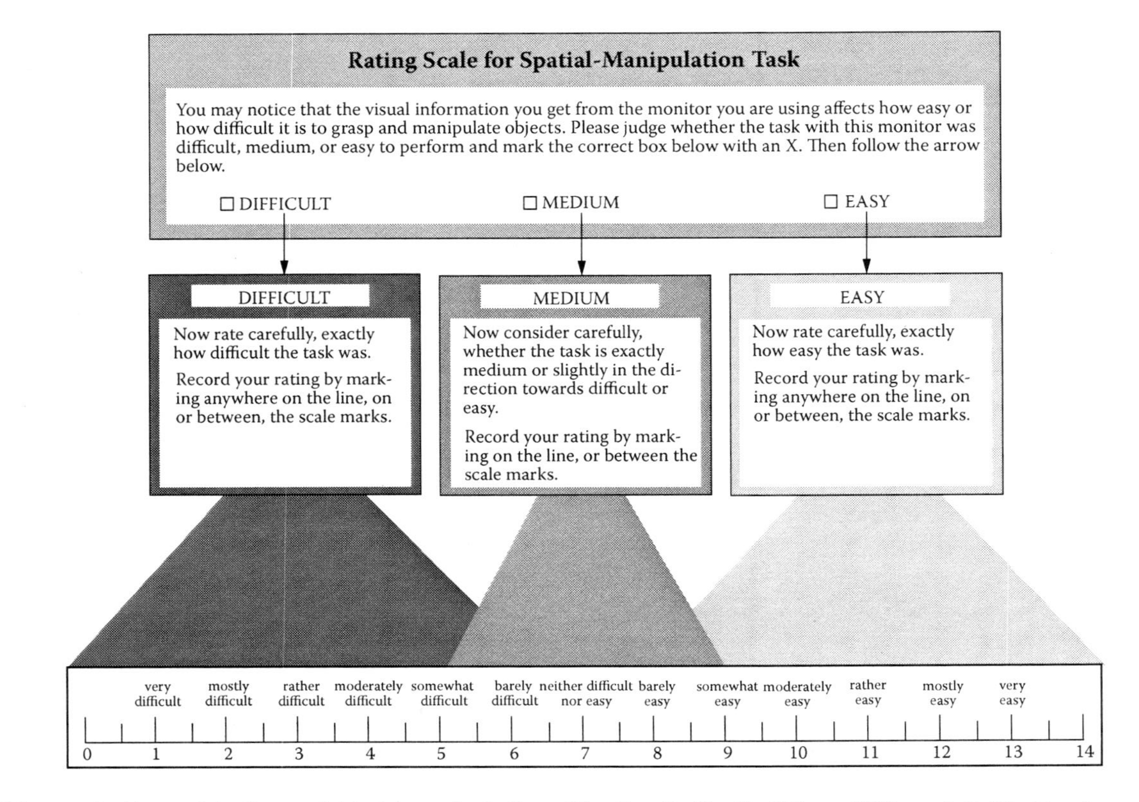

FIGURE 13 Fifteen-point Form of the Sequential Judgment Scale (from Pfendler, C., Pitrella, F.D., and Wiegand, D. *Workload measurement in human engineering test and evaluation*. Forschungsinstitut fur Anthropotechnik, Bericht Number 109, July 1994, p. 31).

Strength and limitations. Kappler, Pitrella, and Godthelp (1988) reported that the Sequential Judgment Scale ratings varied significantly between loaded and unloaded trucks as well as between different models of trucks. Kappler and Godthelp (1989) reported significantly more difficulty in vehicle handling as tire pressure and lane width decreased. The subjects drove on a closed-loop straight lane.

Pitrella (1988) reported that the scale significantly discriminated between 10 difficulty levels in a tracking task. Difficulty was manipulated by varying the amplitude and frequency of the forcing function. Pfendler (1993) reported higher validity estimates for the Sequential Judgment Scale (+0.72) than the German version of the NASA TLX (+0.708) in a color detection task.

Reliabilities have also been high (+0.92 to +0.99, Kappler, Pitrella, and Godthelp, 1988; +0.87 to +0.99, Pitrella, 1989; +0.87, Pfendler, 1993). Although the scale is an interval scale, parametric statistics can be used to analyze the data.

The scale has two disadvantages: "1) if only overall workload is measured, rating results will have low diagnosticity and 2) information on validity of the scale is restricted to psychomotor and perceptual tasks" (Pfendler, Pitrella, and Wiegand, 1994, p. 30).

Data requirements. Subjects mark the scale in pen or pencil. The experimenter then measures the distance from the right end of the scale. This measure is converted to a percentage of the complete scale.

Thresholds. 0 to 100%.

SOURCES

Kappler, W.D. and Godthelp, H. *Design and use of the two-level Sequential Judgment Rating Scale in the identification of vehicle handling criteria: I. Instrumented car experiments on straight lane driving.* Wachtberg: Forschungsinstitut fur Anthropotechnik, FAT Report Number 79, 1989.

Kappler, W.D., Pitrella, F.D., and Godthelp, H. *Psychometric and performance measurement of light weight truck handling qualities.* Wachtberg: Forschungsinstitut fur Anthropotechnik, FAT Report Number 77, 1988.

Pfendler, C. Vergleich der Zwei-Ebenen Intensitats-Skala und des NASA Task Load Index bei de Beanspruchungsbewertung wahrend ternvorgangen. *Z. Arb. Wise* 47 (19 NF) 1993/1, 26–33.

Pfendler, C., Pitrella, F.D., and Wiegand, D. *Workload measurement in human engineering test and evaluation.* Forschungsinstitut fur Anthropotechnik, Bericht Number 109, July 1994.

Pitrella, F.D. A cognitive model of the internal rating process. Wachtberg: Forschungsinstitut fur Anthropotechnik, FAT Report Number 82, 1988.

Pitrella, F.D. and Kappler, W.D. *Identification and evaluation of scale design principles in the development of the sequential judgment, extended range scale.* Wachtberg: Forschungsinstitut fur Anthropotechnik, FAT Report Number 80, 1988.

3.3.3 SET OF SUBSCALES SUBJECTIVE WORKLOAD MEASURES

The final type of subjective workload measure is a set of subscales each of which was designed to measure different aspects of workload. Examples include the following: Crew Status Survey, which separates fatigue and workload (section 3.3.3.1), Finegold Workload Rating Scale (section 3.3.3.2), Flight Workload Questionnaire (section

3.3.3.3), Hart and Hauser Rating Scale (section 3.3.3.4), Multi-Descriptor Scale (section 3.3.3.5), Multidimensional Rating Scale (section 3.3.3.6), Multiple Resources Questionnaire (section 3.3.3.7), NASA Bipolar Rating Scale (section 3.3.3.8), NASA Task Load Index (section 3.3.3.9), Profile of Mood States (section 3.3.3.10), Subjective Workload Assessment Technique (section 3.3.3.11), and Workload/Compensation/Interference/Technical Effectiveness (section 3.3.3.12).

3.3.3.1 Crew Status Survey

General description. The original Crew Status Survey was developed by Pearson and Byars (1956) and contained 20 statements describing fatigue states. The staff of the Air Force School of Aerospace Medicine Crew Performance Branch, principally Storm and Parke, updated the original survey. They selected the statements anchoring the points on the fatigue scale of the survey through iterative presentations of drafts of the survey to aircrew members. The structure of the fatigue scale was somewhat cumbersome, because the dimensions of workload, temporal demand, system demand, system management, danger, and acceptability were combined on one scale. However, the fatigue scale was simple enough to be well received by operational crews. The fatigue scale of the survey was shortened to seven statements and subsequently tested for sensitivity to fatigue as well as for test-retest reliability (Miller and Narvaez, 1986). Finally, a seven-point workload scale was added. The current Crew Status Survey (see figure 14) provides measures of self-reported fatigue and workload as well as space for general comments. Ames and George (1993) modified the workload scale to enhance reliability. Their scale descriptors are the following:

1. Nothing to do; no system demands
2. Light activity; minimum demands
3. Moderate activity; easily managed; considerable spare time
4. Busy; challenging but manageable; adequate time available
5. Very busy; demanding to manage; barely enough time
6. Extremely busy; very difficult; nonessential tasks postponed
7. Overloaded; system unmanageable; important tasks undone; unsafe (p. 4)

Strengths and limitations. These scales have been found to be sensitive to changes in task demand and fatigue but are independent of each other (Courtright, Frankenfeld, and Rokicki, 1986). Storm and Parke (1987) used the Crew Status Survey to assess the effects of temazepam on FB-111A crewmembers. The effect of the drug was not significant. The effect of performing the mission was, however. Specifically, the fatigue ratings were higher at the end than at the beginning of a mission. Gawron et al. (1988) analyzed Crew Status Survey ratings made at four times during each flight. They found a significant segment effect on fatigue and workload. Fatigue ratings increased over the course of the flight (preflight = 1.14, predrop = 1.47, postdrop = 1.43, and postflight = 1.56). Workload ratings were highest around a simulated airdrop (preflight = 1.05, predrop = 2.86, postdrop = 2.52, and postflight = 1.11).

George, Nordeen, and Thurmond (1991) collected workload ratings from Combat Talon II aircrew members during arctic deployment. None of the median ratings

NAME	DATE AND TIME

SUBJECT FATIGUE

(Circle the number of the statement which describes how you feel RIGHT NOW.)

1	Fully Alert, Wide Awake; Extremely Peppy
2	Very Lively; Responsive, But Not at Peak
3	Okay; Somewhat Fresh
4	A Little Tired; Less Than Fresh
5	Moderately Tired; Let Down
6	Extremely Tired; Very Difficult to Concentrate
7	Completely Exhausted; Unable to Function Effectively; Ready to Drop

COMMENTS

WORKLOAD ESTIMATE

(Circle the number of the statement which describes the MAXIMUM workload you experienced during the past work period. Put an X over the number of the statement which best describes the AVERAGE workload you experienced during the past work period)

1	Nothing to do; No System Demands
2	Little to do; Minimum System Demands
3	Active Involvement Required, But Easy to Keep Up
4	Challenging, But Manageable
5	Extremely Busy; Barely Able to Keep Up
6	Too Much to do; Overloaded; Postponing Some Tasks
7	Unmanageable; Potentially Dangerous; Unacceptable

COMMENTS

SAM FORM **202** **CREW STATUS SURVEY**
APR 81

PREVIOUS EDITION WILL BE USED

FIGURE 14 Crew Status Survey.

were greater than four. However, level 5 ratings occurred for navigators during airdrop s and self-contained approach run-ins. These authors also used the Crew Status Survey workload scale during terrain-following training flights on Combat Talon II. Pilots and copilots gave a median rating of 7. The ratings were used to identify major crewstation deficiencies.

However, George and Hollis (1991) reported confusion between adjacent categories at the high-workload end of the Crew Status Survey. They also found adequate ordinal properties for the scale, but very large variance in most order-of-merit tables.

Data requirements. Although the Crew Status Survey is printed on card stock, subjects find it difficult to fill in the rating scale during high-workload periods. Further, sorting (for example, by the time completed) the completed card-stock ratings after the flight is also difficult and not error free. A larger-character-size version of the survey has been included on flight cards at the Air Force Flight Test Center. Verbal ratings prompted by the experimenter work well if (1) subjects can quickly scan a card-stock copy of the rating scale to verify the meaning of a rating and (2) subjects are not performing a conflicting verbal task. Each scale can be used independently.

Thresholds. For subjective fatigue, 1 to 7; for workload, 1 to 7.

SOURCES

Ames, L.L. and George, E.J. *Revision and verification of a seven-point workload estimate scale (AFFTC-TIM-93-01).* Edwards Air Force Base, CA: Air Force Flight Test Center, 1993.

Courtright, J.F., Frankenfeld, C.A., and Rokicki, S.M. The independence of ratings of workload and fatigue. Paper presented at the *Human Factors Society 30th Annual Meeting,* Dayton, Ohio, 1986.

Gawron, V.J., Schiflett, S.G., Miller, J., Ball, J., Slater, T., Parker, F., Lloyd, M., Travale, D., and Spicuzza, R.J. *The effect of pyridostigmine bromide on in-flight aircrew performance (USAFSAM-TR-87-24).* Brooks Air Force Base, TX: School of Aerospace Medicine; January 1988.

George, E. and Hollis, S. *Scale validation in flight test.* Edwards Air Force Base, CA: Flight Test Center; December 1991.

George, E.J., Nordeen, M., and Thurmond, D. *Combat Talon II human factors assessment (AFFTC TR 90-36).* Edwards Air Force Base, CA: Flight Test Center; November 1991.

Miller, J.C. and Narvaez, A. A comparison of two subjective fatigue checklists. *Proceedings of the 10th Psychology in the DoD Symposium.* Colorado Springs, CO: United States Air Force Academy, 514–518, 1986.

Pearson, R.G. and Byars, G.E. *The development and validation of a checklist for measuring subjective fatigue (TR-56-115).* Brooks Air Force Base, TX: School of Aerospace Medicine, 1956.

Storm, W.F. and Parke, R.C. FB-111A aircrew use of temazepam during surge operations. *Proceedings of the NATO Advisory Group for Aerospace Research and Development (AGARD) Biochemical Enhancement of Performance Conference* (Paper number 415, pp. 12-1 to 12-12). Neuilly-sur-Seine, France: AGARD, 1987.

3.3.3.2 Finegold Workload Rating Scale

General description. The Finegold Workload Rating Scale has five subscales (see figure 15). It was developed to evaluate workload at each crew station aboard the AC-130H Gunship.

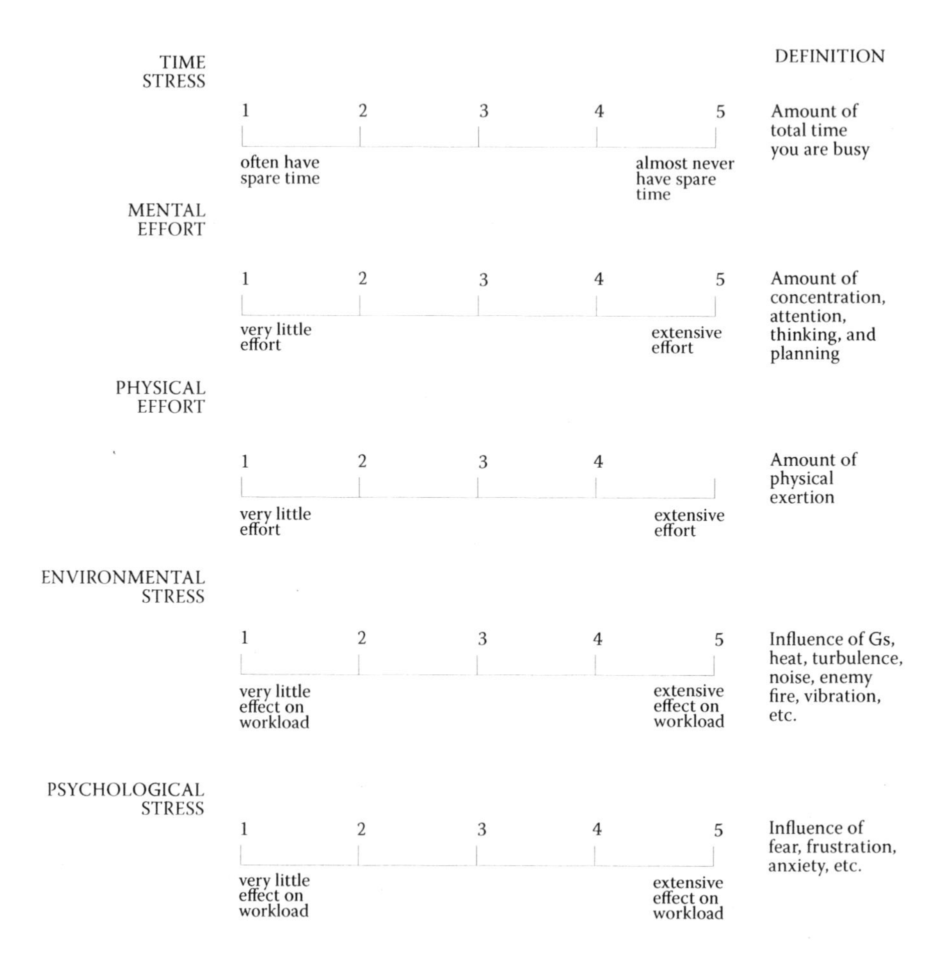

FIGURE 15 Finegold Workload Rating Scale.

Strengths and limitations. Finegold, Lawless, Simons, Dunleavy, and Johnson (1986) reported lower ratings associated with cruise than with engagement or threat segments. Analysis of the subscales indicated that time stress was rated differently at each crew station. Lozano (1989) replicated the Finegold et al. (1986) test using the AC-130H Gunship; again, ratings on subscales varied by crew position. George (1994) is replicated both studies using the current version of the AC-130U Gunship.

Data requirements. Average individual subscales as well as the complete workload Rating Scale scores.

Thresholds. For low workload, 1; for high workload, 5.

SOURCES

Finegold, L.S., Lawless, M.T., Simons, J.L., Dunleavy, A.O., and Johnson, J. *Estimating crew performance in advanced systems. Volume II: Application to future gunships Appendix B: Results of data analysis for AC-130H and hypothetical AC-130H (RP).* Edwards Air Force Base, CA: Air Force Flight Test Center; October 1986.

George, E.J. *AC-130U gunship workload evaluation (C4654-3-501)*. Edwards AFB, CA: Air Force Flight Test Center; April 1994.
Lozano, M.L. *Human engineering test report for the AC-130U gunship (NA-88-1805)*. Los Angeles, CA: Rockwell International; January 1989.

3.3.3.3 Flight Workload Questionnaire

General description. The Flight Workload Questionnaire is a four-item, behaviorally anchored rating scale. The items and the end points of the rating scales are the following: workload category (*low* to very *high*), fraction of time busy (*seldom has much to do* to *fully occupied at all times*), degree of mental effort (*minimal thinking* to *a great deal of thinking*), and feelings (*relaxing* to *very stressful*).

Strengths and limitations. The questionnaire is sensitive to differences in experience and ability. For example, Stein (1984) found significant differences in the flight workload ratings between experienced and novice pilots. Specifically, experienced pilots rated their workload during an air transport flight lower than novice pilots did. However, Stein also found great redundancy in the value of the ratings given for the four questionnaire items. This suggests that the questionnaire may evoke a response bias. The questionnaire provides a measure of overall workload but cannot differentiate between flight segments and/or events.

Data requirements. Not stated.

Thresholds. Not stated.

SOURCE

Stein, E.S. *The measurement of pilot performance: A master-journeyman approach (DOT/FAA/CT-83/15)*. Atlantic City, NJ: Federal Aviation Administration Technical Center; May 1984.

3.3.3.4 Hart and Hauser Rating Scale

General description. Hart and Hauser (1987) used a six-item rating scale (see figure 16) to measure workload during a 9 h flight. The items and their scales were the following: stress (*completely relaxed* to *extremely tense*), mental/sensory effort (*very low* to *very high*), fatigue (*wide-awake* to *worn-out*), time pressure (*none* to *very rushed*), overall workload (*very low* to *very high*), and performance (*completely unsatisfactory* to *completely satisfactory*). Subjects were instructed to mark the scale position corresponding to their experience.

Strengths and limitations. The scale was developed for use in-flight. In the initial study, Hart and Hauser (1987) asked subjects to complete the questionnaire at the end of each of seven flight segments. They reported significant segment effects in the 7 h flight. Specifically, stress, mental/sensory effort, and time pressure were lowest during a data-recording segment. There was a sharp increase in rated fatigue after the start of the data-recording segment. Overall, the aircraft commander rated workload higher than the copilot. Finally, performance received the same ratings throughout the flight.

	Stress	
Completely Relaxed	______	Extremely Tense
	Mental/Sensory Effort	
Very Low	______	Very High
	Fatigue	
Wide Awake	______	Worn Out
	Time Pressure	
None	______	Very Rushed
	Overall Workload	
Very Low	______	Very High
	Performance	
Completely Unsatisfactory	______	Completely Satisfactory

FIGURE 16 Hart and Hauser Rating Scale.

Data requirements. The scale is simple to use but requires a stiff writing surface and minimal turbulence.

Thresholds. Not stated.

SOURCE

Hart, S.G. and Hauser, J.R. Inflight application of three pilot workload measurement techniques. *Aviation, Space, and Environmental Medicine.* 58, 402–410, 1987.

3.3.3.5 Multi-Descriptor Scale

General description. The Multi-Descriptor (MD) Scale is composed of six descriptors: (1) attentional demand, (2) error level, (3) difficulty, (4) task complexity, (5) mental workload, and (6) stress level. Each descriptor is rated after a task. The MD score is the average of the six descriptor ratings.

Strengths and limitations. Wierwille, Rahimi, and Casali (1985) reported that the MD scores were not sensitive to variations in difficulty of mathematical calculations performed during a simulated flight task.

Data requirements. The six rating scales must be presented to the subject after a flight, and the average of the resultant ratings calculated.

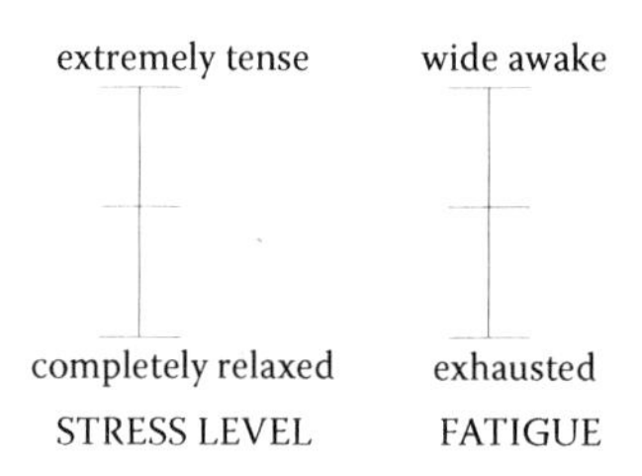

SOURCE

Wierwille, W.W., Rahimi, M., and Casali, J.G. Evaluation of 16 measures of mental workload using a simulated flight task emphasizing mediational activity. *Human Factors.* 27(5), 489–502, 1985.

TABLE 22
Multidimensional Rating Scale

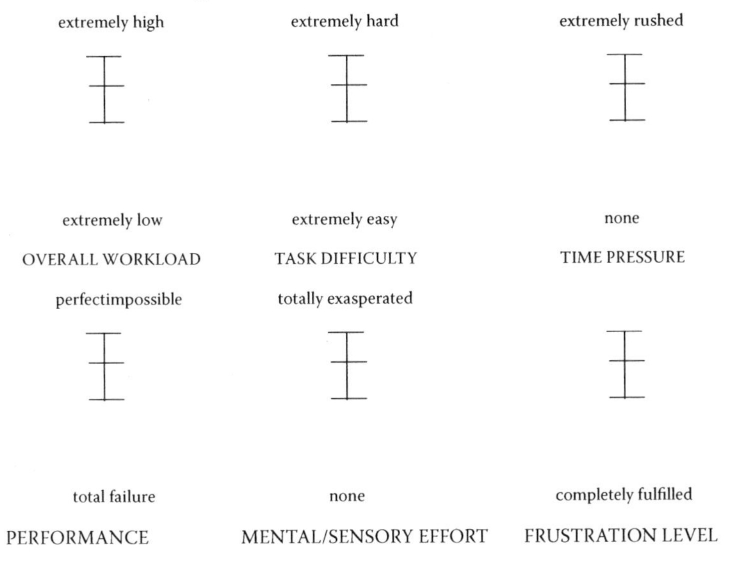

3.3.3.6 Multidimensional Rating Scale

General description. The Multidimensional Rating Scale is composed of eight bipolar scales (see table 22). Subjects are asked to draw a horizontal line on the scale to indicate their rating.

Strengths and limitations. Damos (1985) reported high correlations among several of the subscales (+0.82) as well as between overall workload and task difficulty single-task condition, +0.73 task variations. The time pressure and overall workload scales were also significantly associated with task-by-pacing condition interactions. The mental/sensory effort scale was significantly associated with a task by behavior pattern interaction.

Data requirements. The vertical line in each scale must be 100 mm long. The rater must measure the distance from the bottom of the scale to the subject's horizontal line to determine the rating.

Thresholds. Zero to 100.

SOURCE

Damos, D. The relation between the Type A behavior pattern, pacing, and subjective workload under single- and dual-task conditions. *Human Factors*. 27(6), 675–680, 1985.

3.3.3.7 Multiple Resources Questionnaire

General description. The Multiple Resources Questionnaire (MRQ) uses one rating scale (0 = no usage, 1 = light usage, 2 = moderate usage, 3 = heavy usage, 4 = extreme usage) for 17 dimensions of workload (see table 23).

TABLE 23
Multiple Resources Questionnaire

Multiple Resources Questionnaire
Auditory emotional process
Auditory linguistic process
Facial figural process
Facial motive process
Manual process
Short-term memory process
Spatial attentive process
Spatial categorical process
Spatial concentrative process
Spatial emergent process
Spatial positional process
Spatial quantitative process
Tactile figural process
Visual lexical process
Visual phonetic process
Visual temporal process
Vocal process

Strengths and limitations. Boles and Adair (2001) reported interrater reliabilities of undergraduate students for seven computer games to range from $r = +0.67$ to $r = +0.83$. In a second study using just two computer games, the interrater reliabilities ranged between $r = +0.57$ and $+0.65$.

Data requirements. Ratings on each of the 17 dimensions are used separately and without transformation.

Thresholds. Each rating varies between 0 and 4.

SOURCE

Boles, D.B. and Adair, L.P. The Multiple Resources Questionnaire (MRQ). *Proceedings of the Human Factors and Ergonomics Society 45th Annual Meeting.* 1790–1794, 2001.

3.3.3.8 NASA Bipolar Rating Scale

General description. The NASA Bipolar Rating Scale has 10 subscales. The titles, endpoints, and descriptions of each scale are presented in table 24; the scale itself, in figure 17. If a scale is not relevant to a task, it is given a weight of zero (Hart, Battiste, and Lester, 1984). A weighting procedure is used to enhance intrasubject reliability by 50% (Miller and Hart, 1984).

Strengths and limitations. The scale is sensitive to flight difficulty. For example, Bortolussi, Kantowitz, and Hart (1986) reported significant differences in the bipolar ratings between an easy and a difficult flight scenario. Bortolussi, Hart, and Shively (1987) and Kantowitz, Hart, Bortolussi, Shively, and Kantowitz (1984) reported similar results. However, Haworth, Bivens, and Shively (1986) reported that although the

TABLE 24
NASA Bipolar Rating Scale Descriptions

Title	Endpoints	Descriptions
Overall workload	Low, High	The total workload associated with the task considering all sources and components.
Task difficulty	Low, High	Whether the task was easy, demanding, simple or complex, exacting or forgiving.
Time pressure	None, Rushed	The amount of pressure you felt due to the rate at which the task elements occurred. Whether the task was slow and leisurely or rapid and frantic.
Performance	Perfect, Failure	How successful you think you were in doing what we asked you to do and how satisfied you were with what you accomplished.
Mental/Sensory effort	None, Impossible	The amount of mental and/or perceptual activity that was required (e.g., thinking, deciding, calculating, remembering, looking, searching, etc.).
Physical effort	None, Impossible	The amount of physical activity that was required (e.g., pushing, pulling, turning, controlling, activating, etc.).
Frustration level	Fulfilled, Exasperated	How insecure, discouraged, irritated, and annoyed versus secure, gratified, content, and complacent you felt.
Stress level	Relaxed, Tense	How anxious, worried, uptight, and harassed or calm, tranquil, placid, and relaxed you felt.
Fatigue	Exhausted, Alert	How tired, weary, worn-out, and exhausted or fresh, vigorous, and energetic you felt.
Activity type	Skill based, Rule based, Knowledge based	The degree to which the task required mindless reaction to well-learned routines or required the application of known rules or required problem solving and decision making.

Source: From Lysaght, R.J., Hill, S.G., Dick, A.O., Plamondon, B.D., Linton, P.M., Wierwille, W.W., Zaklad, A.L., Bittner, A.C., and Wherry, R.J. *Operator workload: Comprehensive review and evaluation of operator workload methodologies (Technical Report 851).* Alexandria, VA: Army Research Institute for the Behavioral and Social Sciences; June 1989, p. 91.

scale discriminated between control configurations in a single-pilot configuration, it did not do so in a pilot–copilot configuration.

Biferno (1985) reported a correlation between workload and fatigue ratings for a laboratory study. Bortolussi, Kantowitz, and Hart (1986) and Bortolussi, Hart, and Shively (1987) reported that the bipolar scales discriminated between two levels of difficulty in a motion-based simulator task. Vidulich and Pandit (1986) reported that the bipolar scales discriminated between levels of training in a category search task. Haworth, Bivens, and Shively (1986) reported correlations of +0.79 with Cooper–Harper Rating Scale ratings and +0.67 with SWAT ratings in a helicopter nap-of-the-earth mission. Vidulich and Tsang (1985a, 1985b, 1985c, 1986) reported that the NASA Bipolar Rating Scale were sensitive to task demand, had higher interrater reliability than SWAT, and required less time to complete than SWAT. Vidulich and

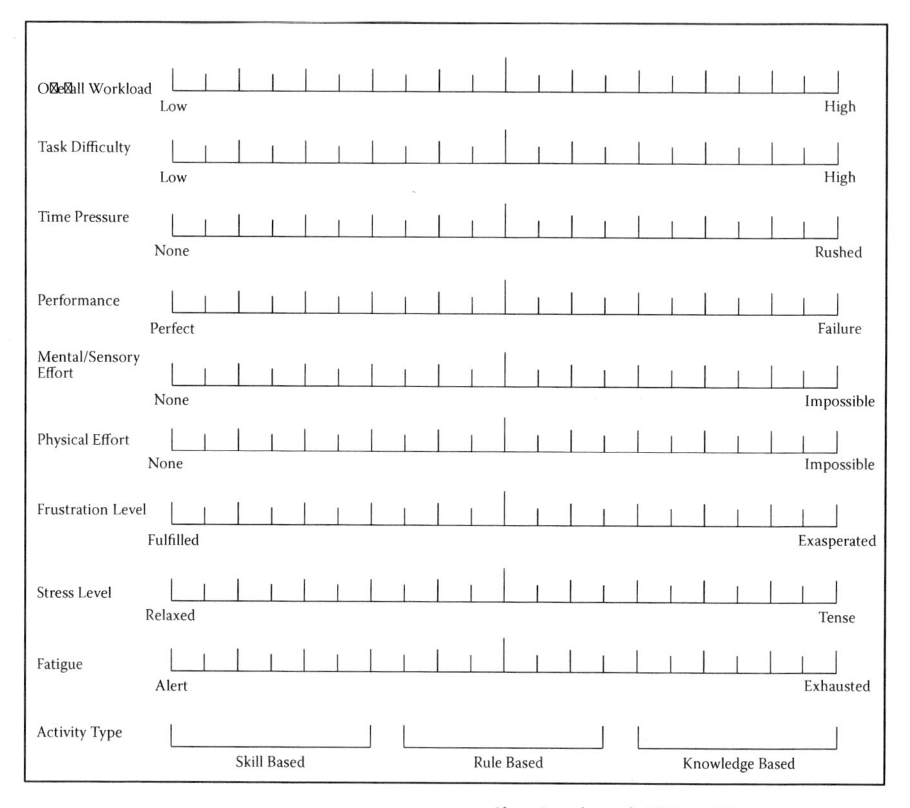

(from Lysaght et al., 1989, p. 92)

FIGURE 17 NASA Bipolar Rating Scale (from Lysaght, R.J., Hill, S.G., Dick, A.O., Plamondon, B.D., Linton, P.M., Wierwille, W.W., Zaklad, A.L., Bittner, A.C., and Wherry, R.J. *Operator workload: Comprehensive review and evaluation of operator workload methodologies (Technical Report 851).* Alexandria, VA: Army Research Institute for the Behavioral and Social Sciences; June 1989, p. 92).

Bortolussi (1988) reported significant increases in the overall workload rating from cruise to combat phase in simulated helicopter. Control configuration had no effect.

Data requirements. The number of times a dimension is selected by a subject is used to weight each scale. These weights are then multiplied by the scale score, summed, and divided by the total weight to obtain a workload score. The minimum workload value is 0; the maximum, 100. The scale provides a measure of overall workload, but is not sensitive to short-term demands. Further, the activity-type dimension must be carefully explained to pilots before use in flight.

Thresholds. Not stated.

SOURCES

Biferno, M.H. *Mental workload measurement: Event-related potentials and ratings of workload and fatigue (NASA CR-177354).* Washington, DC: NASA, 1985.

Bortolussi, M.R., Hart, S.G., and Shively, R.J. Measuring moment-to-moment pilot workload using synchronous presentations of secondary tasks in a motion-base trainer. In Jensen, R.S. (Ed) *Proceedings of the 4th Symposium on Aviation Psychology* (pp. 651–657). Columbus, OH: Ohio State University, 1987.
Bortolussi, M.R., Kantowitz, B.H., and Hart, S.G. Measuring pilot workload in a motion base trainer: A comparison of four techniques. *Applied Ergonomics.* 17, 278–283, 1986.
Hart, S.G., Battiste, V., and Lester, P.T. POPCORN: A supervisory control simulation for workload and performance research (NASA-CP-2341). *Proceedings of the 20th Annual Conference on Manual Control.* (pp. 431–453). Washington, DC: NASA, 1984.
Haworth, L.A., Bivens, C.C., and Shively, R.J. An investigation of single-piloted advanced cockpit and control configuration for nap-of-the-earth helicopter combat mission tasks. *Proceedings of the 42nd Annual Forum of the American Helicopter Society.* 675–671, 1986.
Kantowitz, B.H., Hart, S.G., Bortolussi, M.R., Shively, R.J., and Kantowitz, S.C. *Measuring pilot workload in a moving-base simulator: II. Building levels of workload,* 1984.
Lysaght, R.J., Hill, S.G., Dick, A.O., Plamondon, B.D., Linton, P.M., Wierwille, W.W., Zaklad, A.L., Bittner, A.C., and Wherry, R.J. *Operator workload: Comprehensive review and evaluation of operator workload methodologies (Technical Report 851).* Alexandria, VA: Army Research Institute for the Behavioral and Social Sciences; June 1989.
Miller, R.C. and Hart, S.G. Assessing the subjective workload of directional orientation tasks (NASA-CP-2341). *Proceedings of the 20th Annual Conference on Manual Control.* (pp. 85-95). Washington, DC: NASA, 1984.
Vidulich, M.A. and Bortolussi, M.R. Speech recognition in advanced rotorcraft: Using speech controls to reduce manual control overload. *Proceedings of the National Specialists' Meeting Automation Applications for Rotorcraft,* 1988.
Vidulich, M.A. and Pandit, P. Training and subjective workload in a category search task. *Proceedings of the Human Factors Society 30th Annual Meeting* (pp. 1133–1136). Santa Monica, CA: Human Factors Society, 1986.
Vidulich, M.A. and Tsang, P.S. Assessing subjective workload assessment: A comparison of SWAT and the NASA-Bipolar methods. *Proceedings of the Human Factors Society 29th Annual Meeting.* (pp. 71–75). Santa Monica, CA: Human Factors Society, 1985a.
Vidulich, M.A. and Tsang, P.S. Techniques of subjective workload assessment: A comparison of two methodologies. *Proceedings of the Third Symposium on Aviation Psychology* (pp. 239–246). Columbus, OH: Ohio State University, 1985b.
Vidulich, M.A. and Tsang, P.S. Evaluation of two cognitive abilities tests in a dual-task environment. *Proceedings of the 21st Annual Conference on Manual Control.* (pp. 12.1–12.10). Columbus, OH: Ohio State University, 1985c.
Vidulich, M.A. and Tsang, P.S. Techniques of subjective workload assessment: A comparison of SWAT and NASA-Bipolar methods. *Ergonomics.* 29(11), 1385–1398, 1986.

3.3.3.9 NASA Task Load Index

General description. The NASA Task Load Index (TLX) is a multidimensional subjective workload-rating technique (see figure 18). In NASA TLX, workload is defined as the "cost incurred by human operators to achieve a specific level of performance." The subjective experience of workload is defined as an integration of weighted subjective responses (emotional, cognitive, and physical) and weighted evaluation of behaviors. The behaviors and subjective responses, in turn, are driven by perceptions of task demand. Task demands can be objectively quantified in terms of magnitude and importance. An experimentally based process of elimination led to the identification of six dimensions for the subjective experience of workload:

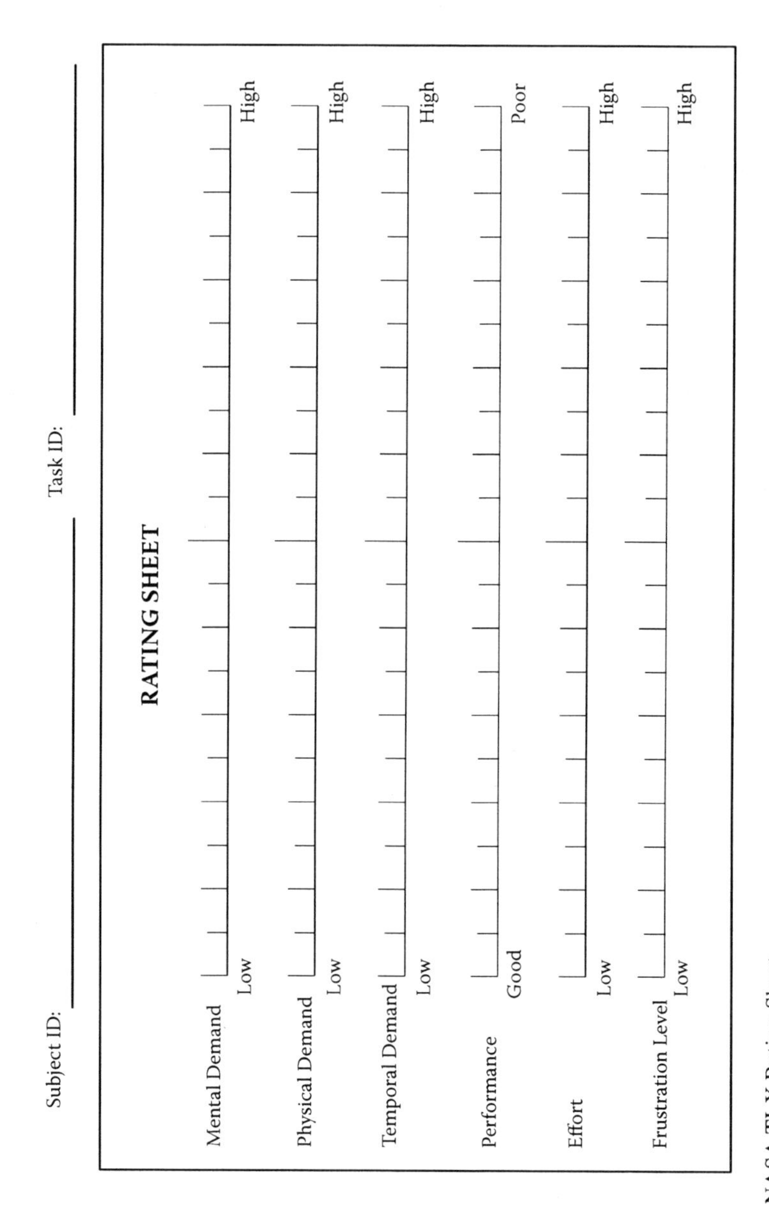

FIGURE 18 NASA TLX Rating Sheet.

TABLE 25
NASA TLX Rating Scale Descriptions

Title	Endpoints	Descriptions
Mental demand	Low, High	How much mental and perceptual activity was required (e.g., thinking, deciding, calculating, remembering, looking, searching, etc.)? Was the task easy or demanding, simple or complex, exacting or forgiving?
Physical demand	Low, High	How much physical activity was required (e.g., pushing, pulling, turning, controlling, activating, etc.)? Was the task easy or demanding, slow or brisk, slack or strenuous, restful or laborious?
Temporal demand	Low, High	How much time pressure did you feel due to the rate or pace at which the tasks or task elements occurred? Was the pace slow and leisurely or rapid and frantic?
Performance	Good, Poor	How successful do you think you were in accomplishing the goals of the task set by the experimenter (or yourself)? How satisfied were you with your performance in accomplishing these goals?
Effort	Low, High	How hard did you have to work (mentally and physically) to accomplish your level of performance?
Frustration level	Low, High	How insecure, discouraged, irritated, stressed, and annoyed versus secure, gratified, content, relaxed, and complacent did you feel during the task? (NASA Task Load Index, p. 13)

mental demand, physical demand, temporal demand, perceived performance, effort, and frustration level. The rating-scale definitions are presented in table 25.

Strengths and limitations. Sixteen investigations were carried out, establishing a database of 3461 entries from 247 subjects (Hart and Staveland, 1987). All dimensions were rated on bipolar scales ranging from 1 to 100, anchored at each end with a single adjective. An overall workload rating was determined from a weighted combination of scores on the six dimensions. The weights were determined from a set of relevance ratings provided by the subjects.

In an early in-flight evaluation in the NASA Kuiper Airborne Observatory, Hart, Hauser, and Lester (1984) reported a significant difference between left and right seat positions as well as in-flight segments related to ratings of overall fatigue, time pressure, stress, mental/sensory effort, fatigue, and performance. The data were from nine NASA test pilots over 11 flights. In another Hart study, Hart and Staveland (1987) concluded that the NASA TLX provides a sensitive indicator of overall workload as it distinguished between tasks of various cognitive and physical demands. They also stated that the weights and magnitudes determined for each NASA TLX dimension provide important diagnostic information about the sources of loading within a task. They reported that the six NASA TLX ratings took less than a minute to acquire and suggested the scale would be useful in operational environments.

Battiste and Bortolussi (1988) reported significant workload effects as well as a test-retest correlation of +0.769. Corwin, Sandry-Garza, Biferno, Boucek, Logan, Jonsson, and Metalis (1989) reported that NASA TLX was a valid and reliable measure of workload.

NASA TLX has been used extensively in the flight environment. Bittner, Byers, Hill, Zaklad, and Christ (1989), Byers, Bittner, Hill, Zaklad, and Christ (1988), Hill, Byers, Zaklad, and Christ (1989), Hill, Zaklad, Bittner, Byers, and Christ (1988), and Shively, Battiste, Matsumoto, Pepitone, Bortolussi, and Hart (1987), based on in-flight data, stated that NASA TLX ratings significantly discriminated between flight segments.

Nataupsky and Abbott (1987) successfully applied NASA TLX to a multitask environment. Vidulich and Bortolussi (1988) replicated the significant flight-segment effect but reported no significant differences in NASA TLX ratings between control configurations or between combat countermeasure conditions. In a later study, Tsang and Johnson (1989) reported reliable increases in NASA TLX ratings when target acquisition and engine failure tasks were added to the primary flight task. Vidulich and Bortolussi (1988) reported significant increases in NASA TLX ratings from the cruise to the combat phase during a simulated helicopter mission. Control configuration, however, had no effect. In a display study, Stark, Comstock, Prinzel, Burdette, and Scerbo (2001) reported a significant decrease in workload when pilots in a fixed-base simulator had a tunnel or pathway in the sky display or a smaller rather than a larger display.

Nygren (1991) reported that NASA TLX is a measure of general workload experienced by aircrews. Selcon, Taylor, and Koritsas (1991) concluded from pilot ratings of an air combat flight simulation that NASA TLX was sensitive to difficulty but not the pilot experience. Hancock, Williams, Manning, and Miyake (1995) reported that the NASA TLX score was highly correlated with difficulty of a simulated flight task.

Hendy, Hamilton, and Landry (1993) examined one-dimensional and multidimensional measures of workload in a series of four experiments (low-level helicopter operations, peripheral version display evaluation, flight simulator fidelity, and aircraft-landing task). They concluded that if an overall measure of workload is required, then a univariate measure is as sensitive as an estimate derived from multivariate data. If a univariate measure is not available, then a simple unweighted additive method can be used to combine ratings into an overall workload estimate.

Byers, Bittner, and Hill (1989) suggested using raw NASA TLX scores. Moroney, Biers, Eggemeier, and Mitchell (1992) reported that the prerating weighting scheme is unnecessary, because the correlation between weighted and unweighted scores was +0.94. Further, delays of 15 min did not affect the workload ratings; delays of 48 h, however, did. After 48 h, ratings no longer discriminate between workload conditions. Moroney, Biers, and Eggemeier (1995) concluded from a review of relevant studies that 15 min delays do not affect NASA TLX.

NASA TLX has been applied to other environments. Hill, Lavecchia, Byers, Bittner, Zaklad, and Christ (1992) reported that the NASA TLX was sensitive to different levels of workload and high in user acceptance. Their subjects were army operators. Jordan and Johnson (1993) concluded from an on-road evaluation of a car stereo that NASA TLX was a useful measure of mental workload.

Hancock and Caird (1993) reported a significant increase in the overall workload rating scale of the NASA TLX as shrink rate of a target decreased. The highest ratings were on paths with four steps rather than 2, 8, or 16 steps from cursor to target. Dember, Warm, Nelson, Simons, and Hancock (1993) used the NASA TLX to measure workload on a visual vigilance task performed for 10, 20, 30, 40, or 50 min and at either an easy or difficult level of discrimination. They reported linearly increase in workload with time and a decrease with salience. NASA TLX scores significantly increased as ambient noise increased (Becker, Warm, Dember, and Hancock, 1995).

Harris, Hancock, Arthur, and Caird (1995) reported significantly higher ratings on five (mental demand, temporal demand, effort, frustration, and physical demand) of the six NASA-TLX scales for manual than automatic tracking. In another automation study, Heers and Casper (1998) reported higher workload ratings without automatic terrain avoidance, missile warning receiver, and laser-guided rockets than with these advanced technologies. The data were collected in a Scout Helicopter simulator. The subjects were eight U.S. Army helicopter pilots. In another simulator study, Bustamante, Fallon, Bliss, Bailey, and Anderson (2005) reported a significantly higher time pressure ratings at 20 nm (nautical mile) to weather than 160 nm to weather. Their subjects were 24 commercial airline pilots using a desktop flight simulator.

Aretz, Shacklett, Acquaro, and Miller (1995) reported that the number of concurrent tasks had the greatest impact on NASA TLX ratings followed by subject's flight experience. Their data were collected in a fixed-base simulator flown by 15 U.S. Air Force Academy cadets with 0 to 15.9 flight hours. Grubb, Warm, Dember, and Berch (1995), on the basis of vigilance data from 144 observers, reported increased workload as display uncertainty increased. Period of watch had no significant effect. (10, 20, 30, or 40 min). Temple, Warm, Dember, Jones, LaGrange, and Matthews (2000) used NASA TLX to verify the high workload induced by a 12 min computerized vigilance task. Szalma, Warm, Matthews, Dember, Weiler, Meier, and Eggemeier (2004) found that modality (auditory versus visual task) or time (four 10 min vigilance periods) had no significant effects on the global workload score of the NASA TLX. Also related to vigilance, Metzger and Parasuraman (2005) reported significant differences in NASA TLX as a function of decision-aid reliability in an air traffic scenario.

Vidulich and Tsang (1985) compared the SWAT and TLX. They stated that the collection of ratings is simpler with SWAT. However, the SWAT card sort is more tedious and time consuming. Battiste and Bortolussi (1988) reported that no significant correlation between SWAT and NASA TLX in a simulated B-727 flight. Jordan, Farmer, and Belyavin (1995) reported that the patterns of results for NASA-TLX and the Prediction of Performance (POP) workload measure was the same with the NASA-TLX having significant differences in all levels of workload conditions on a desk-top flight simulator. Hancock (1996) stated that NASA TLX and SWAT "were essentially equivalent in terms of their sensitivity to task manipulations." The task was tracking.

Tsang and Johnson (1987) reported good correlations between NASA TLX and a one dimensional workload scale. Vidulich and Tsang (1987) replicated the Tsang and Johnson finding as well as reported a good correlation between NASA TLX and the AHP. Leggatt and Noyes (1997) reported that there was no significant difference in NASA TLX workload ratings between others and own ratings of a subject's workload. There was, however, an interaction in that subordinates rated the leader's workload higher than the leader rated his or her own workload. The subjects were armored fighting vehicle drivers. Riley, Lyall, and Wiener (1994) compared results from NASA TLX with 22 methods of measuring workload. All methods gave approximately the same results for airline pilots completing Line Oriented Flight Training exercises. Moroney, Reising, Biers, and Eggemeier (1993) reported that NASA TLX ratings were not significantly affected by the workload in a previous simulated flight. Workload was manipulated by crosswind level being simulated.

Endsley and Kaber (1999) used NASA TLX to measure workload associated with varying levels of automation: manual control, action support, batch processing, shared control, decision support, blended decision making, rigid system, automated decision making, supervisory control, and full automation. Blended decision making, automated decision making, supervisory control, and full automation had significantly lower workload. Their subjects were 30 undergraduate students.

Vidulich and Pandit (1987) reported only three significant correlations between NASA TLX and seven personality tests (Jenkins Activity Survey, Rotter's Locus of Control, Cognitive Failures Questionnaire, Cognitive Interference Questionnaire, Thought Occurrence Questionnaire, California Q-Sort, and the Myers–Briggs Type Indicator): the speed scale of the Jenkins Activity Survey and the physical demand scale of the NASA TLX ($r = -0.23$), Locus of control and physical demand ($r = +0.21$) and, finally, locus of control and effort ($r = +0.23$). In a more recent study, Szalma (2002) reported that stress-coping strategy had no effect on NASA TLX after completion of 24 min vigilance task. The subjects were 48 male and 48 female undergraduate students. Similarly, Brown and Galster (2004) reported no significant effect of imposed workload on NASA TLX rating. The imposed workload was differences in reliability of automation in a simulated flight.

In a different domain, Levin, France, Hemphill, Jones, Chen, Rickard, Makowski, and Aronsky (2006) collected NASA TLX data from emergency room physicians after 180 min of tasking in an actual emergency room. High workloads were associated with exchanging patient information, directing patient care, completing phone calls and consults, and charting. The highest workload dimension was temporal demand. Lio, Bailey, Carswell, Seales, Clarke, and Payton (2006) reported significant increases in NASA TLX as the precision requirements in a laparoscopic task increased.

In yet another domain, Jerome, Witner, and Mouloua (2006) used NASA TLX to assess augmented reality cues. They reported a significant decrease in workload with the presence of haptic cues. Riley and Strater (2006) used NASA TLX to compare four robot-control modes and reported significant differences in workload. Alm, Kovordanyi, and Ohlsson (2006) used NASA TLX to assess night-vision systems in passenger vehicles. NASA TLX ratings of mental demands and effort were significantly lower in the presence of the night-vision system than when the system was absent. Finomore, Warm, Matthews, Riley, Dember, Shaw, Ungar, and Scerbo (2006) reported significant increase in NASA TLX for detecting the absence of a feature in a target than for detecting the presence of a feature in a target.

Windell, Wiebe, Converse-Lane, and Beith (2006) compared NASA TLX scores with a Short Subjective Instrument (SSI) (i.e., a single question for overall workload). The SSI showed significant differences between modules designed to vary in workload, whereas the NASA TLX mental demands scale ratings did not. Liu (1996) asked subjects to perform a pursuit tracking task with a decision-making task. Visual scanning resulted in increased workload as measured by the NASA TLX.

Svensson, Angelborg-Thanderz, Sjoberg, and Olsson (1997) reported the reliability of the NASA TLX to be +0.77 among 18 pilots flying simulated low-level, high-speed missions. The correlation with the Bedford Workload Scale was +0.826 and with the SWAT was +0.735.

Data requirements. Use of the TLX requires two steps. First, subjects rate each task performed on each of the six subscales. Hart suggests that subjects practice using the rating scales in a training session. Second, subjects must perform 15 pair-wise comparisons of six workload scales. The number of times each scale is rated as contributing more to the workload of a task is used as the weight for that scale. Separate weights should be derived for diverse tasks; the same weights can be used for similar tasks. Note that a set of PC compatible programs has been written to gather ratings and weights, and to compute the weighted workload scores. The programs are available from the Human Factors Division at NASA Ames Research Center, Moffett Field, California.

Thresholds. Knapp and Hall (1990) used NASA TLX to evaluate a highly automated communication system. Using 40 as a high workload threshold, the system was judged to impose high workload and difficult cognitive effort on operators.

SOURCES

Alm, T., Kovordanyi, R., and Ohlsson, K. Continuous versus situation-dependent night vision presentation in automotive applications. *Proceedings of the Human Factors and Ergonomics Society 50th Annual Meeting* (pp. 2033–2037), 2006.

Aretz, A.J., Shacklett, S.F., Acquaro, P.L., and Miller, D. The prediction of pilot subjective workload ratings. *Proceedings of the Human Factors and Ergonomics Society 39th Annual Meeting.* 1: 94–97, 1995.

Battiste, V. and Bortolussi, M.R. Transport pilot workload: A comparison of two objective techniques. *Proceedings of the Human Factors Society 32nd Annual Meeting.* 150–154, 1988.

Becker, A.B., Warm, J.S., Dember, W.N., and Hancock, P.A. Effects of jet engine noise and performance feedback on perceived workload in a monitoring task. *International Journal of Aviation Psychology.* 5(1), 49–62, 1995.

Bittner, A.C., Byers, J.C., Hill, S.G., Zaklad, A.L., and Christ, R.E. Generic workload ratings of a mobile air defense system. *Proceedings of the Human Factors Society 33rd Annual Meeting* (pp. 1476–1480). Santa Monica, CA: Human Factors Society, 1989.

Brown, R.D. and Galster, S.M. Effects of reliable and unreliable automation on subjective measures of mental workload, situation awareness, trust and confidence in a dynamics flight task. *Proceedings of the Human Factors and Ergonomics Society 48th Annual Meeting.* 147–151, 2004.

Bustamante, E.A., Fallon, C.K., Bliss, J.P., Bailey, W.R., and Anderson, B.L. Pilots' workload, situation awareness, and trust during weather events as a function of time pressure, role assignment, pilots' rank, weather display, and weather system. *International Journal of Applied Aviation Studies.* 5(2): 348–368, 2005.

Byers, J.C., Bittner, A.C., and Hill, S.G. Traditional and raw Task Load Index (TLX) correlations: Are paired comparisons necessary? In *Advances in industrial ergonomics and safety.* London: Taylor and Frances, 1989.

Byers, J.C., Bittner, A.C., Hill, S.G., Zaklad, A.L., and Christ, R.E. Workload assessment of a remotely piloted vehicle (RPV) system. *Proceedings of the Human Factors Society 32nd Annual Meeting.* (pp. 1145–1149). Santa Monica, CA: Human Factors Society, 1988.

Corwin, W.H., Sandry-Garza D.L., Biferno, M.H., Boucek, G.P., Logan, A.L., Jonsson, J.E., and Metalis, S.A. *Assessment of crew workload measurement methods, techniques, and procedures. Volume I—Process, methods, and results (WRDC-TR-89-7006).* Wright-Patterson Air Force Base, OH, 1989.

Dember, W.N., Warm, J.S., Nelson, W.T., Simons, K.G., and Hancock, P.A. The rate of gain of perceived workload in sustained operations. *Proceedings of the Human Factors and Ergonomics Society 37th Annual Meeting.* 2, 1388–1392, 1993.

Endsley, M.R. and Kaber, D.B. Level of automation effects on performance, situation awareness, and workload in a dynamic control task. *Ergonomics.* 42(3): 462–492, 1999.

Finomore, V.S., Warm, J.S., Matthews, G., Riley, M.A., Dember, W.N., Shaw, T.H., Ungar, N.R., and Scerbo, M.W. Measuring the workload of sustained attention. *Proceedings of the Human Factors and Ergonomics Society 50th Annual Meeting.* (pp. 1614–1618), 2006.

Grubb, P.L., Warm, J.S., Dember, W.N., and Berch, D.B. Effects of multiple-signal discrimination on vigilance performance and perceived workload. *Proceedings of the Human Factors and Ergonomics Society 39th Annual Meeting.* 2, 1360–1364, 1995.

Hancock, P.A. Effects of control order, augmented feedback, input device, and practice on tracking performance and perceived workload. *Ergonomics.* 39(9), 1146–1162, 1996.

Hancock, P.A. and Caird, J.K. Experimental evaluation of a model of mental workload. *Human Factors.* 35(3), 413–419, 1993.

Hancock, P.A., William G., Manning, C.M., and Miyake, S. Influence of task demand characteristics on workload and performance. *International Journal of Aviation Psychology.* 5(1), 63–86, 1995.

Harris, W.C., Hancock, P.A., Arthur, E.J., and Caird, J.K. Performance, workload, and fatigue changes associated with automation. *International Journal of Aviation Psychology.* 5(2), 169–185, 1995.

Hart, S.G., Hauser, J.R., and Lester, P.T. Inflight evaluation of four measures of pilot workload. *Proceedings of the Human Factors Society 28th Annual Meeting.* 2, 945–949, 1984.

Hart, S.G. and Staveland, L.E. Development of NASA-TLX (Task Load Index): Results of empirical and theoretical research. In Hancock, P.A. and Meshkati, N. (Eds.) *Human mental workload.* Amsterdam: Elsevier, 1987.

Heers, S.T. and Casper, P.A. Subjective measurement assessment in a full mission scout-attack helicopter simulation. *Proceedings of the Human Factors and Ergonomics Society 42nd Annual Meeting.* 1, 26–30, 1998.

Hendy, K.C., Hamilton, K.M., and Landry, L.N. Measuring subjective workload: When is one scale better than many? *Human Factors.* 35(4), 579–601, 1993.

Hill, S.G., Byers, J.C., Zaklad, A.L., and Christ, R.E. Subjective workload assessment during 48 continuous hours of LOS-F-H operations. *Proceedings of the Human Factors Society 33rd Annual Meeting.* (pp. 1129–1133). Santa Monica, CA: Human Factors Society, 1989.

Hill, S.G., Lavecchia, H.P., Byers, J.C., Bittner, A.C., Zaklad, A.L., and Christ, R.E. Comparison of four subjective workload rating scales. *Human Factors.* 34, 429–439, 1992.

Hill, S.G., Zaklad, A.L., Bittner, A.C., Byers, J.C., and Christ, R.E. Workload assessment of a mobile air defense system. *Proceedings of the Human Factors Society 32nd Annual Meeting.* (pp. 1068–1072). Santa Monica, CA: Human Factors Society, 1988.

Jerome, C.J., Witner, B., and Mouloua, M. Attention orienting in augmented reality environments: Effects of multimodal cues. *Proceedings of the Human Factors and Ergonomics Society 50th Annual Meeting.* (pp. 2114–2118). Santa Monica, CA: Human Factors and Ergonomics Society, 2006.

Jordan, C.S., Farmer, E.W., and Belyavin, A.J. The DRA Workload Scales (DRAWS): A validated workload assessment technique. *Proceedings of the Eighth International Symposium on Aviation Psychology.* 2, 1013–1018, 1995.

Jordan, P.W. and Johnson, G.L. Exploring mental workload via TLX: The case of operating a car stereo whilst driving. In A.G. Gale, I.D. Brown, C.M. Haslegrave, H.W. Kruysse, and S.P. Taylor (Eds.). *Vision in vehicles—IV.* Amsterdam: North-Holland, 1993.

Knapp, B.G. and Hall, B.J. *High performance concerns for the TRACKWOLF system (ARI Research Note 91-14).* Alexandria, VA, 1990.

Leggatt, A. and Noyes, J. Workload judgments: Self-assessment versus assessment of others. In Harris, D. (Ed.) *Engineering psychology and cognitive ergonomics Volume one Transportation systems.* Aldershot, UK: Ashgate, 443–449, 1997.

Levin, S., France, D.J., Hemphill, R., Jones, I., Chen, K.Y., Rickard, D., Makowski, R., and Aronsky, D. Tracking workload in the emergency department. *Human Factors.* 48(3): 526–539, 2006.

Lio, C.H., Bailey, K., Carswell, C.M., Seales, W.B., Clarke, D., and Payton, G.M. Time estimation as a measure of mental workload during the training of laparoscopic skills. *Proceedings of the Human Factors and Ergonomics Society 50th Annual Meeting.* (pp. 1910–1913), 2006.

Liu, Y. Quantitative assessment of effects of visual scanning on concurrent task performance. *Ergonomics.* 39(3): 382–399, 1996.

Metzger, U. and Parasuraman, R. Automation in future air traffic management: Effects of decision aid reliability on controller performance and mental workload. *Human Factors.* 47(1), 35–49, 2005.

Moroney, W.F., Biers, D.W., and Eggemeier, F.T. Some measurement and methodological considerations in the application of subjective workload measurement techniques. *International Journal of Aviation Psychology.* 5(1), 87–106, 1995.

Moroney, W.E., Biers, D.W., Eggemeier, F.T., and Mitchell, J.A. A comparison of two scoring procedures with the NASA Task Load Index in a simulated flight tasks. *NAECON Proceedings.* (pp. 734–740). Dayton, OH, 1992.

Moroney, W.F., Reising, J., Biers, D.W., and Eggemeier, D.W. The effect of previous level of workload on the NASA Task Load Index (TLX) in a simulated flight environment. *Proceedings of the Seventh International Symposium on Aviation Psychology.* 2, 882–890, 1993.

Nataupsky, M. and Abbott, T.S. Comparison of workload measures on computer-generated primary flight displays. *Proceedings of the Human Factors Society 31st Annual Meeting.* (pp. 548–552). Santa Monica, CA: Human Factors Society, 1987.

Nygren, T.E. Psychometric properties of subjective workload measurement techniques: Implications for their use in the assessment of perceived mental workload. *Human Factors.* 33 (1), 17–33, 1991.

Riley, J.M. and Strater, L.D. Effects of robot control mode on situational awareness and performance in a navigation task. *Proceedings of the Human Factors and Ergonomics Society 50th Annual Meeting.* (pp. 540–544), 2006.

Riley, V., Lyall, E., and Wiener, E. Analytic workload models for flight deck design and evaluation. *Proceedings of the Human Factors and Ergonomics Society 38th Annual Meeting.* 1, 81–84, 1994.

Selcon, S.J., Taylor, R.M., and Koritsas, E. Workload or situational awareness?: TLX vs. SART for aerospace systems design evaluation. *Proceedings of the Human Factors Society 35th Annual Meeting.* 62–66, 1991.

Shively, R.J., Battiste, V., Matsumoto, J.H., Pepitone, D.D., Bortolussi, M.R., and Hart, S.G. Inflight evaluation of pilot workload measures for rotorcraft research. In R.S. *Jensen Proceedings of the 4th Symposium on Aviation Psychology.* (pp. 637–643). Columbus, OH: Ohio State University, 1987.

Stark, J.M., Comstock, J.R., Prinzel, L.J., Burdette, D.W., and Scerbo, M.W. A preliminary examination of situation awareness and pilot performance in a synthetic vision environment. *Proceedings of the Human Factors and Ergonomics Society 45th Annual Meeting.* 40–43, 2001.

Svensson, E., Angelborg-Thanderz, M., Sjoberg, L., and Olsson, S. Information complexity–mental workload and performance in combat aircraft. *Ergonomics*. 40(3), 362–380, 1997.
Szalma, J.L. Individual difference in the stress and workload of sustained attention. *Proceedings of the Human Factors and Ergonomics Society 46th Annual Meeting*. 1002–1006, 2002.
Szalma, J.L., Warm, J.S., Matthews, G., Dember, W.N., Weiler, E.M., Meier, A., and Eggemeier, F. T. Effects of sensory modality and task duration on performance, workload, and stress in sustained attention. *Human Factors*. 46(2), 219–233, 2004.
Temple, J.G., Warm, J.S., Dember, W.N., Jones, K.S., LaGrange, C.M., and Matthews, G. The effects of signal salience and caffeine on performance, workload, and stress in an abbreviated vigilance task. *Human Factors*. 42(2), 183–194, 2000.
Tsang, P.S. and Johnson, W. Automation: Changes in cognitive demands and mental workload. *Proceedings of the Fourth Symposium on Aviation Psychology*. Columbus, OH: Ohio State University, 1987.
Tsang, P.S. and Johnson, W.W. Cognitive demands in automation. *Aviation, Space, and Environmental Medicine*. 60, 130–135, 1989.
Vidulich, M.A. and Bortolussi, M.R. Control configuration study. *Proceedings of the American Helicopter Society National Specialist's Meeting: Automation Application for Rotorcraft*, 1988.
Vidulich, M.A. and Bortolussi, M.R. Speech recognition in advanced rotorcraft: Using speech controls to reduce manual control overload. *Proceedings of the National Specialists' Meeting Automation Applications for Rotorcraft*, 1988.
Vidulich, M.A. and Pandit, P. Individual differences and subjective workload assessment: Comparing pilots to nonpilots. *Proceedings of the International Symposium on Aviation Psychology*. 630–636, 1987.
Vidulich, M.A. and Tsang, P.S. Assessing subjective workload assessment: A comparison of SWAT and the NASA-bipolar methods. *Proceedings of the Human Factors Society 29th Annual Meeting*. (pp. 71–75). Santa Monica, CA: Human Factors Society, 1985.
Vidulich, M.A. and Tsang, P.S. Absolute magnitude estimation and relative judgment approaches to subjective workload assessment. *Proceedings of the Human Factors Society 31st Annual Meeting* (pp. 1057–1061). Santa Monica, CA: Human Factors Society, 1987.
Windell, D., Wiebe, E., Converse-Lane, S., and Beith, B. A comparison of two mental workload instruments in multimedia instruction. *Proceedings of the Human Factors and Ergonomics Society 50th Annual Meeting*. (1764–1768). Santa Monica, CA: Human Factors and Ergonomics Society, 2006.

3.3.3.10 Profile of Mood States

General description. The shortened version of the Profile of Mood States (POMS) scale (Shachem, 1983) provides measures of self-rated tension, depression, anger, vigor, fatigue, and confusion.

Strengths and limitations. Reliability and validation testing of the POMS has been extensive. For example, McNair and Lorr (1964) reported test-retest reliabilities of 0.61 to 0.69 for the six factors. Reviews of the sensitivity and reliability of the POMS have been favorable (Norcross, Guadagnoli, and Prochaska, 1984). Constantini, Braun, Davis, and Iervolino (1971) reported significant positive correlations between POMS and the Psychological Screening Inventory, thus yielding consensual validation. Pollock, Cho, Reker, and Volavka (1979) correlated POMS scales and physiological measures from eight healthy males. The tension and depression scores were significantly correlated with heart rate (+0.75 and +0.76, respectively) and diastolic blood pressure (+0.71 and +0.72, respectively). Heart rate was also significantly correlated with the anger score (+0.70).

The POMS has been used extensively in psychotherapy research (e.g., Haskell, Pugatch, and McNair, 1969; Lorr, McNair, Weinstein, Michaux, and Raskin, 1961; McNair, Goldstein, Lorr, Cibelli, and Roth, 1965; Pugatch, Haskell, and McNair, 1969) and drug research (e.g., Mirin, Shapiro, Meyer, Pillard, and Fisher, 1971; Nathan, Titler, Lowenstein, Solomon, and Rossi, 1970; Nathan, Zare, Ferneau, and Lowenstein, 1970; Pillard and Fisher, 1970).

Storm and Parke (1987) used the POMS to assess the mood effects of a sleep-inducing drug (temazepam) for EF-111 aircrews. As anticipated, there were no significant drug effects on any of the six subscales. Gawron et al. (1988) asked subjects to complete the POMS after a 1.75 h flight. There were no significant crew position effects on rated vigor or fatigue. There was a significant order effect on fatigue, however. Subjects who had been pilots first had higher ratings (2.7) than those who had been copilots first (1.3).

Harris, Hancock, Arthur, and Caird (1995) did not find a significant difference in the fatigue rating between a manual and an automatic tracking group.

Data requirements. The POMS takes about 10 min to complete and requires a stiff writing surface. The POMS is available from the Educational and Industrial Testing Service, San Diego, California.

Thresholds. Not stated.

SOURCES

Costantini, A.F., Braun, J.R., Davis, J.E., and Iervolino, A. The life change inventory: A device for quantifying psychological magnitude of changes experienced by college students. *Psychological Reports,* 34 (3, Pt. 1), 991–1000, June 1971.

Gawron, V.J., Schiflett, S., Miller, J., Ball, J., Slater, T., Parker, F., Lloyd, M., Travale, D., and Spicuzza, R.J. *The effect of pyridostigmine bromide on in-flight aircrew performance (USAFSAM-TR-87-24).* Brooks Air Force Base, TX: School of Aerospace Medicine; January 1988.

Harris, W.C., Hancock, P.A., Arthur, E.J., and Caird, J.K. Performance, workload, and fatigue changes associated with automation. *International Journal of Aviation Psychology.* 5(2), 169–185, 1995.

Haskell, D.H., Pugatch, D. and McNair, D.M. Time-limited psychotherapy for whom? *Archives of General Psychiatry.* 21, 546–552, 1969.

Lorr, M., McNair, D.M., Weinstein, G.J., Michaux, W.W., and Raskin, A. Meprobromate and chlorpromazine in psychotherapy. *Archives of General Psychiatry.* 4, 381–389, 1961.

McNair, D.M., Goldstein, A.P., Lorr, M., Cibelli, L.A., and Roth, I. Some effects of chlordiazepoxide and meprobromate with psychiatric outpatients. *Psychopharmacologia.* 7, 256–265, 1965.

McNair, D.M., and Lorr, M. An analysis of mood in neurotics. *Journal of Abnormal Psychology.* 69, 620–627, 1964.

Mirin, S.M., Shapiro, L.M., Meyer, R.E., Pillard, R.C. and Fisher, S. Casual versus heavy use of marijuana: A redefinition of the marijuana problem. *American Journal of Psychiatry.* 172, 1134–1140, 1971.

Nathan, P.F., Titler, N.A., Lowenstein, L.M., Solomon, P., and Rossi, A.M. Behavioral analyses of chronic alcoholism: Interaction of alcohol and human contact. *Archives of General Psychiatry.* 22, 419–430, 1970.

Nathan, P.F., Zare, N.C., Ferneau, E.W. and Lowenstein, L.M. Effects of congener differences in alcohol beverages on the behavior of alcoholics. *Quarterly Journal on Studies of Alcohol. Supplement No. 5.* 87–100, 1970.

Norcross, J.C., Guadagnoli, E., and Prochaska, J.O. Factor structure of the profile of mood states (POMS): Two partial replications. *Journal of Clinical Psychology.* 40, 1270–1277, 1984.

Pillard, R.C. and Fisher, S. Aspects of anxiety in dental clinic patients. *Journal of the American Dental Association.* 80, 1331–1334, 1970.

Pollock, V., Cho, D.W., Reker, D., and Volavka, J. Profile of mood states: The factors and their correlates. *Journal of Nervous Mental Disorders.* 167, 612–614, 1979.

Pugatch, D., Haskell, D.H., and McNair, D.M. *Predictors and patterns of change associated with the course of time limited psychotherapy (Mimeo Report),* 1969.

Shachem, A. A shortened version of the profile of mood states. *Journal of Personality Assessment.* 47, 305–306, 1983.

Storm, W.F., and Parke, R.C. FB-111A aircrew use of temazepam during surge operations. *Proceedings of NATO Advisory Group for Aerospace Research and Development (AGARD).* Biochemical Enhancement of Performance Conference (Paper No. 415, pp. 12–1 to 12–12). Neuilly-sur-Seine, France: AGARD, 1987.

3.3.3.11 Subjective Workload Assessment Technique

General description. The Subjective Workload Assessment Technique (SWAT) combines ratings of three different scales (see table 26) to produce an interval scale of mental workload. These scales are the following: (1) time load, which reflects

TABLE 26
SWAT Scales

Time Load

1. Often have spare time. Interruptions or overlap among activities occur infrequently or not at all.
2. Occasionally have spare time. Interruptions or overlap among activities occur frequently.
3. Almost never have spare time. Interruptions or overlap among activities are frequent or occur all the time.

Mental Effort Load

1. Very little conscious mental effort or concentration required. Activity is almost automatic, requiring little or no attention.
2. Moderate conscious mental effort or concentration required. Complexity of activity is moderately high owing to uncertainty, unpredictability, or unfamiliarity. Considerable attention required.
3. Extensive mental effort and concentration are necessary. Very complex activity requiring total attention.

Psychological Stress Load

1. Little confusion, risk, frustration, or anxiety exists and can be easily accommodated.
2. Moderate stress due to confusion, frustration, or anxiety noticeably adds to workload. Significant compensation is required to maintain adequate performance.
3. High to very intense stress due to confusion, frustration, or anxiety. High-to-extreme determination and self-control required. (Potter, S.S. and Bressler, J.R. *Subjective workload assessment technique (SWAT): A user's guide.* Wright-Patterson Air Force Base, OH: Armstrong Aerospace Medical Research Laboratory; July 1989, pp. 12–14).

the amount of spare time available in planning, executing, and monitoring a task; (2) mental effort load, which assesses how much conscious mental effort and planning are required to perform a task; and (3) psychological stress load, which measures the amounts of risk, confusion, frustration, and anxiety associated with task performance. A more complete description is given by Reid and Nygren (1988). A description of the initial conjoint measurement model for SWAT is described by Nygren (1982, 1983).

Strengths and limitations. SWAT has been found to be a valid (Albery, Repperger, Reid, Goodyear, and Roe, 1987; Haworth, Bivens, and Shively, 1986; Masline, 1986; Reid, Shingledecker, and Eggemeier, 1981; Reid, Shingledecker, Nygren, and Eggemeier, 1981; Vidulich and Tsang, 1985, 1987; Warr, Colle, and Reid, M.G, 1986), sensitive (Eggemeier, Crabtree, Zingg, Reid, and Shingledecker, 1982), reliable (Corwin, Sandry-Garza, Biferno, Boucek, Logan, Jonsson, and Metalis, 1989; Gidcomb, 1985), and relatively unobtrusive (Crabtree, Bateman, and Acton, 1984; Courtright and Kuperman, 1984; and Eggemeier, 1988) measure of workload. Further, SWAT ratings are not affected by delays of up to 30 min (Eggemeier, Crabtree, and LaPointe, 1983), nor by intervening tasks of all but difficult tasks (Eggemeier, Melville, and Crabtree, 1984; Lutmer and Eggemeier, 1990). Moroney, Biers, and Eggemeier (1995) concur. Also, Eggleston (1984) found a significant correlation between projected SWAT ratings made during system concept evaluation and those made during ground-based simulation of the same system.

Warr (1986) reported that SWAT ratings were less variable than Modified Cooper–Harper Rating Scale ratings. Kilmer et al. (1988) reported that SWAT was more sensitive to changes in difficulty of a tracking task than the Modified Cooper–Harper Rating Scale was. Finally, Nygren (1991) stated that SWAT provides a good cognitive model of workload, sensitive to individual differences. Anthony and Biers (1997), however, found no difference between Overall Workload Scale and SWAT ratings. Their subjects were 48 introductory psychology students performing a memory recall task.

SWAT has been used in diverse environments, for example, test aircraft (Papa and Stoliker, 1988); a high-G centrifuge (Albery, Ward, and Gill, 1985; Albery, 1989); command, control, and communications centers (Crabtree, Bateman, and Acton, 1984); nuclear power plants (Beare and Dorris, 1984); domed flight simulators (Reid, Eggemeier, and Shingledecker, 1982; Skelly and Simons, 1983); tank simulators (Whitaker, Peters, and Garinther, 1989); and the benign laboratory setting (Graham and Cook, 1984; Kilmer, Knapp, Burdsal, Borresen, Bateman, and Malzahn, 1988). In the laboratory, SWAT has been used to assess the workload associated with critical tracking and communication tasks (Reid, Shingledecker, and Eggemeier, 1981), memory tasks (Eggemeier, Crabtree, Zingg, Reid, and Shingledecker, 1982; Eggemeier and Stadler, 1984; Potter and Acton, 1985), and monitoring tasks (Notestine, 1984). Hancock and Caird (1993) reported significant increases in SWAT rating as the shrink rate of the target decreased and as the number of steps from the cursor to the target increased.

Usage in simulated flight has also been extensive (Haworth, Bivens, and Shively, 1986; Nataupsky and Abbott, 1987; Schick and Hann, 1987; Skelly and Purvis, 1985; Skelly, Reid, and Wilson, 1983; Thiessen, Lay, and Stern, 1986; Ward and

Hassoun, 1990). For example, Bateman and Thompson (1986) reported that SWAT ratings increased as task difficulty increased. Their data were collected in an aircraft simulator during a tactical mission. Vickroy (1988), also using an aircraft simulator, reported that SWAT ratings increased as the amount of air turbulence increased. Fracker and Davis (1990) reported significant increases in SWAT as the number of simulated enemy aircraft increased from 1 to 3. Hankey and Dingus (1990) reported that SWAT was sensitive to changes in time on task and fatigue. Hancock, Williams, Manning, and Miyake (1995) reported that SWAT was highly correlated with the difficulty of a simulated flight task. However, See and Vidulich (1997) reported significant effects of target type and threat status on SWAT scores in a combat aircraft simulator. There were no significant correlations of SWAT with overall workload, but two subscales correlated with peak workload (effort, $r = +0.78$; stress, $r = +0.76$).

Usage in actual flight has been extensive. For example, Pollack (1985) used SWAT to assess differences in workload between flight segments. She reported that C-130 pilots had the highest SWAT scores during the approach segment of the mission. She also reported higher SWAT ratings during the preflight segments of tactical, rather than proficiency, missions. Haskell and Reid (1987) found significant difference in SWAT ratings between flight maneuvers, and also between successfully completed maneuvers and those that were not successfully completed. Gawron et al. (1988) analyzed SWAT ratings made by the pilot and copilot four times during each familiarization and data flight: (1) during the taxi out to the runway, (2) just before a simulated drop, (3) just after a simulated drop, and (4) during the taxi back to the hangar. There were significant segment effects. Specifically, SWAT ratings were highest before the drop and lowest for preflight. The ratings during postdrop and postflight were both moderate.

Experience with SWAT has not been all positive, however. For example, Boyd (1983) reported that there were significant positive correlations between the three workload scales in a text-editing task. This suggests that the three dimensions of workload are not independent. This in turn poses a problem for use of conjoint measurement techniques. Further, Derrick (1983) and Hart (1986) suggest that three scales may not be adequate for assessing workload. In examining the three scales, Biers and Masline (1987) compared three alternative analysis methods for SWAT: conjoint analysis, simple sum of the three subscales, and a weighted linear combination. They reported that the individual scales were differentially sensitive to different task demands. Masline and Biers (1987) also reported greater correspondence between projective and posttask workload ratings using magnitude estimation than either SWAT or equal-appearing intervals. Further, Battiste and Bortolussi (1988) reported a test-retest correlation of +0.751 but also stated that of the 144 SWAT ratings reported during a simulated B-727 flight, 59 were zero.

Corwin (1989) reported no difference between in-flight and postflight ratings of SWAT in only two of three flight conditions. Gidcomb (1985) reported casual card sorts and urged emphasizing the importance of the card sort to SWAT raters. A computerized version of the traditional card sort was developed at the Air Force School of Aerospace Medicine. This version eliminates the tedium and dramatically reduces the time to complete the SWAT card sort.

Haworth, Bivens, and Shively (1986) reported that although the SWAT was able to discriminate between control configuration conditions in a single-pilot configuration, it could not discriminate between these same conditions in a pilot–copilot configuration. Wilson, Hughes, and Hassoun (1990) reported no significant differences in SWAT ratings among display formats, in contrast to pilot comments. van de Graaff (1987) reported considerable (60 points) intersubject variability in SWAT ratings during an in-flight approach task. Hill, Iavecchia, Byers, Bittner, Zaklad, and Christ (1992) reported that SWAT was not as sensitive to workload or as accepted by army operators as NASA TLX and the Overall Workload Scale.

Vidulich and Tsang (1986) reported that SWAT failed to detect resource competition effects in dual-task performance of tracking and a directional transformation task. Vidulich (1991) reported test-retest reliability of +0.606 in SWAT ratings for tracking, choice RT, and Sternberg tasks. In addition, Rueb, Vidulich, and Hassoun (1992) reported that only one of three difficult simulated aerial refueling missions had SWAT scores above the 40 redline.

Vidulich and Pandit (1987) concluded that SWAT was not an effective measure of individual differences. This conclusion was based on insignificant correlations of SWAT with any of the scales on the Jenkins Activity Survey, Rotter's Locus of Control, the Cognitive Failures Questionnaire, the Cognitive Interference Questionnaire, the Thought Occurrence Questionnaire, the California Q-Sort, and the Myers–Briggs Type Indicator.

Arbak, Shew, and Simons (1984) applied SWAT in a reflective manner based on mission performance of B-52 pilots and copilots. The authors concluded that this reflective manner was useful when a two-on-one interview technique was applied, the original situation was described in detail, segment boundaries are well-identified, and instances reviewed impact performance. Kuperman and Wilson (1985) applied SWAT projectively early in system design.

Reid (1985) warned that SWAT was most sensitive when workload was moderate to high. Acton and Rokicki (1986) surveyed SWAT users in the Air Force Test and Evaluation community and suggested the development of a user's guide to help train raters. Additionally, they suggested the development of guidelines for task selection and methods to handle small data sets. In addition, Nygren, Schnipke, and Reid (1998) reported that how individuals weighted the SWAT dimensions affected their workload ratings. Their subjects were 124 introductory psychology students who were categorized into one of six groups based on their SWAT dimension weightings.

Svensson, Angelborg–Thanderz, Sjoberg, and Olsson (1997) reported the reliability of SWAT to be +0.74 among 18 pilots flying simulated low-level, high-speed missions. The correlation with the Bedford Workload Scale was +0.687 and with the NASA TLX was +0.735.

Thresholds. The minimum value is 0 and the maximum value is 100. High workload is associated with the maximum value. In addition, ratings of the time, effort, and stress scales may be individually examined as workload components (Eggemeier, McGhee, and Reid, 1983). Colle and Reid (2005) reported a redline value of 41.1, which was within the 40 +/–10 suggested by Reid and Colle (1988).

Data requirements. SWAT requires two steps to use: scale development and event scoring. Scale development requires subjects to rank, from lowest to highest workload, 27 combinations of three levels of the three workload subscales. The levels of each subscale are presented in table 26. Reid, Eggemeier, and Nygren (1982) describe their individual differences approach to scale development. Programs to calculate the SWAT score for every combination of ratings on the three subscales are available from the Air Force Research Laboratory at Wright-Patterson Air Force Base. A user's manual is also available from the same source.

During event scoring, the subject is asked to provide a rating (1, 2, 3) for each subscale. The experimenter then maps the set of ratings to the SWAT score (1 to 100) calculated during the scale development step. Haskell and Reid (1987) suggests that the tasks to be rated be meaningful to the subjects and, further, that the ratings not interfere with performance of the task. Acton and Colle (1984) reported that the order in which the subscale ratings are presented does not affect the SWAT score. However, it is suggested that the order remain constant to minimize confusion. Eggleston and Quinn (1984) recommended developing a detailed system and operating environment description for prospective ratings. Finally, Biers and McInerney (1988) reported that the card sort did not affect the task workload ratings, and therefore may not be necessary.

SOURCES

Acton, W. and Colle, H. The effect of task type and stimulus pacing rate on subjective mental workload ratings. *Proceedings of the IEEE 1984 National Aerospace and Electronics Conference* (pp. 818–823). Dayton, OH: IEEE, 1984.

Acton, W.H. and Rokicki, S.M. Survey of SWAT use in operational test and evaluation. *Proceedings of the Human Factors Society 30th Annual Meeting*. 2, 1221–1224, 1986.

Albery, W. B. The effect of sustained acceleration and noise on workload in human operators. *Aviation, Space, and Environmental Medicine*. 6(10), 943–948, 1989.

Albery, W., Repperger, D., Reid, G., Goodyear, C., and Roe, M. Effect of noise on a dual task: Subjective and objective workload correlates. *Proceedings of the National Aerospace and Electronics Conference*. Dayton, OH: IEEE, 1987.

Albery, W.B., Ward, S.L., and Gill, R.T. *Effect of acceleration stress on human workload (Technical Report AMRL-TR-85-039)*. Wright-Patterson Air Force Base, OH: Aerospace Medical Research Laboratory, May 1985.

Anthony, C.R. and Biers, D.W. Unidimensional versus multidimensional workload scales and the effect of number of rating scale categories. *Proceedings of the Human Factors and Ergonomics Society 41st Annual Meeting*. 2, 1084–1088, 1997.

Arbak, C.J., Shew, R.L., and Simons, J.C. The use of reflective SWAT for workload assessment. *Proceedings of the Human Factors Society 28th Annual Meeting*. 2, 959–962, 1984.

Bateman, R.P and Thompson, M.W. Correlation of predicted workload with actual workload using the subjective workload assessment technique. *Proceedings of the SAE AeroTech Conference*, 1986.

Battiste, V. and Bortolussi, M.R. Transport pilot workload: A comparison of two subjective techniques. *Proceedings of the Human Factors Society 32nd Annual Meeting*. 150–154, 1988.

Beare, A. and Dorris, R. The effects of supervisor experience and the presence of a shift technical advisor on the performance of two-man crews in a nuclear power plant simulator. *Proceedings of the Human Factors Society 28th Annual Meeting*. 242–246. Santa Monica, CA: Human Factors Society, 1984.

Biers, D.W. and Masline, P.J. Alternative approaches to analyzing SWAT data. *Proceedings of the Human Factors Society 31st Annual Meeting.* 1, 63–66, 1987.

Biers, D.W. and McInerney, P. An alternative to measuring subjective workload: Use of SWAT without the card sort. *Proceedings of the Human Factors Society 32nd Annual Meeting.* 2, 1136–1139, 1988.

Boyd, S.P. Assessing the validity of SWAT as a workload measurement instrument. *Proceedings of the Human Factors Society 27th Annual Meeting.* 124–128, 1983.

Colle, H. A. and Reid, G.B. Estimating a mental workload redline in a simulated air-to-ground combat mission. *The International Journal of Aviation Psychology.* 15(4), 303–319, 2005.

Corwin, W.H. In-flight and post-flight assessment of pilot workload in commercial transport aircraft using SWAT. *Proceedings of the Fifth Symposium on Aviation Psychology.* 808–813, 1989.

Corwin, W.H., Sandry-Garza, D.L., Biferno, M.H., Boucek, G.P., Logan, A.L., Jonsson, J.E., and Metalis, S.A. *Assessment of crew workload measurement methods, techniques, and procedures. Vol. I—Process methods and results (WRDC-TR-89-7006).* Wright-Patterson Air Force Base, OH; September, 1989.

Courtright J.F. and Kuperman, G. Use of SWAT in USAF system T&E. *Proceedings of the Human Factors Society 28th Annual Meeting.* (pp. 700–703). Santa Monica, CA: Human Factors Society, 1984.

Crabtree, M.A. Bateman, R.P., and Acton, W. Benefits of using objective and subjective workload measures. *Proceedings of the Human Factors Society 28th Annual Meeting.* (pp. 950–953). Santa Monica, CA: Human Factors Society, 1984.

Derrick W.L. Examination of workload measures with subjective task clusters. *Proceedings of the Human Factors Society 27th Annual Meeting.* (pp. 134–138). Santa Monica, CA: Human Factors Society, 1983.

Eggemeier, F.T. Properties of workload assessment techniques. In Hancock, P.A. and Meshtaki, N. (Eds.) *Human mental workload.* (pp. 41–62). Amsterdam: North-Holland, 1988.

Eggemeier, F.T., Crabtree, M.S., and LaPointe, P. The effect of delayed report on subjective ratings of mental workload. *Proceedings of the Human Factors Society 27th Annual Meeting.* (pp. 139–143). Santa Monica, CA: Human Factors Society, 1983.

Eggemeier, F.T., Crabtree, M.S., Zingg, J.J., Reid, G.B., and Shingledecker, C.A. Subjective workload assessment in a memory update task. *Proceedings of the Human Factors Society 26th Annual Meeting.* Santa Monica, CA: Human Factors Society. 643–647, 1982.

Eggemeier, F.T., McGhee, J.Z., and Reid, G.B. The effects of variations in task loading on subjective workload scales. *Proceedings of the IEEE 1983 National Aerospace and Electronics Conference.* (pp. 1099–1106). Dayton, OH: IEEE, 1983.

Eggemeier, F.T., Melville, B., and Crabtree, M. The effect of intervening task performance on subjective workload ratings. *Proceedings of the Human Factors Society 28th Annual Meetings.* (pp. 954–958). Santa Monica, CA: Human Factors Society, 1984.

Eggemeier, F.T. and Stadler, M. Subjective workload assessment in a spatial memory task. *Proceedings of the Human Factors Society 28th Annual Meeting.* (pp. 680–684). Santa Monica, CA: Human Factors Society, 1984.

Eggleston, R.G. A comparison of projected and measured workload ratings using the subjective workload assessment technique (SWAT). *Proceedings of the National Aerospace and Electronics Conference. Vol.* 2, 827–831, 1984.

Eggleston, R.G. and Quinn, T.J. A preliminary evaluation of a projective workload assessment procedure. *Proceedings of the Human Factors Society 28th Annual Meeting* (pp. 695–699). Santa Monica, CA: Human Factors Society, 1984.

Fracker, M.L. and Davis, S.A. Measuring operator situation awareness and mental workload. *Proceedings of the 5th Mid-Central Ergonomics/Human Factors Conference*, Dayton, OH, 23–25 May 1990.

Gawron, V.J., Schiflett, S., Miller, J., Ball, J., Slater, T., Parker, F., Lloyd, M., Travale, D., and Spicuzza, R.J. *The effect of pyridostigmine bromide on in-flight aircrew performance (USAFSAM-TR-87-24)*. Brooks Air Force Base, TX: School of Aerospace Medicine, January 1988.

Gidcomb, C. *Survey of SWAT use in flight test (BDM/A-85-0630-7R.)* Albuquerque, NM: BDM Corporation, 1985.

Graham, C.H. and Cook, M.R. *Effects of pyridostigmine on psychomotor and visual performance (TR-84-052);* September 1984.

Hancock, P.A. and Caird, J.K. Experimental evaluation of a model of mental workload. *Human Factors*. 35(3), 413–419, 1993.

Hancock, P.A., Williams, G., Manning, C.M., and Miyake, S. Influence of task demand characteristics on workload and performance. *International Journal of Aviation Psychology*. 5(1), 63–86, 1995.

Hankey, J.M. and Dingus, T.A. A validation of SWAT as a measure of workload induced by changes in operator capacity. *Proceedings of the Human Factors Society 34th Annual Meeting*. 1, 113–115, 1990.

Hart, S.G. Theory and measurement of human workload. In Seidner, J. (Ed.) *Human productivity enhancement, Vol. 1* (pp. 396–455). New York: Praeger, 1986.

Haskell, B.E., and Reid, G.B. The subjective perception of workload in low-time private pilots: A preliminary study. *Aviation, Space, and Environmental Medicine*. 58, 1230–1232, 1987.

Haworth, L.A., Bivens, C.C., and Shively, R.J. An investigation of single-piloted advanced cockpit and control configuration for nap-of-the-earth helicopter mission tasks. *Proceedings of the 42nd Annual Forum of the American Helicopter Society*. 657–671, 1986.

Hill, S.G., Iavecchia, H.P., Byers, J.C., Bittner, A.C., Zaklad, A.L., and Christ, R.E. Comparison of four subjective workload rating scales. *Human Factors*. 34, 429–439, 1992.

Kilmer, K.J., Knapp, R., Burdsal, C., Borresen, R., Bateman, R.P., and Malzahn, D. A comparison of the SWAT and modified Cooper–Harper scales. *Proceedings of the Human Factors 32nd Annual Meeting*. 155–159, 1988.

Kuperman, G.G. and Wilson, D.L. A workload analysis for strategic conventional standoff capability missions. *Proceedings of the Human Factors Society 29th Annual Meeting*. 2, 635–639, 1985.

Lutmer, P.A. and Eggemeier, F.T. The effect of intervening task performance and multiple ratings on subjective ratings of mental workload. Paper presented at the *5th Mid-Central Ergonomics Conference*, University of Dayton, Dayton, OH, 1990.

Masline, P.J. A comparison of the sensitivity of interval scale psychometric techniques in the assessment of subjective mental workload. Unpublished master's thesis, University of Dayton, Dayton, OH, 1986.

Masline, P.J. and Biers, D.W. An examination of projective versus post-task subjective workload ratings for three psychometric scaling techniques. *Proceedings of the Human Factors Society 31st Annual Meeting*. 1, 77–80, 1987.

Moroney, W.F., Biers, D.W., and Eggemeier, F.T. Some measurement and methodological considerations in the application of subjective workload measurement techniques. *International Journal of Aviation Psychology*. 5(1), 87–106, 1995.

Nataupsky, M. and Abbott, T.S. Comparison of workload measures on computer-generated primary flight displays. *Proceedings of the Human Factors Society 31st Annual Meeting* (pp. 548–552). Santa Monica, CA: Human Factors Society, 1987.

Notestine, J. Subjective workload assessment and effect of delayed ratings in a probability monitoring task. *Proceedings of the Human Factors Society 28th Annual Meeting* (pp. 685–690). Santa Monica, CA: Human Factors Society, 1984.

Nygren, T.E. *Conjoint measurement and conjoint scaling: A user's guide (AFAMRL-TR-82-22).* Wright-Patterson Air Force Base, OH: Aerospace Medical Research Laboratory; April 1982.

Nygren, T.E. *Investigation of an error theory for conjoint measurement methodology (763025/714404).* Columbus, OH: Ohio State University Research Foundation, May 1983.

Nygren, T.E. Psychometric properties of subjective workload measurement techniques: Implications for their use in the assessment of perceived mental workload. *Human Factors.* 33, 17–33, 1991.

Nygren, T.E., Schnipke, S., and Reid, G. Individual differences in perceived importance of SWAT workload dimensions: Effects on judgment and performance in a virtual high workload environment. *Proceedings of the Human Factors and Ergonomics Society 42nd Annual Meeting.* 1, 816–820, 1998.

Papa, R.M. and Stoliker, J.R. *Pilot workload assessment: A flight test approach.* Washington, DC: American Institute of Aeronautics and Astronautics, 88–2105, 1988.

Pollack, J. *Project report: An investigation of Air Force reserve pilots' workload.* Dayton, OH: Systems Research Laboratory; November 1985.

Potter, S.S. and Acton, W. Relative contributions of SWAT dimensions to overall subjective workload ratings. *Proceedings of Third Symposium on Aviation Psychology.* Columbus, OH: Ohio State University, 1985.

Potter, S.S. and Bressler, J.R. *Subjective workload assessment technique (SWAT): A user's guide.* Wright-Patterson Air Force Base, OH: Armstrong Aerospace Medical Research Laboratory; July 1989.

Reid, G.B. Current status of the development of the Subjective Workload Assessment Technique. *Proceedings of the Human Factors Society 29th Annual Meeting.* 1, 220–223, 1985.

Reid, G.B. and Colle, H. A. Critical SWAT values for predicting operator workload. *Proceedings of the Human Factors Society 32nd annual meeting.* 1414–1418, 1988.

Reid, G.B., Eggemeier, F., and Nygren, T. An individual differences approach to SWAT scale development. *Proceedings of the Human Factors Society 26th Annual Meeting.* (pp. 639–642). Santa Monica, CA: Human Factors Society, 1982.

Reid, G.B., Eggemeier, F.T., and Shingledecker, C.A. In Frazier, M.L. and Crombie, R.B. (Eds.) *Proceedings of the workshop on flight testing to identify pilot workload and pilot dynamics, AFFTC-TR-82-5* (pp. 281–288). Edwards AFB, CA: May 1982.

Reid, G.B. and Nygren, T.E. The subjective workload assessment technique: A scaling procedure for measuring mental workload. In Hancock, P.A. and Mehtaki, N. (Eds.) *Human mental workload* (pp. 185–218). Amsterdam: North Holland, 1988.

Reid, G.B., Shingledecker, C.A., and Eggemeier, F.T. Application of conjoint measurement to workload scale development. *Proceedings of the Human Factors Society 25th Annual Meeting.* 522–526, 1981.

Reid, G.B., Shingledecker, C.A., Nygren, T.E., and Eggemeier, F.T. Development of multidimensional subjective measures of workload. *Proceedings of the IEEE International Conference on Cybernetics and Society.* 403–406, 1981.

Rueb, J., Vidulich, M., and Hassoun, J.A. Establishing workload acceptability: An evaluation of a proposed KC-135 cockpit redesign. *Proceedings of the Human Factors Society 36th Annual Meeting.* 17–21, 1992.

Schick, F.V. and Hann, R.L. The use of subjective workload assessment technique in a complex flight task. In Roscoe, A.H. (Ed.) *The practical assessment of pilot workload,* AGARDograph No. 282 (pp. 37–41). Neuilly-sur-Seine, France: AGARD, 1987.

See, J.E. and Vidulich, M.A. Assessment of computer modeling of operator mental workload during target acquisition. *Proceedings of the Human Factors and Ergonomics Society 41st Annual Meeting.* 1303–1307, 1997.

Skelly, J.J. and Purvis, B. *B-52 wartime mission simulation: Scientific precision in workload assessment.* Paper presented at the 1985 Air Force Conference on Technology in Training and Education. Colorado Springs, CO; April 1985.

Skelly, J.J., Reid, G.B., and Wilson, G.R. *B-52 full mission simulation: Subjective and physiological workload applications.* Paper presented at the Second Aerospace Behavioral Engineering Technology Conference, 1983.

Skelly, J.J. and Simons, J.C. Selecting performance and workload measures for full-mission simulation. *Proceedings of the IEEE 198 National Aerospace and Electronics Conference.* 1082–1085, 1983.

Svensson, E., Angelborg-Thanderz, M., Sjoberg, L., and Olsson, S. Information complexity–mental workload and performance in combat aircraft. *Ergonomics.* 40(3), 362–380, 1997.

Thiessen, M.S., Lay, J.E., and Stern, J.A. *Neuropsychological workload test battery validation study (FZM 7446).* Fort Worth, TX: General Dynamics, 1986.

van de Graaff, R.C. *An in-flight investigation of workload assessment techniques for civil aircraft operations (NLR-TR-87119 U).* Amsterdam, the Netherlands: National Aerospace Laboratory, 1987.

Vickroy, S.C. *Workload prediction validation study: The verification of CRAWL predictions.* Wichita, KS: Boeing Military Airplane Company, 1988.

Vidulich, M.A. The Bedford Scale: Does it measure spare capacity? *Proceedings of the 6th International Symposium on Aviation Psychology.* 2, 1136–1141, 1991.

Vidulich, M.A. and Pandit, P. Individual differences and subjective workload assessment: Comparing pilots to nonpilots. *Proceedings of the International Symposium on Aviation Psychology.* 630–636, 1987.

Vidulich, M.A. and Tsang, P.S. Techniques of subjective workload assessment: A comparison of two methodologies. *Proceedings of the Third Symposium on Aviation Psychology.* 239–246, 1985.

Vidulich, M.A. and Tsang, P.S. Techniques of subjective workload assessment: A comparison of SWAT and NASA-Bipolar methods. *Ergonomics.* 29(11), 1385–1398, 1986.

Vidulich, M.A. and Tsang, P.S. Absolute magnitude estimation and relative judgment approaches to subjective workload assessment. *Proceedings of the Human Factors Society 31st Annual Meeting.* 1057–1061, 1987.

Ward, G.F. and Hassoun, J.A. *The effects of head-up display (HUD) pitch ladder articulation, pitch number location and horizon line length on unusual altitude recoveries for the F-16 (ASD-TR-90-5008).* Wright-Patterson Air Force Base, OH: Crew Station Evaluation Facility; July 1990.

Warr, D.T. *A comparative evaluation of two subjective workload university measures: The subjective assessment technique and the modified Cooper-Harper Rating.* Masters thesis, Dayton, OH: Wright State, 1986.

Warr, D., Colle, H., and Reid, G.B. *A comparative evaluation of two subjective workload measures: The subjective workload assessment technique and the Modified Cooper–Harper Scale.* Paper presented at the Symposium on Psychology in the Department of Defense. USAFA, Colorado Springs, CO., 1986.

Whitaker, L.A., Peters, L., and Garinther, G. Tank crew performance: Effects of speech intelligibility on target acquisition and subjective workload assessment. *Proceedings of the Human Factors Society 33rd Annual Meeting.* 1411–1413, 1989.

Wilson, G.F., Hughes, E., and Hassoun, J.A Physiological and subjective evaluation of a new aircraft display. *Proceedings of the Human Factors Society 34th Annual Meeting.* 1441–1443, 1990.

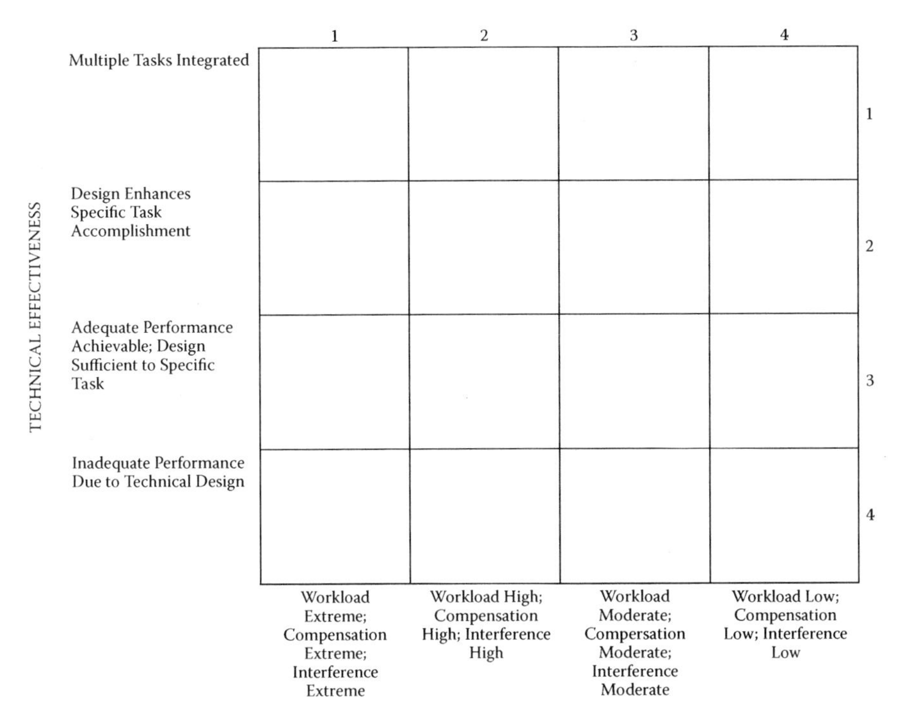

FIGURE 19 WCI/TE Scale Matrix (from Lysaght, R.J., Hill, S.G., Dick, A.O., Plamondon, B.D., Linton, P.M., Wierwille, W.W., Zaklad, A.L., Bittner, A.C., and Wherry, R.J. *Operator workload: Comprehensive review and evaluation of operator workload methodologies (Technical Report 851).* Alexandria, VA: Army Research Institute for the Behavioral and Social Sciences; June 1989, p. 110).

3.3.3.12 Workload/Compensation/Interference/Technical Effectiveness

General description. The Workload/Compensation/Interference/Technical Effectiveness (WCI/TE) rating scale (see figure 19) requires subjects to rank the 16 matrix cells and then rate specific tasks. The ratings are converted by conjoint scaling techniques to values of 0 to 100.

Strengths and limitations. Wierwille and Connor (1983) reported sensitivity of WCI/TE ratings to three levels of task difficulty in a simulated flight task. Wierwille, Casali, Connor, and Rahimi (1985) reported sensitivity to changes in difficulty in psychomotor, perceptual, and mediational tasks. Wierwille, Rahimi, and Casali (1985) reported that WCI/TE was sensitive to variations in the difficulty of a secondary mathematical task during a simulated flight task. However, O'Donnell and Eggemeier (1986) suggest that the WCI/TE not be used as a direct measure of workload.

Data requirements. Subjects must rank the 16 matrix cells and then rate specific tasks. Complex mathematical processing is required to convert the ratings to WCI/TE values.

Thresholds. The minimum workload is 0, and the maximum workload is 100.

SOURCES

Lysaght, R.J., Hill, S.G., Dick, A.O., Plamondon, B.D., Linton, P.M., Wierwille, W.W., Zaklad, A.L., Bittner, A.C., and Wherry, R.J. *Operator workload: Comprehensive review and evaluation of operator workload methodologies (Technical Report 851).* Alexandria, VA. Army Research Institute for the Behavioral and Social Sciences, June 1989.

O'Donnell, R.D. and Eggemeier, F.T. Workload assessment methodology. In Boff, K.R., Kaufman, L., and Thomas, J.P. (Eds.) *Handbook of perception and human performance. Volume 2, Cognitive processes and performance.* New York: Wiley, 1986.

Wierwille, W.W., Casali, J.G., Connor, S.A., and Rahimi, M. Evaluation of the sensitivity and intrusion of mental workload estimation techniques. In Roner, W. (Ed.) *Advances in man-machine systems research. Vol. 2* (pp. 51–127). Greenwich, CT: J.A.I. Press, 1985.

Wierwille, W.W., Rahimi, M., and Casali, J.G. Evaluation of 16 measures of mental workload using a simulated flight task emphasizing mediational activity. *Human Factors.* 27(5), 489–502, 1985.

Wierwille, W.W. and Connor, S.A. Evaluation of twenty workload assessment measures using a psychomotor task in a motion-base aircraft simulation. *Human Factors.* 25, 1–16, 1983.

3.3.4 Single-Number Subjective Workload Measures

As the name implies, these subjective workload measures require the subject to give only one number to rate the workload. Examples include the following: Continuous Subjective Assessment of Workload (section 3.3.4.1), Dynamic Workload Scale (section 3.3.4.2), Equal-Appearing Intervals (section 3.3.4.3), Hart and Bortolussi Rating Scale (section 3.3.4.4), Instantaneous Self Assessment (section 3.3.4.5), McDonnell Rating Scale (section 3.3.4.6), Overall Workload Scale (section 3.3.4.7), and Utilization (section 3.3.4.8).

3.3.4.1 Continuous Subjective Assessment of Workload

General description. The Continuous Subjective Assessment of Workload (C-SAW) requires subjects to provide ratings of 1 to 10 (corresponding to the Bedford Workload Scale descriptors) while viewing a videotape of their flight immediately after landing. Computer prompts ratings at rates up to 3 s. A bar chart or graph against the timeline is the output.

Strengths and limitations. Jensen (1995) stated that subjects could reliably provide ratings every 3 s. He reported that C-SAW was sensitive to differences between a HUD and a head-down display. C-SAW has high face validity but has not been formally validated.

Thresholds. The minimum value is zero.

SOURCE

Jensen, S.E. Developing a flight workload profile using Continuous Subjective Assessment of Workload (C-SAW). *Proceedings of the 21st Conference of the European Association for Aviation Psychology.* Chapter 46, 1995.

Workload Assessment		CRITERIA			Appreciation
		Reserve Capacity	Interruptions	Effort or Stress	
Light	2	Ample			Very Acceptable
Moderate	3	Adequate	Some		Well Acceptable
Fair	4	Sufficient	Recurring	Not Undue	Acceptable
High	5	Reduced	Repetitive	Marked	High but Acceptable
Heavy	6	Little	Frequent	Significant	Just Acceptable
Extreme	7	None	Continuous	Acute	Not Acceptable Continuously
Supreme	8	Impairment	Impairment	Impairment	Not Acceptable Instantaneously

from Lysaght, et al. (1989) p. 108

FIGURE 20 Dynamic Workload Scale.

3.3.4.2 Dynamic Workload Scale

General description. The Dynamic Workload Scale is a seven-point workload scale (see figure 20) developed as a tool for aircraft certification. It has been used extensively by Airbus Industries.

Strengths and limitations. Speyer, Fort, Fouillot, and Bloomberg (1987) reported a high concordance between pilot and observer ratings as well as sensitivity to workload increases.

Data requirements. Dynamic Workload Scale ratings must be given by both a pilot and an observer-pilot. The pilot is cued to make a rating; the observer gives a rating whenever workload changes or 5 min have passed.

Thresholds. Two is the minimum workload, and eight is the maximum workload.

SOURCE

Speyer, J., Fort, A., Fouillot, J., and Bloomberg, R. Assessing pilot workload for minimum crew certification. In A.H. Roscoe (Ed.) *The practical assessment of pilot workload.* AGARDograph Number 282 (pp. 90–115). Neuilly-sur-Seine, France: AGARD, 1987.

3.3.4.3 Equal-Appearing Intervals

General description. Subjects rate the workload in one of several categories using the assumption that each category is equidistant from adjacent categories.

Strengths and limitations. Hicks and Wierwille (1979) reported sensitivity to task difficulty in a driving simulator. Masline (1986) reported comparable results with the magnitude estimates and SWAT ratings but greater ease of administration. Masline, however, warned of rater bias.

Data requirements. Equal intervals must be clearly defined.

Thresholds. Not stated.

SOURCES

Hicks, T.G. and Wierwille, W.W. Comparison of five mental workload assessment procedures in a moving-base driving simulator. *Human Factors.* 21, 129–143, 1979.

Masline, P.J. A comparison of the sensitivity of interval scale psychometric techniques in the assessment of subjective mental workload. Unpublished master's thesis, University of Dayton, Dayton, OH, 1986.

3.3.4.4 Hart and Bortolussi Rating Scale

General description. Hart and Bortolussi (1984) used a single rating scale to estimate workload. The scale units were 1 to 100, with 1 being low workload and 100 being high workload.

Strengths and limitations. The workload ratings significantly varied across flight segments, with takeoff and landing having higher workload than climb or cruise. The workload ratings were significantly correlated to ratings of stress (+0.75) and effort (+0.68). These results were based on data from 12 instrument-rated pilots reviewing a list of 163 events.

Moray, Dessouky, Kijowski, and Adapathya (1991) used the same rating scale but numbered the scale from 1 to 10 rather than from 1 to 100. This measure was significantly related to time pressure but not to knowledge or their interaction.

Data requirements. The subjects need only the end points of the scale.

Thresholds. Low workload equals 1; high workload equals 100.

SOURCES

Hart, S.A. and Bortolussi, M.R. Pilot errors as a source of workload. *Human Factors*. 25(5), 545–556, 1984.

Moray, N., Dessouky, M.I., Kijowski, B.A., and Adapathya, R.S. Strategic behavior, workload, and performance in task scheduling. *Human Factors*. 33(6), 607–629, 1991.

3.3.4.5 Instantaneous Self Assessment (ISA)

General description. The Instantaneous Self Assessment (ISA) is a five-point rating scale (see table 27) that was originally developed in the United Kingdom to evaluate workload of air traffic controllers. ISA has since been applied to evaluating workload of Joint Strike Fighter pilots. On-line access to workload ratings was added to ISA and the resultant system renamed Eurocontrol Recording and Graphical display On-line (ERGO) (Hering and Coatleven, 1994).

TABLE 27
Instantaneous Self Assessment

ISA Button Number	Color	Legend	Definition
5	Red	VERY HIGH	Workload level is too demanding and unsustainable, even for a short period of time.
4	Yellow	HIGH	Workload level is uncomfortably high, although it can be sustained for a short period of time.
3	White	FAIR	Workload level is sustainable and comfortable.
2	Green	LOW	Workload level is low, with occasional periods of inactivity. Operator has considerable spare capacity and is relaxed.
1	Blue	VERY LOW	Workload level is too low. Operator is resting or not contributing to crew tasks.

Source From Hering, H. and Coatleven, G. ERGO (version 1) for Instantaneous Self Assessment of Workload (EEC Note No. 24/94. Brussels, Belgium: EUROCONTROL Agency, April 1994. (With kind permission of The European Organization for the Safety of Air Navigation (EUROCONTROL)© 1994 EUROCONTROL. All rights reserved.)

Strengths and limitations. Hering and Coatleven (1996) stated that the ISA has been used in ATC simulations since 1993. Castle and Leggatt (2002) performed a laboratory experiment to compare the workload estimates from three rating scales: ISA, NASA TLX, and the Bedford Workload Scale. They asked 16 pilots and 16 nonpilots to rate their workload using each of these three workload scales while performing the Multiple Attribute Task Battery. As a control, subjects also performed the task battery without rating their workload. Finally, subjects were asked to complete a face validity questionnaire. Average ratings for the 11 scales on the questionnaire were between 4 and 6 on a scale of 1 to 7 (7 being the highest positive rating). This was comparable to the other two workload measures.

There were, however, significant differences between the two groups. The nonpilots rated the ISA to be significantly more professional in appearance, and the pilots rated the ISA to be significantly more reliable. ISA was not sensitive to differences between pilots and nonpilots in the performance of a task battery designed to simulate flying a fixed-wing aircraft. The correlation between ISA and the Bedford Workload Scale was +0.49 and the NASA TLX was +0.55. The correlation with ratings of observers with ISA ratings was +0.80. The correlation with task loading on the Multiple Attribute Task Battery was highest for the ISA (+0.82), and lower for the NASA TLX (+0.57) and the Bedford Workload Scale (+0.53). There were no significant correlations between ISA rating and performance. Nor were there significant effects on performance whether or not the ISA rating was given. That was also true for the NASA TLX and Bedford Workload Scale scales. Internal consistency as measured by Cronbach's alpha varied between 0.43 and 0.78 for subjects and 0.64 to 0.81 for observers. Retest reliability for the same task performed 2 weeks later was +0.84. Instant ratings were reported to be more consistent than ratings made 2 min after the task.

Tattersall and Foord (1996) in a laboratory study using a tracking task reported that tracking task performance decreased when ISA responses were made, and therefore warned of its intrusiveness on primary task performance.

Lamoureux (1999) compared 81 categories of aircraft relationships in ATC and predicted versus ISA subjective workload ratings. The predictions were 73% accurate.

Data requirements. Use of the standard rating scale.

Thresholds. These vary from 1 to 5.

SOURCES

Castle, H. and Leggatt, H. *Instantaneous Self Assessment (ISA) – Validity & Reliability (JS 14865 Issue 1).* Bristol, United Kingdom: BAE Systems, November 2002.

Hering, H. and Coatleven, G. *ERGO (version 1) for Instantaneous Self Assessment of Workload (EEC Note No. 24/94).* Brussels, Belgium: EROCONTROL Agency, April 1994.

Hering, H. and Coatleven, G. *ERGO (version 2) for Instantaneous Self Assessment of Workload in a real-time ATC simulation environment (EEC Report No. 10/96).* Bruxelles, Belgium: EROCONTROL Agency, April 1996.

Lamoureux, T. The influence of aircraft proximity data on the subjective mental workload of controllers in the air traffic control task. *Ergonomics.* 42(11), 1482–1491, 1999.

Tattersall, A.J. and Foord, P.S. An experimental evaluation of instantaneous self-assessment as a measure of workload. *Ergonomics.* 39(5), 740–748, 1996.

3.3.4.6 McDonnell Rating Scale

General description. The McDonnell rating scale (see figure 21) is a ten-point scale requiring a pilot to rate workload based on the attentional demands of a task.

<table>
<tr><td rowspan="9">Controllable
Capable of being controlled or managed in context of mission, with available pilot attention</td><td rowspan="6">Acceptable
May have deficiencies which warrant improvement, but adequate for mission. Pilot compensation, If required to achieve acceptable performance, is feasible.</td><td rowspan="3">Satisfactory
Meets all requirements and expectations; good enough without improvement. Clearly adequate for mission.</td><td>Excellent, Highly desirable</td><td>A1</td></tr>
<tr><td>Good, pleasant, well behaved</td><td>A2</td></tr>
<tr><td>Fair. Some mildly unpleasant characteristics. Good enough for mission without improvement.</td><td>A3</td></tr>
<tr><td rowspan="3">Unsatisfactory
Reluctantly acceptable. Deficiencies which warrant improvement. Performance adequate for mission with feasible pilot compensation.</td><td>Some minor but annoying deficiencies. Improvement is requested. Effect on performance is easily compensated for by pilot.</td><td>A4</td></tr>
<tr><td>Moderately objectionable deficiencies. Improvement is needed. Reasonable performance requires considerable pilot compensation.</td><td>A5</td></tr>
<tr><td>Very objectionable deficiencies. Major improvements are needed. Requires best availble pilot compensation to achieve acceptable performance.</td><td>A6</td></tr>
<tr><td rowspan="3">Unacceptable
Deficiencies which require mandatory improvement. Inadequate performance for mission, even with maximum feasible pilot compensation.</td><td rowspan="3"></td><td>Major deficiencies which require mandatory improvement for acceptance. Controllable. Performance inadequate for mission, or pilot compensation required for minimum acceptable performance in mission is too high.</td><td>U7</td></tr>
<tr><td>Controllable with difficulty. Requires substantial pilot skill and attention to retain control and continue mission.</td><td>U8</td></tr>
<tr><td>Marginally controllable in mission. Requires maximum available pilot skill and attention to retain control.</td><td>U9</td></tr>
<tr><td colspan="3">Uncontrollable
Control will be lost during some portion of the mission.</td><td>Uncontrollable in Mission</td><td>U10</td></tr>
</table>

FIGURE 21 McDonnell Rating Scale (from McDonnell, J.D. *Pilot rating techniques for the estimation and evaluation of handling qualities (AFFDL-TR-68-76).* Wright-Patterson Air Force Base, TX: Air Force Flight Dynamics Laboratory, 1968, p. 7).

Strengths and limitations. van de Graaff (1987) reported significant differences in workload among various flight approach segments and crew conditions. Intersubject variability between McDonnell ratings was less than that between SWAT ratings.

Data requirements. Not stated.

Thresholds. Not stated.

SOURCES

McDonnell, J.D. *Pilot rating techniques for the estimation and evaluation of handling qualities (AFFDL-TR-68-76).* Wright-Patterson Air Force Base, TX: Air Force Flight Dynamics Laboratory, 1968.

van de Graaff, R.C. *An in-flight investigation of workload assessment techniques for civil aircraft operations, (NLR-TR-87119U).* Amsterdam, the Netherlands: National Aerospace Laboratory, 1987.

3.3.4.7 Overall Workload Scale

General description. The Overall Workload (OW) Scale is a bipolar scale (*low* on the left; *high* on the right) requiring subjects to provide a single workload rating on a horizontal line divided into 20 equal intervals.

Strengths and limitations. The OW scale is easy to use but is less valid and reliable than NASA TLX or Analytical Hierarchy Process (AHP) ratings (Vidulich and Tsang, 1987). Hill, Iavecchia, Byers, Bittner, Zaklad, and Christ (1992) reported that OW was consistently more sensitive to workload and had greater operator acceptance than the Modified Cooper–Harper rating scale or the SWAT. Anthony and Biers (1997), however, found no difference between OW and SWAT ratings. Their subjects were 48 introductory psychology students performing a memory recall task. Harris, Hill, Lysaght, and Christ (1992) reported that the OW Scale has been sensitive across tasks, systems, and environments.

The scale can be used retrospectively or prospectively (Eggleston and Quinn, 1984). It has been used in assessing workload in mobile air defense missile system (Hill, Zaklad, Bittner, Byers, and Christ 1988), remotely piloted vehicle systems (Byers, Bittner, Hill, Zaklad, and Christ, 1988), helicopter simulators (Iavecchia, Linton, and Byers, 1989); and laboratories (Harris, Hancock, Arthur, and Caird, 1995).

Data requirements. Not stated.

Thresholds. Not stated.

SOURCES

Anthony, C.R. and Biers, D.W. Unidimensional versus multidimensional workload scales and effect of number of rating scale categories. *Proceedings of the Human Factors and Ergonomics Society 41st Annual Meeting.* 2, 1084–1088, 1997.

Byers, J.C., Bittner, A.C., Hill, S.G., Zaklad, A.L., and Christ, R.E. Workload assessment of a remotely piloted vehicle (RPV) system. *Proceedings of the Human Factors Society 32nd Annual Meeting* (pp. 1145–1149). Santa Monica, CA: Human Factors Society, 1988.

Eggleston, R.G. and Quinn, T.J. A preliminary evaluation of a projective workload assessment procedure. *Proceedings of the Human Factors Society 28th Annual Meeting* (pp. 695–699). Santa Monica, CA: Human Factors Society, 1984.

Harris, W.C., Hancock, P.A., Arthur, E.J., and Caird, J.K. Performance, workload, and fatigue changes associated with automation. *International Journal of Aviation Psychology*. 5(2), 169–185, 1995.

Harris, R.M., Hill, S.G., Lysaght, R.J., and Christ, R.E. *Handbook for operating the OWL-KNEST technology (ARI Research note 92-49)*. Alexandria, VA: United States Army Research Institute for the Behavioral and Social Sciences, 1992.

Hill, S.G., Iavecchia, H.P., Byers, J.C., Bittner, A.C., Zaklad, A.L., and Christ, R.E. Comparison of four subjective workload rating scales. *Human Factors*. 34, 429–439, 1992.

Hill, S.G., Zaklad, A.L., Bittner, A.C., Byers, J.C., and Christ, R.E. Workload assessment of a mobile air defense missile system. *Proceedings of the Human Factors Society 32nd Annual Meeting* (pp. 1068–1072). Santa Monica, CA: Human Factors Society, 1988.

Iavecchia, H.P., Linton, P.M., and Byers, J.C. Operator workload in the UH-60A Black Hawk crew results vs. TAWL model predictions. *Proceedings of the Human Factors Society 33rd Annual Meeting* (pp. 1481–1481). Santa Monica, CA: Human Factors Society, 1989.

Vidulich, M.A. and Tsang, P.S. Absolute magnitude estimation and relative judgment approaches to subjective workload assessment. *Proceedings of the Human Factors Society 31st Annual Meeting*. 1057–1061, 1987.

3.3.4.8 Pilot Objective/Subjective Workload Assessment Technique

General description. The Pilot Objective/Subjective Workload Assessment Technique (POSWAT) is a ten-point subjective scale (see table 28) developed at the Federal Aviation Administration's Technical Center (Stein, 1984). The scale is a modified Cooper–Harper Rating Scale, but does not include the characteristic binary decision tree. It does, however, divide workload into five categories: low, minimal, moderate, considerable, and excessive. Similar to the Cooper–Harper Rating Scale, the lowest three levels (1 through 3) are grouped into a low category. A similar scale, the Air Traffic Workload Input Technique (ATWIT), has been developed for Air Traffic Controllers (Porterfield, 1997).

Strengths and limitations. The immediate predecessor of POSWAT was the Workload Rating System. It consisted of a workload entry device with an array of 10 push buttons. Each push button corresponded to a rating from 1 (very easy) to 10 (very hard). The scale was sensitive to changes in flight-control stability (Rehmann, Stein, and Rosenberg, 1983). It also generally decreased as experience increased (Mallery, 1987).

Stein (1984) reported that POSWAT ratings significantly differentiated between experienced and novice pilots, and high (initial and final approach) and low (en route) flight segments. There was also a significant learning effect: workload ratings were significantly higher on the first than on the second flight. Although the POSWAT scale was sensitive to manipulations of pilot experience level for flights in a light aircraft and in a simulator, the scale was cumbersome. Seven dimensions (workload, communications, control inputs, planning, "deviations," error, and pilot complement) are combined on one scale. Further, the number of ranks on the ordinal scale is confusing, because there are both 5 and 10 levels.

TABLE 28
POSWAT

Pilot Workload	Workload	Characteristics
1	Little or none	Any tasks completed immediately. Nominal control inputs, no direct communications. No required planning. No chance of any deviations.
2	Minimal	All tasks easily accomplished. No chance of deviation.
3	Minimal	All tasks accomplished. Minimal chance of deviation.
4	Moderate	All tasks accomplished. Tasks are prioritized. Minimal chance of deviation.
5	Moderate	All tasks are accomplished. Tasks are prioritized. Moderate chance of deviation.
6	Moderate	All tasks are accomplished. Tasks are prioritized. Considerable chance of deviation.
7	Considerable	Almost all tasks accomplished. Lowest priority tasks dropped. Considerable chance of deviation.
8	Considerable	Most tasks accomplished. Lower-priority tasks dropped. Considerable chance of deviation.
9	Excessive	Only high-priority tasks accomplished. Chance of error or major deviations. Second pilot desired for flight.
10	Excessive	Only highest-priority tasks (safety of flight) tasks accomplished. Errors or frequent deviations occur. Second pilot needed for safe flight.

Source: From Mallery, C.J. and Maresh, J.L. Comparison of POSWAT ratings for aircraft and simulator workload. *Proceedings of the Fourth International Symposium on Aviation Psychology.* 644–650, 1987, p. 655.

Rehman, Stein, and Rosenberg (1983) obtained POSWAT ratings once per minute. These investigators found that pilots reliably reported workload differences in a tracking task on a simple ten-point nonadjectival scale. Therefore, the cumbersome structure of the POSWAT scale may not be necessary.

Data requirements. Stein (1984) suggested not analyzing POSWAT ratings for short flight segments if the ratings are given at 1 min intervals.

Thresholds. Not stated.

SOURCES

Mallery, C.J. The effect of experience on subjective ratings for aircraft and simulator workload during IFR flight. *Proceedings of the Human Factors Society 31st Annual Meeting.* 2, 838–841, 1987.

Mallery, C.J. and Maresh, J.L. Comparison of POSWAT ratings for aircraft and simulator workload. *Proceedings of the Fourth International Symposium on Aviation Psychology.* 644–650, 1987.

Porterfield, D.H. Evaluating controller communication time as a measure of workload. *The International Journal of Aviation Psychology.* 7(2), 171–182, 1997.

Rehman, J.T., Stein, E.S., and Rosenberg, B.L. Subjective pilot workload assessment. *Human Factors*. 25(3), 297–307, 1983.
Stein, E.S. *The measurement of pilot performance: A master-journeyman approach (DOT/FAA/CT-83/15)*. Atlantic City, NJ: Federal Aviation Administration Technical Center; May 1984.

3.3.4.9 Utilization

General description. Utilization (p) is the probability of the operator being in a busy status (Her and Hwang, 1989).

Strengths and limitations. Utilization has been a useful measure of workload in continuous process tasks (e.g., milling, drilling, system controlling, loading, and equipment setting). It accounts for both arrival time of work in a queue and service time on that work.

Data requirements. A queuing process must be in place.

Thresholds. The minimum value is 0, and the maximum value is 1. High workload is associated with the maximum value.

SOURCE

Her, C. and Hwang, S. Application of queuing theory to quantify information workload in supervisory control systems. *International Journal of Industrial Ergonomics*. 4, 51–60, 1989.

3.3.5 Task-Analysis Based Subjective Workload Measures

These measures break the tasks into subtasks and subtask requirements for workload evaluation. Examples include the following: Arbeitswissenshaftliches Erhebungsverfahren zur Tatigkeitsanalyze (section 3.3.5.1), Computerized Rapid Analysis of Workload (section 3.3.5.2), McCracken–Aldrich Technique (section 3.3.5.3), Task Analysis Workload (section 3.3.5.4), and Zachary/Zaklad Cognitive Analysis (section 3.3.5.3).

3.3.5.1 Arbeitswissenshaftliches Erhebungsverfahren zur Tatigkeitsanalyze

General description. Arbeitswissenschaftliches Erhebungsverfahren zur Tatigkeitsanalyze (AET) was developed in Germany to measure workload. AET has the following three parts: (1) work system analysis, which rates the "type and properties of work objects, the equipment to be used as well as physical social and organizational work environment" (North and Klaus, 1980, p. 788) on both nominal and ordinal scales; (2) task analysis, which uses a 31-item ordinal scale to rate "material work objects, abstract (immaterial) work objects, and man-related tasks" (p. 788); and (3) job-demand analysis, which is used to evaluate the conditions under which the job is performed. "The 216 items of the AET are rated on nominal or ordinal scales using five codes as indicated for each item: frequency, importance, duration, alternative and special (intensity) code" (p. 790).

Strengths and limitations. AET has been used in over 2000 analyses of both manufacturing and management jobs.

Data requirements. Profile analysis is used to analyze the job workload. Cluster analysis is used to identify elements of jobs that "have a high degree of natural association among one another" (p. 790). Multivariate statistics are used for "placement, training, and job classification" (p. 790).

Thresholds. Not stated.

SOURCE

North, R.A. and Klaus, J. Ergonomics methodology: An obstacle or promoter for the implementation of ergonomics in industrial practice? *Ergonomics*. 23 (8), 781–795, 1980.

3.3.5.2 Computerized Rapid Analysis of Workload

General description. The Computerized Rapid Analysis of Workload (CRAWL) is a computer program that helps designers predict workload in systems being designed. CRAWL inputs are mission timelines and task descriptions. Tasks are described in terms of cognitive, psychomotor, auditory, and visual demands.

Strengths and limitations. Bateman and Thompson (1986) reported increases in CRAWL ratings as task difficulty increased. Vickroy (1988) reported similar results as air turbulence increased.

Data requirements. The mission timeline must provide detailed second-by-second descriptions of the aircraft status.

Thresholds. Not stated.

SOURCES

Bateman, R.P. and Thompson, M.W. Correlation of predicted workload with actual workload measured using the Subjective Workload Assessment Technique. *Proceedings of the SAE AeroTech Conference,* 1986.

Vickroy, S.C. Workload prediction validation study: The verification of CRAWL predictions; 1988.

3.3.5.3 McCracken–Aldrich Technique

General description. The McCracken–Aldrich Technique was developed to identify workload associated with flight control, flight support, and mission-related activities (McCracken and Aldrich, 1984).

Strength and limitations. The technique may require months of preparation to use. It has been useful in assessing workload in early system design stages.

Data requirements. A mission must be decomposed into segments, functions, and performance elements (e.g., tasks). Subject matter experts rate workload (from 1 to 7) for each performance element. A FORTRAN programmer is required to generate the resulting scenario timeline.

Thresholds. Not stated.

SOURCE

McCracken, J.H. and Aldrich, T.B. *Analysis of selected LHX mission functions: Implications for operator workload and system automation goals (TNA ASI 479-24-84).* Fort Rucker, AL: Anacapa Sciences, 1984.

3.3.5.4 Task Analysis Workload

General description. The Task Analysis/Workload (TAWL) technique requires missions to be decomposed into phases, segments, functions, and tasks. For each task, a subject matter expert rates the workload on a scale of 1 to 7. The tasks are combined into a scenario timeline and workload estimated for each point on the timeline.

Strengths and limitations. TAWL is sensitive to task workload but requires about 6 months to develop (Harris, Hill, Lysaght, and Christ, 1992). It has been used to identify workload in helicopters (Szabo and Bierbaum, 1986).

$$p = b_0 + b_1N + b_2S$$

where

p = utilization
b = intercept determined from regression analysis
b_1 = slope determined from regression analysis
N = number of information types
S = quantity of information in an information type.

Hamilton and Cross (1993), based on seven task conditions, performance of 20 AH-64 aviators, and two analysts, reported significant correlations (+0.89 and +0.99) between the measures predicted by the TAWL model and the actual data.

Data requirements. A detailed task analysis is required. Then subject matter experts must rate the workload of each task on six channels (auditory, cognitive, kinesthetic, psychomotor, visual, and visual-aided). A PC compatible system is required to run the TAWL software. A user's guide (Hamilton, Bierbaum, and Fulford, 1991) is available.

Thresholds. Not stated.

SOURCES

Hamilton, D.B., Bierbaum, C.R., and Fulford, L.A. *Task Analysis/Workload (TAWL) User's guide—version 4.0 (ASI 690-330-90).* Fort Rucker, AL: Anacapa Sciences, 1991.

Hamilton, D.B. and Cross, K.C. *Preliminary validation of the task analysis/workload methodology (ARI RN92-18).* Alexandria, VA: Army Research Institute for the Behavioral and Social Sciences, 1993.

Harris, R.M., Hill, S.G., Lysaght, R.J., and Christ, R.E. *Handbook for operating the OWL & NEST technology (ARI Research Note 92-49).* Alexandria, VA: United States Army Research Institute for the Behavioral and Social Sciences, 1992.

Szabo, S.M. and Bierbaum, C.R. *A comprehensive task analysis of the AH-64 mission with crew workload estimates and preliminary decision rules for developing an AH-64 workload prediction model (ASI 678-204-86[B]).* Fort Rucker, AL: Anacapa Sciences, 1986.

3.3.5.5 Zachary/Zaklad Cognitive Analysis

General Description. The Zachary/Zaklad Cognitive Analysis Technique requires both operational subject matter experts and cognitive scientists to identify operator strategies for performing all tasks listed in a detailed cognitive mission task analysis. A second group of subject matter experts then rates, using 13 subscales, workload associated with performing each task.

Strengths and limitations. The method has only been applied in two evaluations, one for the P-3 aircraft (Zaklad, Deimler, Iavecchia, and Stokes, 1982), the other for F/A-18 aircraft (Zachary, Zaklad, and Davis, 1987; Zaklad, Zachary, and Davis, 1987).

Data requirements. A detailed cognitive mission timeline must be constructed. Two separate groups of subject matter experts are required, one to develop the timeline, the other to rate the associated workload.

Thresholds. Not stated.

SOURCES

Zaklad, A.L., Deimler, J.D., Iavecchia, H.P., and Stokes, J. *Multisensor correlation and TACCO workload in representative ASW and ASUW environments (TR-1753A).* Analytics, 1982.

Zachary, W., Zaklad, A., and Davis, D. A cognitive approach to multisensor correlation in an advanced tactical environment. *Proceedings of the 1987 Tri-Service Data Fusion Symposium.* Fourel, MD; Johns Hopkins University, 1987.

Zaklad, A., Zachary, W., and Davis, D. A cognitive model of multisensor correlation in an advanced aircraft environment. *Proceedings of the Fourth Mid-Central Ergonomics/Human Factors Conference,* Urbana IL, 1987.

3.4 SIMULATION OF WORKLOAD

Several digital models have been used to evaluate workload. These include the following: (1) Null Operation System Simulation (NOSS), (2) SAINT (Buck 1979), Modified Petri Nets (MPN) (White, MacKinnon, and Lyman, 1986), (3) task analyses (Bierbaum and Hamilton, 1990), and (4) Workload Differential Model (Ntuen and Watson, 1996).

In addition, Riley (1989) described the Workload Index (W/INDEX), a simulation that enables designers to compare alternate physical layouts and interface technologies, use of automation, and sequence of tasking. The designer inputs the system design concept and assigns each to an interface channel (i.e., visual, auditory, manual, and verbal) and a task timeline. W/INDEX then predicts workload.

A Micro Saint model of the workload in a simulated air-to-ground combat mission did not predict workload as measured by SWAT (See and Vidulich, 1997).

SOURCES

Bierbaum, C.R. and Hamilton, D.B. *Task analysis and workload prediction model of the MH-60K mission and a comparison with UH-60A workload predictions; Volume III: Appendices H through N. (ARI Research Note 91-02).* Alexandria, VA: U.S. Army Research Institute for the Behavioral and Social Sciences, October 1990.

Buck, J. Workload estimation through simulation. Paper presented at the *Workload Program of the Indiana Chapter of the Human Factors Society.* Crawfordsville, Indiana, March 31, 1979.

Ntuen, C.A. and Watson, A.R. *Workload prediction as a function of system complexity.* IEEE 0-8186-7493, 96–100, August 1996.

Riley, V. W/INDEX: A crew workload prediction tool. *Proceedings of the Fifth International Symposium on Aviation Psychology.* 2, 832–837, 1989.

See, J.E. and Vidulich, M.A Assessment of computer modeling of operator mental workload during target acquisition. *Proceedings of the Human Factors and Ergonomics Society 41st Annual Meeting.* 2, 1303–1307, 1997.

White, S.A., MacKinnon, D.P., and Lyman, J. *Modified Petri Net Modal sensitivity to workload manipulations (NASA-CR-177030).* Moffett Field, CA: NASA Ames Research Center, 1986.

3.5 DISSOCIATION OF WORKLOAD AND PERFORMANCE

Wickens and Yeh (1983) presented a theory of dissociation between subjective measures of workload and performance. The theory proposes that the subjective measures are determined by the number of tasks, and performance by task difficulty. In subsequent work, Yeh and Wickens (1985, 1988) identified five conditions in which the relationship between performance and subjective measures of workload imply different effects on workload. In the first condition, termed *motivation,* performance improves and subjective ratings of workload increase (see figure 22). In the second condition, which Yeh and Wickens termed *underload,* as demand increases, performance remains the same but subjective measures of workload increase. In this condition, performance implies that workload remains constant, whereas the subjective measures imply increased workload. In the third condition, *resource-limited tasks,* as the amount of invested resources increases, performance degrades and subjective workload measures increase. However, the proportion of the change in subjective ratings is greater than the proportion of the change in performance. In this condition, performance implies that workload increases somewhat, whereas subjective workload measures imply that workload increases greatly. In the fourth condition, comparison of *dual-task configurations with different degrees of competition for common resources,* as demand for common resources increases, performance degrades and subjective workload measures increase. This time, however, the change in performance is greater than the change in subjective ratings. In this condition, performance implies that workload increases greatly, whereas subjective workload measures suggest that workload increases slightly. In the fifth condition, which Yeh and Wickens termed *overload,* as demand increases, performance degrades, whereas subjective measures remain unchanged. In this condition, performance implies that workload increases, whereas the subjective measures imply that workload remains the same.

Yeh and Wickens defined these differences in implied workload as *dissociation,* and they suggested that dissociation occurs because in these five conditions different factors determine performance and subjective workload measures. These factors are listed in table 29. As can be seen from table 29, there is only one factor common to both performance and subjective measures of workload: amount of invested resources. However, there are different resource dichotomies. For example, Yeh and Wickens (1988) defined four types of resources: (1) perceptual/central versus response stage, (2) verbal versus spatial code, (3) visual versus auditory input modality, and (4) manual versus speech output modality.

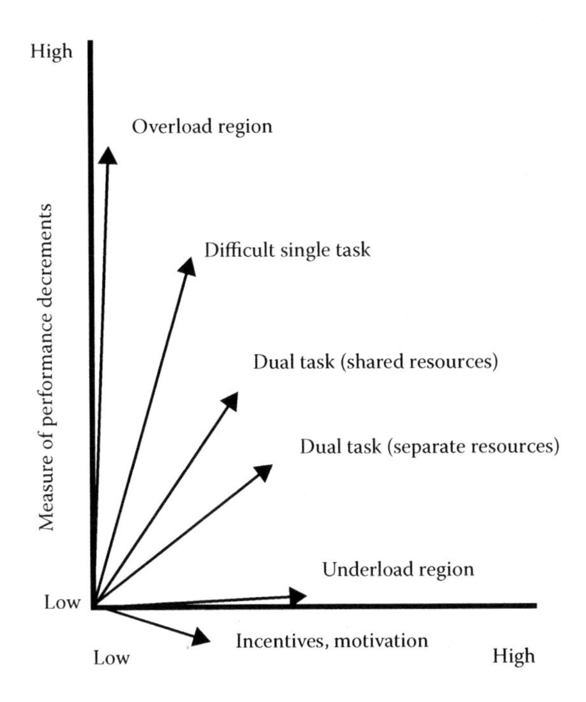

FIGURE 22 Dissociations between performance and subjective measures of workload as predicted by theory (adapted from Yeh, Y. and Wickens, C.D. (1988). Dissociation of performance and subjective measures of workload. *Human Factors*. 30, p. 115).

The importance of these dichotomies has been supported in many experiments. For example, Wickens and Liu (1988) reported greater tracking error when a one-dimensional compensatory tracking task was time-shared with a spatial decision task (same code: spatial) than with a verbal decision task (different codes: spatial and verbal). Tracking error was also greater when the response to the time-shared decision task was manual (same modality: manual) rather than verbal (different modality: manual versus speech).

Derrick (1985) analyzed performance on 18 computer-based tasks and a global subjective estimate for workload on each task. The tasks were performed in four configurations: (1) single easy, (2) single hard, (3) dual with the same task, and (4) dual with different tasks. He concluded: "If a task was increased in difficulty by adding to the perceptual and central processing resource load (and thus performance declined), people rated the workload as higher than for the same task with lower perceptual and central processing resource demands. However, tasks made more difficult by increasing the resource demands of responding (leading to worse performance) did not produce increased workload ratings" (p. 1022). In addition, "performance in the Dual With the Same Task configuration was worse than in the Dual with Different Tasks configuration, but the workload ratings were essentially equivalent for these two conditions" (p. 1022).

Derrick summarized his findings as well as those of Yeh and Wickens (1985) in table 30. The values in the cells indicate the weight associated with each portion of the dissociation effect.

TABLE 29
Determinants of Performance and Subjective Measures of Workload

Measure	Primary Determinant	Secondary Determinant
Single-task performance	Amount of invested resources	Task difficulty
		Subject's motivation
		Subjective criteria of optimal performance
	Resource efficiency	Task difficulty
		Data quality
		Practice
Dual-task performance	Amount of invested resources	Task difficulty
		Subject's motivation
		Subjective criteria of optimal performance
	Resource efficiency	Task difficulty and/or complexity
		Data quality
		Practice
Subjective workload	Amount of invested resources	Task difficulty
		Subject's motivation
		Subjective criteria of optimal performance
	Demands on working memory	Amount of time sharing between tasks
		Amount of information held in working memory
		Demand on perceptual and/or central processing resources

Source: Adapted from Yeh, Y. and Wickens, C.D. Dissociation of performance and subjective measures of workload. *Human Factors*. 30, 111–120, 1988.

TABLE 30
A Theory of Dissociation

	Sources	Performance Decreases	Subjective Difficulty Increases
1	Increased single-task difficulty	4	3
	Perceptual/cognitive	2	2
	Response	2	1
2	Concurrent task demand	3	4
	Same resources	2	2
	Different resources	1	2

The relationship between the factors that determine performance and subjective measures of workload and the resource dichotomies is complex. Greater resource supply improves performance of resource-limited tasks but not of data-limited tasks. For dual tasks, performance degrades when the tasks compete for common resources.

The importance of accurate performance and workload measurement during system evaluation cannot be overstated. As Derrick (1985) stated: "A workload practitioner who relies on these [subjective] ratings [of workload] rather than performance will be biased to choose a nonoptimal system that requires operators to perform just one task rather than the system that demands dual task performance" (p. 1023). "The practitioner who relies solely on subjective data may likely favor the system that has serious performance limitations, especially under emergency conditions" (p. 1024). "A system design option that increases control demands, and ultimately degrades performance under high workload conditions, will not be rated as a problem in the normal single task workload evaluations" (p. 1024). Therefore, the potential for inaccurate interpretation of the results is worrisome.

Gawron made an extensive search was made of Defense Technical Information Center (DTIC) reports and refereed journals to identify publications in which both task performance and subjective measures of workload were reported. Only 11 publications meeting this criterion were found (see table 31). The publications spanned 11 years from 1978 to 1996. Eight were performed in flight simulators, one in a centrifuge, and two in a laboratory using a tracking task. The type of facility, mission, duration, number of trials, and type of subjects in each of these publications are also described in table 31. As can be seen from the table, missions varied from airdrop (1), to research (2) to combat (3), with half being transport missions (5). The duration of trials also varied widely, from 1 to 200 min with up to 100 replications. Each publication reported data based on at least 6 subjects, with two publications reporting data from 48 subjects. Half the studies collected data from only male subjects.

The high- and low-workload conditions, task, performance metric, and subjective metric for each data point gleaned from these publications is presented in table 32. The workload conditions covered the entire range from physical (flight acceleration forces), mental effort (inexperience, different responses to the same cues, difficult navigation problems, dual task, and no autopilot), psychological stress (occurrence of malfunctions, high winds), and time stress (high target rates). Performance metrics included both speed (1 publication) and error (10 publications) measures. The following subjective workload metrics were also diverse: four publications reported a workload rating, three the Modified Cooper–Harper Rating Scale, one AHP, one NASA TLX, one POSWAT, and two SWAT. Three publications reported *z*-scores; seven did not. The publications were from the following four laboratories: Air Force Research Laboratory (3 publications/22 data points), FAA Tech Center (1/8), University of Illinois (2/4), University of Minnesota (1/4), and VPI (3/19).

All but one (point 21, in which workload decreased) of the 47 data points showed increased ratings of workload between the low- and high-workload conditions. All but six of the data points showed decreased performance (either increase in errors or time) in the high-workload condition. The six points in which performance improved while workload increased are presented in table 33. The points come from three different studies by three different research teams. They use different workload rating scales and performance metrics (although five of the six are response time). What is common is that each of the data points involves auditory stimuli and/or responses. (Note that there was no dissociation with other performance measures for the same independent variable). These points are further categorized in table 34. Five of the six

TABLE 31
Description of Studies Reviewed

Study	Reference	Facility	Mission	Duration	Trials	Subjects	Laboratory
1	Albery (1989)	Dynamic Environment Simulator (centrifuge)	Combat	60 s	24 or 12 per subject	9 male military personnel	Air Force Research Laboratory
2	Casali and Wierwille (1983)	Singer/Link GAT-1B simulator	Transport	Not stated	3 per subject	29 male and 1 female civilian pilot	Virginia Polytechnic Institute and State University
3	Casali and Wierwille (1984)	Singer/Link GAT-1B simulator	Transport	12 min	3 per subject	48 male pilots	Virginia Polytechnic Institute and State University
4	Hancock (1996)	1D compensatory tracking	Research	2 min	100 per subject	6 right-handed male university volunteers	University of Minnesota
5	Kramer, Sirevaag, and Braune (1987)	ILLIMAC fixed-based flight simulator	Transport	45 min	4 per subject	7 right-handed male student pilots	University of Illinois
6	Madero, Sexton, Gunning, and Moss (1979)	Air Force Flight Dynamics Laboratory multicrew simulator	Airdrop	Not stated	3 per crew	8 C-130 crews made up of pilot, copilot, and loadmaster	Air Force Research Laboratory
7	Stein (1984)	Singer/Link General Aviation Trainer (simulated Cessna 421)	Transport	35 min	2 per subject	12 air transport commercial pilots and 12 recently qualified instrument pilots	FAA Technical Center
8	Vidulich and Bortolussi (1988)	NASA Ames 1-Cab fixed-base simulator	Combat	1 to 1.5 h	4	12 male army AH-64 helicopter pilots male RAF helicopter test pilot male retired army helicopter pilot	Air Force Research Laboratory
9	Vidulich and Wickens (1985, 1986)	2D compensatory tracking and Sternberg task	Research	Not stated	14 per subject	40 students	University of Illinois
10	Wierwille, Rahimi, and Casali (1985)	Singer/Link GAT-1B simulator	Transport	Not stated	3 per subject	48 male pilots	Virginia Polytechnic Institute and State University
11	Wolf (1978)	F-4 simulator	Combat	2 min	120 total	7 RF-4B aircraft pilots from Air National Guard	Air Force Research Laboratory

TABLE 32

Summary of Research Reporting Both Performance and Subjective Measures of Workload

Study	Data Point	High Workload	Low Workload	Task	Performance Metric	Subjective Workload Metric
1	1	3.75G acceleration	1.4G acceleration	Compensatory tracking	Tracking error	SWAT
1	2	100 dBA noise	40 dBA noise	Compensatory tracking	Tracking error	SWAT
2	3	One call sign every 2 s on average; nontarget call signs were permutations	One call sign every 12 s on average; nontarget call signs were permutations	Control aircraft	*z*-Score errors of omission in communications	Modified Cooper–Harper Rating Scale
2	4	One call sign every 2 s on average; nontarget call signs were permutations	One call sign every 12 s on average; nontarget call signs were permutations	Control aircraft	*z*-Score errors of commission in communications	Modified Cooper–Harper Rating Scale
2	5	One call sign every 2 s on average; nontarget call signs were permutations	One call sign every 12 s on average; nontarget call signs were permutations	Control aircraft	*z*-Score communications response time	Modified Cooper–Harper Rating Scale
2	6	One call sign every 2 s on average; nontarget call signs were permutations	One call sign every 12 s on average; nontarget call signs were permutations	Control aircraft	*z*-Score errors of omission in communications	Multidescriptor Scale
2	7	One call sign every 2 s on average; nontarget call signs were permutations	One call sign every 12 s on average; nontarget call signs were permutations	Control aircraft	*z*-Score errors of commission in communications	Multidescriptor Scale
2	8	One call sign every 2 s on average; nontarget call signs were permutations	One call sign every 12 s on average; nontarget call signs were permutations	Control aircraft	*z*-Score communications response time	Multidescriptor Scale

Continued

TABLE 32 (Continued)

Summary of Research Reporting Both Performance and Subjective Measures of Workload

Study	Data Point	High Workload	Low Workload	Task	Performance Metric	Subjective Workload Metric
3	9	High icing, potential malfunction in any engine or fuel gauges, average rate of malfunctions 5 s	Low icing, average rate of icing hazard 1 every 50 s	Control aircraft	Pitch high-pass mean square	Modified Cooper–Harper Rating Scale
3	10	High icing, potential malfunction in any engine or fuel gauges, average rate of malfunctions 5 s	Low icing, average rate of icing hazard 1 every 50 s	Control aircraft	Roll high-pass mean square	Modified Cooper–Harper Rating Scale
3	11	High icing, potential malfunction in any engine or fuel gauges, average rate of malfunctions 5 s	Low icing, average rate of icing hazard 1 every 50 s	Control aircraft	Response time to hazard	Modified Cooper–Harper Rating Scale
3	12	High icing, potential malfunction in any engine or fuel gauges, average rate of malfunctions 5 s	Low icing, average rate of icing hazard 1 every 50 s	Control aircraft	Pitch high-pass mean square	Multidescriptor Scale
3	13	High icing, potential malfunction in any engine or fuel gauges, average rate of malfunctions 5 s	Low icing, average rate of icing hazard 1 every 50 s	Control aircraft	Roll high-pass mean square	Multidescriptor Scale
3	14	High icing, potential malfunction in any engine or fuel gauges, average rate of malfunctions 5 s	Low icing, average rate of icing hazard 1 every 50 s	Control aircraft	Response time to hazard	Multidescriptor Scale

3	15	High icing, potential malfunction in any engine or fuel gauges, average rate of malfunctions 5 s	Low icing, average rate of icing hazard 1 every 50 s	Control aircraft	Pitch high-pass mean square	Workload/Compensation/Interference/Technical Effectiveness Scale
3	16	High icing, potential malfunction in any engine or fuel gauges, average rate of malfunctions 5 s	Low icing, average rate of icing hazard 1 every 50 s	Control aircraft	Roll high-pass mean square	Workload/Compensation/Interference/Technical Effectiveness Scale
3	17	High icing, potential malfunction in any engine or fuel gauges, average rate of malfunctions 5 s	Low icing, average rate of icing hazard 1 every 50 s	Control aircraft	Response time to hazard	Workload/Compensation/Interference/Technical Effectiveness Scale
4	18	Initial 10 trials	Final 10 trials	1D compensatory tracking	Tracking error	NASA TLX
4	19	Initial 10 trials	Final 10 trials	1D compensatory tracking	Tracking error	SWAT
4	20	Initial 10 trials 30 d later	Final 10 trials 30 d later	1D compensatory tracking	Tracking error	NASA TLX
4	21	Initial 10 trials 30 d later	Final 10 trials 30 d later	1D compensatory tracking	Tracking error	SWAT
5	22	30 mph wind from 270°, moderate turbulence, and partial suction failure in heading indicator during approach	No wind, no turbulence, and no system failure	Control aircraft	Heading deviation	Subjective Workload Rating
5	23	30 mph wind from 270°, moderate turbulence, and partial suction failure in heading indicator during approach	No wind, no turbulence, and no system failure	Control aircraft	Altitude deviation	Subjective Workload Rating

Continued

TABLE 32 (Continued)

Summary of Research Reporting Both Performance and Subjective Measures of Workload

Study	Data Point	High Workload	Low Workload	Task	Performance Metric	Subjective Workload Metric
5	24	30 mph wind from 270°, moderate turbulence, and partial suction failure in heading indicator during approach	No wind, no turbulence, and no system failure	Control aircraft	Glideslope deviation	Subjective Workload Rating
6	25	Without autopilot, without bulk data storage	With autopilot, with bulk data storage	Control aircraft	Course error during cruise segment (feet)	Subjective Workload Rating
6	26	Without autopilot, without bulk data storage	With autopilot, with bulk data storage	Control aircraft	Course error during CARP segment (feet)	Subjective Workload Rating
7	27	Journeymen (recently qualified instrument pilots)	Masters (professional, high-time pilots)	Control aircraft	Automated performance score enroute flight 1	POSWAT Rating
7	28	Journeymen	Masters	Control aircraft	Automated performance score descent flight 1	POSWAT Rating
7	29	Journeymen	Masters	Control aircraft	Automated performance score initial approach flight 1	POSWAT Rating
7	30	Journeymen	Masters	Control aircraft	Automated performance score final approach flight 1	POSWAT Rating
7	31	Journeymen	Masters	Control aircraft	Automated performance score enroute flight 2	POSWAT Rating
7	32	Journeymen	Masters	Control aircraft	Automated performance score descent flight 2	POSWAT Rating

7	33	Journeymen	Masters	Control aircraft	Automated performance score initial approach flight 2	POSWAT Rating
7	34	Journeymen	Masters	Control aircraft	Automated performance score final approach flight 2	POSWAT Rating
8	35	Secondary task required speech response	Secondary task required manual response	Respond to surface-to-air missile	Response time decrement during cruise	AHP Rating
8	36	Secondary task required speech response	Secondary task required manual response	Respond to surface-to-air missile	Response time decrement during hover	AHP Rating
8	37	Secondary task required speech response	Secondary task required manual response	Respond to surface-to-air missile	Response time decrement during combat	AHP Rating
9	38	Inconsistent Sternberg letters	Consistent Sternberg letters	Compensatory tracking with Sternberg	z-Score reaction time	z-Score task difficulty rating
10	39	Difficult navigation question, large number of numerical calculations performed, and large degree of rotation of reference triangle	Easy navigation question, small number of calculations performed, and small degree of rotation of reference triangle	Control aircraft	z-Score error rate	Modified Cooper–Harper Rating Scale
10	40	Difficult navigation question, large number of numerical calculations performed, and large degree of rotation of reference triangle	Easy navigation question, small number of calculations performed, and small degree of rotation of reference triangle	Control aircraft	z-Score reaction time	Modified Cooper–Harper Rating Scale
10	41	Difficult navigation question, large number of numerical calculations performed, and large degree of rotation of reference triangle	Easy navigation question, small number of calculations performed, and small degree of rotation of reference triangle	Control aircraft	z-Score error rate	Workload/Compensation/Interference/Technical Effectiveness Scale

Continued

TABLE 32 (Continued)

Summary of Research Reporting Both Performance and Subjective Measures of Workload

Study	Data Point	High Workload	Low Workload	Task	Performance Metric	Subjective Workload Metric
10	42	Difficult navigation question, large number of numerical calculations performed, and large degree of rotation of reference triangle	Easy navigation question, small number of calculations performed, and small degree of rotation of reference triangle	Control aircraft	*z*-Score reaction time	Workload/Compensation/ Interference/Technical Effectiveness Scale
11	43	Wind gusts to 30 kn, 0.05 rad/s rate limit on aileron and stabilator deflection	No gusts, 0.5 rad/s rate limit on aileron and stabilator deflection	Control aircraft	*z*-Score lateral path root-mean-square error	Workload Rating
11	44	Wind gusts to 30 kn, 0.05 rad/s rate limit on aileron and stabilator deflection	No gusts, 0.5 rad/s rate limit on aileron and stabilator deflection	Control aircraft	*z*-Score speed root-mean-square error	Workload Rating
11	45	Wind gusts to 30 kn, 0.05 rad/s rate limit on aileron and stabilator deflection	No gusts, 0.5 rad/s rate limit on aileron and stabilator deflection	Control aircraft	*z*-Score pitch attitude root-mean-square error	Workload Rating
11	46	Wind gusts to 30 kn, 0.05 rad/s rate limit on aileron and stabilator deflection	No gusts, 0.5 rad/s rate limit on aileron and stabilator deflection	Control aircraft	*z*-Score vertical path root-mean-square error	Workload Rating
11	47	Wind gusts to 30 kn, 0.05 rad/s rate limit on aileron and stabilator deflection	No gusts, 0.5 rad/s rate limit on aileron and stabilator deflection	Control aircraft	*z*-Score roll attitude root-mean-square error	Workload Rating

TABLE 33
Points at Which Workload Increased and Performance Improved

Data Point	High Workload	Low Workload	Task	Performance Metric	Subjective Workload Metric
2	100 dBA	40 dBA	Compensatory tracking	Tracking error	SWAT
5	One call sign every 2 s on average; nontarget call signs were permutations	One call sign every 12 s on average; nontarget call signs were permutations	Control aircraft	*z*-Score communications response time	Modified Cooper–Harper Rating Scale
8	One call sign every 2 s on average; nontarget call signs were permutations	One call sign every 12 s on average; nontarget call signs were permutations	Control aircraft	*z*-Score communications response time	Multidescriptor Scale
21	Initial 10 trials 30 d later	Final 10 trials 30 d later	1D compensatory tracking	Tracking error	SWAT
35	Secondary task required speech response	Secondary task required manual response	Respond to surface-to-air missile	Response time decrement during cruise	AHP Rating
36	Secondary task required speech response	Secondary task required manual response	Respond to surface-to-air missile	Response time decrement during hover	AHP Rating
37	Secondary task required speech response	Secondary task required manual response	Respond to surface-to-air missile	Response time decrement during combat	AHP Rating

TABLE 34
Categorization of Above Points

Data	Task		Resources	
Point	Primary	Secondary	Shared	Separate
2	Tracking			
5	Tracking	Communicating		X
8	Tracking	Communicating		X
21	Tracking			
35	Tracking	Reaction time		X
36	Tracking	Reaction time		X
37	Tracking	Reaction time		X

points require dual-task performance in which one task requires a manual response and the other a speech response. The sixth point, data point 2, requires subjects to ignore an auditory stimulus to concentrate on performing a manual task.

Is the use of separate resources associated then with dissociation between performance and workload? There are two ways to test this: first, against the results from similar points and, second, against the theoretical plot in figure 22. There are communication requirements in data points 3 through 8 and 35 through 37. Why did dissociation not occur in data points 3, 4, 6, and 7? Perhaps because the performance metrics were errors, not time. In each case in which the performance metric was time and speech was required, dissociation occurred. In examining figure 22, these data points would fall in the incentives or motivation region, in which performance is enhanced but workload increased. Avoiding a missile is certainly motivating, and this may have resulted in the dissociation between performance and workload for data points 34, 36, and 37. For points 5 and 8, subjects were instructed to "strive to maintain adequate (specified) performance on all aspects of the primary task" (p. 630). These words may have motivated the subjects.

Dissociation between performance and subjective measures of workload occurred in 7 out of 47 data points reviewed in the preceding text. In five points, performance time decreased, whereas subjective workload ratings increased. These points came from three different experiments, resulted from three different types of workload (flight acceleration forces, communications difficulty, input mode, time), used three different tasks (tracking, controlling aircraft, responding to a surface-to-air missile), and four different subjective workload metrics (SWAT, Modified Cooper–Harper Rating Scale, Multidescriptor scale, and AHP). This diversity supports the existence of the dissociation phenomenon and suggests the need for guidelines in the interpretation of workload data. As a start, Yeh and Wickens (1988) suggest that performance measures be used to provide a "direct index of the benefit of operating a system" (p. 118). Further, subjective workload measures should be used to "indicate potential performance problems that could emerge if additional demands are imposed" (p. 118).

King, Hamerman-Matsumoto, and Hart (1989) tested the multiple resource model by examining RT and errors on four simulated flight tasks. Workload was measured using NASA-TLX. They reported mixed results: the significant decrement associated with single-to-dual task performance was paired with increased overall workload ratings. However, the discrete tasks were unaffected by the single-to-dual tasking, but workload ratings increased. Tracking degraded in the dual-task condition, and workload also decreased.

Bateman, Acton, and Crabtree (1984) examined the relationship between error occurrence and SWAT ratings. They concluded that errors are dependent on task structure and workload on task difficulty and situational stress.

Thornton (1985) concluded based on a hovercraft simulation that dissociation was most pronounced when workload increased at the beginning of the mission.

SOURCES

Albery, W.B. The effect of sustained acceleration and noise on workload in human operators. *Aviation, Space, and Environmental Medicine*. 60(10), 943–948, 1989.

Bateman, R. P., Acton, W. H., and Crabtree, M.S. Workload and performance: orthogonal measures. *Proceedings of the Human Factors Society 28th Annual Meeting*. 2, 678–679, 1984.

Casali, J.G. and Wierwille, W.W. A comparison of rating scale, secondary task, physiological, and primary-task workload estimation techniques in a simulated flight task emphasizing communications load. *Human Factors*. 25, 623–642, 1983.

Casali, J.G. and Wierwille, W.W. On the comparison of pilot perceptual workload: A comparison of assessment techniques addressing sensitivity and intrusion issues. *Ergonomics*, 27, 1033–1050, 1984.

Derrick, W.A. The dissociation between subjective and performance-based measures of operator workload. *National Aerospace and Electronics Conference*. 1020–1025, 1985.

Hancock, P.A. Effects of control order, augmented feedback, input device and practice on tracking performance and perceived workload. *Ergonomics*, 39(9), 1146–1162, 1996.

King, T., Hamerman-Matsumoto, J., and Hart, S.G. Dissociation revisited: Workload and performance in a simulated flight task. *Proceedings of the Fifth International Symposium on Aviation Psychology*. 2, 796–801, 1989.

Kramer, A.F., Sirevaag, E.J., and Braune, R. A psychophysiological assessment of operator workload during simulated flight missions. *Human Factors*. 29, 145–160, 1987.

Madero, R.P., Sexton, G.A., Gunning, D., and Moss, R. *Total aircrew workload study for the AMST (AFFDL-TR-79-3080, Volume 1)*. Wright-Patterson Air Force Base, OH: Air Force Flight Dynamics Laboratory, February 1979.

Stein, E.S. *The measurement of pilot performance; A master-journeyman approach (DOT/FAA/CT-83/15)*. Atlantic City, NJ: Federal Aviation Administration Technical Center, May 1984.

Thornton, D.C. An investigation of the "Von Restorff" phenomenon. *Proceedings of the Human Factors 29th Annual Meeting*. 2, 760–764, 1985.

Vidulich, M.A. and Bortolussi, M.R. A dissociation of objective and subjective workload measures in assessing the impact of speech controls in advanced helicopters. *Proceedings of the 32nd Annual Meeting of the Human Factors Society*, 1988, 1471–1475.

Vidulich, M.A. and Wickens, C.D. Causes of dissociation between subjective workload measures and performance: Caveats for the use of subjective assessments. *Proceedings of the 3rd Symposium on Aviation Psychology*. 223–230, 1985.

Vidulich, M.A. and Wickens, C.D. Causes of dissociation between subjective workload measures and performance. *Applied Ergonomics*. 17(4), 291–296, 1986.

Wickens, C.D. and Liu, Y. Codes and modalities in multiple resources: A success and a qualification. *Human Factors*. 30, 599–616, 1988.

Wickens, C.D. and Yeh, Y. The dissociation between subjective workload and performance: A multiple resource approach. *Proceedings of the Human Factors Society 27th Annual Meeting*. 244–248, 1983.

Wierwille, W.W., Rahimi, M., and Casali, J.G. Evaluation of 16 measures of mental workload using a simulated flight task emphasizing mediational activity. *Human Factors*. 27, 489–502, 1985.

Wolf, J.D. *Crew workload assessment: Development of a measure of operator workload (AFFLD-TR-78-165)*. Wright-Patterson Air Force Base, OH: Air Force Flight Dynamics Laboratory, December 1978.

Yeh, Y. and Wickens, C.D. The effect of varying task difficulty on subjective workload. *Proceedings of the Human Factors Society 29th Annual Meeting*. 2, 765–769, 1985.

Yeh, Y. and Wickens, C.D. Dissociation of performance and subjective measures of workload. *Human Factors*. 30, 111–120, 1988.

4 Measures of Situational Awareness

Situational awareness (SA) is knowledge relevant to the task being performed. For example, pilots must know the state of their aircraft, the environment through which they are flying, and relationships between them, such as the fact that thunderstorms are associated with turbulence. It is a critical component of decision making and has been included in several models of decision making (e.g., Dorfel and Distelmaier model, 1997; see figure 23).

SA has three levels (Endsley, 1991): level 1, perception of the elements in the environment; level 2, comprehension of the current situation; and level 3, projection of future status.

There are four types of SA measures: performance (also known as query methods; Durso and Gronlund, 1999), subjective ratings, simulation, and physiological measures. Individual descriptions of the first three types of measures of SA are provided in the following sections. Articles describing physiological measures of SA were written by French, Clark, Pomeroy, Clarke, and Seymour (2003) and Vidulich, Stratton, Crabtree, and Wilson (1994). A flow chart to help select the most appropriate measure is given in figure 24.

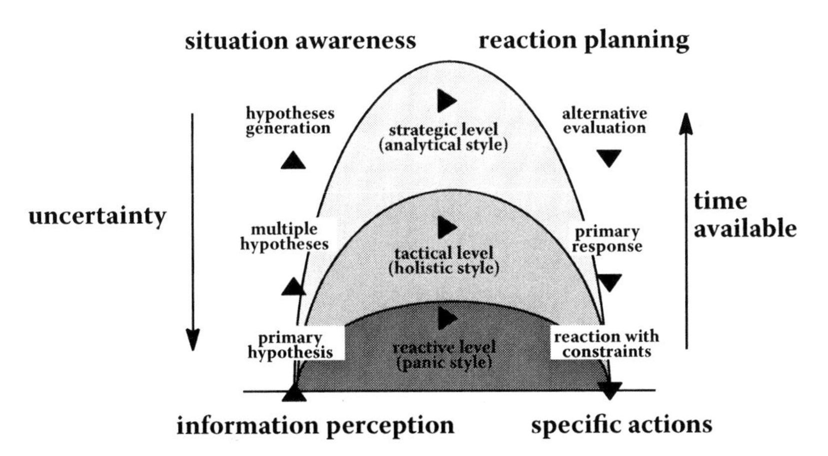

FIGURE 23 Decision making under uncertainty and time pressure. (From Dorfel, G. and Distelmaier, H. Enhancing Situational Awareness by knowledge-based user interfaces. *Proceedings of the 2nd Annual Symposium and Exhibition on Situational Awareness in the Tactical Air Environment.* Patuxent River, MD: Naval Air Warfare Center, 1997. p. 2. With permission.)

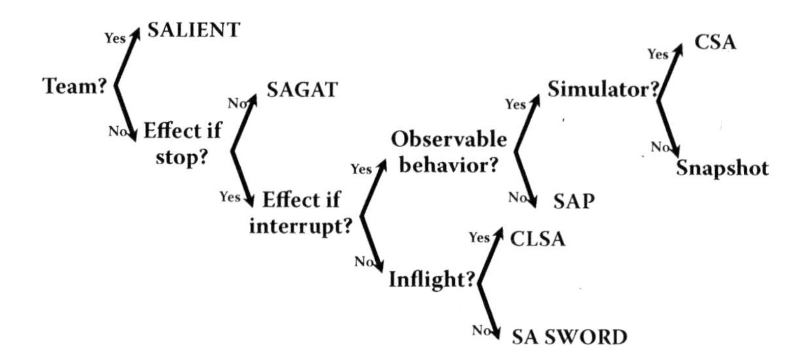

FIGURE 24 Guide to Selecting an SA measure.

SOURCES

Dorfel, G. and Distelmaier, H. Enhancing Situational Awareness by knowledge-based user interfaces. *Proceedings of the 2nd Annual Symposium and Exhibition on Situational Awareness in the Tactical Air Environment.* Patuxent River, MD: Naval Air Warfare Center, 1997.

Durso, F.T. and Gronlund, S.D. Situation Awareness. In F.T. Durso, R.S. Nickerson, R.W. Schvaneveldt, S.T. Dumais, D.S. Lindsay, and M.T.H. Chi (Eds.). *Handbook of applied cognition.* New York: John Wiley and Sons, 1999.

Endsley, M.R. Situation Awareness in dynamic systems. In R.M. Taylor (Ed.) *Situational awareness in dynamic systems (IAM Report 708).* Farnborough, UK: Royal Air Force Institute of Aviation Medicine, 1991.

French, H.T., Clark, E., Pomeroy, D., Clarke, C.R., and Seymour, M. Psycho-physiological measures of Situation Awareness. *Proceedings of the Human Factors of Decision Making in Complex Systems,* Part 6: Assessment and Measurement: Chapter 27, 291–298, August 2003.

Vidulich, M.A., Stratton, M., Crabtree, M., and Wilson, G. Performance-based and physiological measures of situational awareness. *Aviation, Space, and Environmental Medicine.* 65(5), A7–A12, 1994.

4.1 PERFORMANCE MEASURES OF SA

4.1.1 Situational Awareness Global Assessment Technique

General description. The most well-known measure of SA is the Situational Awareness Global Assessment Technique (SAGAT) (Endsley, 1988a). SAGAT was designed around real-time, human-in-the-loop simulation of a military cockpit but could be generalized to other systems. Using SAGAT, the simulation is stopped at random times, and the operators are asked questions to determine their SA at that particular point in time. Subjects' answers are compared with the correct answers that have been simultaneously collected in the computer database. "The comparison of the real and perceived situation provides an objective measure of SA" (Endsley, 1988b, p. 101).

This same technique could be used with any complex system that is simulated, be it a nuclear power plant control room or the engine room of a ship. In addition, if an operational system is properly instrumented, SAGAT is also applicable in this environment. SAGAT uses a graphical computer program for the rapid presentation of queries and data collection. In addition to possessing a high degree of face validity, the SAGAT technique has been tested in several studies, which demonstrated: (1) empirical validity (Endsley, 1989, 1990b)—the technique of freezing the simulation did not impact subject performance, and subjects were able to reliably report SA knowledge for up to 6 min after a freeze without memory decay problems; (2) predictive validity (Endsley, 1990b)—linking SAGAT scores to subject performance; and (3) content validity (Endsley, 1990a)—showing appropriateness of the queries used (for an air-to-air fighter cockpit).

Other researchers have simply asked subjects to indicate locations of objects on maps. This technique is sometimes referred to as mini-SAGAT or snapshots. It has been used in a wide variety of military and civilian tasks. For example, firefighters have been asked to mark the location of a fire on a map. SA is measured as the deviation between the actual fire and the indicated location on the map (Artman, 1999).

Strengths and limitations. SAGAT provides objective measures of SA across all of the operators' SA requirements that can be computed in terms of errors or percent correct and can be treated accordingly. However, Sarter and Woods (1991) suggest that SAGAT does not measure SA but, rather, measures what pilots can recall. Further, Fracker and Vidulich (1991) identified two major problems with the use of explicit measures of SA, such as SAGAT: (1) decay of information and (2) inaccurate beliefs.

Fracker (1989) stated from a simulated air threat study that "the present data encourage the use of memory probes to measure situation awareness."Fracker (1991) evaluated SA measures in a simulated combat task. Test-retest correlations were significant for identity accuracy, identity latency, envelope sensitivity, and kill probability but not for location error and avoidance failure. Only identity latency was significantly correlated with kill probability. This correlation was used to assess criterion validity. Vidulich, McCoy, and Crabtree (1995) used a SAGAT-like memory probe to measure SA on a PC-based flight simulator performance. The SAGAT-like measure was sensitive to differences in information presented during the simulation but not to difficulty, thus enabling the differentiation of SA and workload.

Construct validity was evaluated from correlations of SA metrics with avoidance failure. Three correlations were significant: (1) identity latency, (2) envelope sensitivity, and (3) kill probability. Two correlations were not significant: (1) location error and (2) identity accuracy. There were three significant correlations among SA metrics: (1) identity accuracy and location error, (2) identity accuracy and identity latency, and (3) envelope sensitivity and latency. A 0.10 alpha was used to determine significance. Three correlations were not significant: (1) identity latency and location error, (2) envelope sensitivity and location error, and (3) envelope sensitivity and identify accuracy. Endsley (1995) did not find any significant difference in SAGAT scores as a function of time (0 to 400 s) between event and response nor did she see any performance decrements in a piloting task between stopping the simulation and not stopping the simulation. However, Endsley and Rodgers (1996) evaluated SAGAT scores associated with 15 air traffic scenarios associated with operational errors. Their subjects were 20 air traffic control specialists. Low percent correct on recall items was associated with assigned clearances complete (23.2%), speed (28%), turn (35.1%), and call sign numeric (38.4%). In a study in the same domain, Jones and Endsley (2004) reported a significant correlation between SAGAT and real-times probes. Their subjects were air traffic controllers. In another domain, Huf and French (2004) reported that Situational Awareness Rating Technique (SART) indicated low understanding of the situation in a virtual submarine environment, whereas SAGAT indicated high understanding as evidences by correct responses.

Bolstad (1991) correlated SAGAT with scores from potential SA predictor tests. The subjects were Northrop employees with tactical flight experience. The top correlations with SAGAT were attention-sharing tracking level (+0.717), immediate or delayed memory total errors (+0.547), encoding speed (+0.547), dot estimation (+0.415), and Group Embedded Figures Test number correct (+0.385). In a study by the same lead author, Bolstad, Endsley, Howell, and Costello (2003) and (2002) used SAGAT to measure the effectiveness of time-sharing training on enhancing SA. The subjects were 24 pilots. The data were collected using the Microsoft Flight Sim2000. There was a significant increase in SA after the training. Those with training went from 22% correct awareness of wind direction before training to 55% after training. Those without training went from 58% preflight to 36% postflight. Bolstad and Endsley (2003) used Goal-Directed Cognitive Task Analysis to develop SAGAT queries. Jones, Endsley, Bolstad, and Estes (2004) extended this work to build a toolkit (Designer's Situation Awareness Toolkit) to help develop SAGAT queries.

Hogg, Folleso, Strand-Volden, and Torralba (1995) adapted SAGAT to develop the Situation Awareness Control Room Inventory (SACRI) to evaluate nuclear power plant operator's SA. They completed four simulator studies using SACRI and concluded that it was useful in design assessments when used with time and accuracy measures.

Crooks, Hu, and Mahan (2001) asked 165 undergraduate students to rate their SA using SAGAT for a simulated aircraft identification task. The results were mixed. There were significant differences in SAGAT for corridor, size, and direction but not for speed, range, angle, and identify friend or foe. Further, for range and angle, there was a significant difference in the direction opposite to the direction hypothesized. In addition, Snow and Reising (2000) found no significant correlation between SAGAT and SA-SWORD and, further, only SA-SWORD showed statistically significant effects of visibility and synthetic terrain type in a simulated flight. Similarly, Riley and Strater (2006) used SAGAT to compare SA in four robot control modes and reported no significant differences.

Prince, Ellis, Brannick, and Salas (2007) used a SAGAT-like approach to measure team situation awareness. In a high-fidelity flight simulator, team situation awareness was measured by two instructors who had completed a 3 d training course. The mean correlation between the two instructors on the 48 situation awareness items was r = + 0.88. There were also significant correlations between team situation awareness and ratings of problem solving for icing, boost pump, and fire emergencies. In a low-fidelity simulator, team simulation awareness was measured by responses to flight status. Questions included current altitude and heading, controlling agency, last ATC call, weather at destination, emergency airport, airport bearing and range, and distance to, clock position of, and altitude of traffic. The data were from 41 military aircrews. There were significant correlations between responses to flight status questions and ratings of problem solving for boost pump and fire emergencies, but not for icing.

Data requirements. The proper queries must be identified before the start of the experiment.

Thresholds. Tolerance limits for acceptable deviance of perceptions from real values on each parameter should be identified before the start of the experiment.

SOURCES

Artman, H. Situation awareness and co-operation within and between hierarchical units in dynamic decision making. *Ergonomics*, 42(11), 1404–1417, 1999.

Bolstad, C.A. Individual pilot differences related to situation awareness. *Proceedings of the Human Factors Society*. 1, 52–56, 1991.

Bolstad, C.A. Measuring shared and team situation awareness in the Army's future objective force. *Proceedings of the Human Factors and Ergonomics Society 47th Annual Meeting*. 369–373, 2003.

Bolstad, C.A., Endsley, M.R., Howell, C.D., and Costello, A.M. The effect of time-sharing training on pilot situation awareness. *Proceedings of the 12th International Symposium on Aviation Psychology*. 140–145, 2003.

Bolstad, C.A., Endsley, M.R., Howell, C.D., and Costello, A.M. General aviation pilot training for situation awareness: an evaluation. *Proceedings of the Human Factors and Ergonomics Society 46th Annual Meeting*. 21–25, 2002.

Crooks, C.L., Hu, C.Y., and Mahan, R.P. Cue utilization and situation awareness during simulated experience. *Proceedings of the Human Factors and Ergonomics Society 45th Annual Meeting*. 1563–1567, 2001.

Endsley, M.R. Situational awareness global assessment technique (SAGAT). *Proceedings of the National Aerospace and Electronics Conference*. 789–795, 1988a.

Endsley, M.R. Design and evaluation for situation awareness enhancement. *Proceedings of the 32nd Annual Meeting of the Human Factors Society*. 97–101, 1988b.

Endsley, M.R. A methodology for the objective measurement of pilot situation awareness. Presented at the *AGARD Symposium on Situation Awareness in Aerospace Operations*. Copenhagen, Denmark, October 1989.

Endsley, M.R. *Situation awareness in dynamic human decision making: Theory and measurement (NORDOC 90-49)*. Hawthorne, CA: Northrop Corporation, 1990a.

Endsley, M.R. Predictive utility of an objective measure of situation awareness. *Proceedings of the Human Factors Society 34th Annual Meeting* (pp. 41–45), Santa Monica, CA: Human Factors Society, 1990b.

Endsley, M.R. Toward a theory of situational awareness in dynamic systems. *Human Factors*. 37(1), 32–64, 1995.

Endsley, M.R. and Rodgers, M.D. Attention distribution and Situation Awareness in Air Traffic Control. *Proceedings of the Human Factors and Ergonomics Society 40th Annual Meeting*. 1, 82–85, 1996.

Fracker, M.L. Attention allocation in situation awareness. *Proceedings of the Human Factors Society 33rd Annual Meeting*, 1396–1400, 1989.

Fracker, M.L. *Measures of Situation Awareness: An experimental evaluation (AL-TR-1991-0127)*. Wright-Patterson Air Force Base, OH: October 1991.

Fracker, M.L. and Vidulich, M.A. Measurement of situation awareness: a brief review. In R.M. Taylor (Ed.) *Situational awareness in dynamic systems (IAM Report 708)*. Farnborough, UK: Royal Air Force Institute of Aviation Medicine, 1991.

Hogg, D.N., Folleso, K., Strand-Volden, F., and Torralba, B. Development of a situation awareness measure to evaluate advanced alarm systems in nuclear power plant control rooms. *Ergonomics*. 38(11), 2394–2413, 1995.
Huf, S. and French, H.T. Situation awareness in a networked virtual submarine. *Proceedings of the Human Factors and Ergonomics Society 48th Annual Meeting*. 663–667, 2004.
Jones, D.G. and Endsley, M.R. Use of real-time probes for measuring situation awareness. *International Journal of Aviation Psychology*. 14(4), 343–367, 2004.
Jones, D.G., Endsley, M.R., Bolstad, C.A., and Estes, G. The designer's situation awareness toolkit: Support for user-centered design. *Proceedings of the Human Factors and Ergonomics Society 48th Annual Meeting*. 653–657, 2004.
Prince, C., Ellis, J.E., Brannick, M.T., and Salas, E. Measurement of team situation awareness in low experience level aviators. *International Journal of Aviation Psychology*, 17(1), 41–57, 2007.
Riley, J.M. and Strater, L.D. Effects of robot control mode on situational awareness and performance in a navigation task. *Proceedings of the Human Factors and Ergonomics Society 50th Annual Meeting* 540–544, 2006.
Sarter, N.B. and Woods, D.D. Situational awareness: A critical but ill-defined phenomenon. *International Journal of Aviation Psychology*. 1(1), 45–57, 1991.
Snow, M.P. and Reising, J. M. Comparison of two situation awareness metrics: SAGAT and SA-SWORD. *Proceedings of the IEA 2000/HFES 2000 Congress*. 3, 49–52, 2000.
Vidulich, M.A., McCoy, A.L., and Crabtree, M.S. The effect of a situation display on memory probe and subjective situational awareness metrics. *Proceedings of the Eighth International Symposium on Aviation Psychology*. 2, 765–768, 1995.

4.1.2 Situational Awareness-Linked Instances Adapted to Novel Tasks

General description. The Situational Awareness-Linked Instances Adapted to Novel Tasks (SALIANT) was developed to measure team SA. The SALIANT methodology requires five phases: (1) identify team SA behaviors (see table 35), (2) develop scenarios, (3) define acceptable responses, (4) write a script, and (5) create a structured form with columns for scenario and responses.

Strengths and limitations. SALIANT has been validated using 20 undergraduate students in a 4 h tabletop helicopter simulation. Interrater reliability was r = +0.94. There were significant correlations between SALIENT score and communication frequency (r = +0.74), between SALIENT score and performance (r = +0.63). There were no significant correlations between SALIANT score and the teams' shared mental model (r = −0.04). Additional validation data are available in Muniz, Salas, Stout, and Bowers (1998).

Data requirements. Although the generic behaviors in table 35 can be used, scenarios, responses, scripts, and report forms must be developed for each team task.

Thresholds. Not stated.

TABLE 35
Generic Behavioral Indicators of Team SA

Demonstrated Awareness of Surrounding Environment

Monitored environment for changes, trends, and abnormal conditions
Demonstrated awareness of where he or she was

Recognized Problems

Reported problems
Located potential sources of problem
Demonstrated knowledge of problem consequences
Resolved discrepancies
Noted deviations

Anticipated a Need for Action

Recognized a need for action
Anticipated consequences of actions and decisions
Informed others of actions taken
Monitored actions

Demonstrated Knowledge of Tasks

Demonstrated knowledge of tasks
Exhibited skill time-sharing attention among tasks
Monitored workload
Shared workload within station
Answered questions promptly

Demonstrated Awareness of Information

Communicated important information
Confirmed information when possible
Challenged information when doubtful
Rechecked old information
Provided information in advance
Obtained information of what was happening
Demonstrated understanding of complex relationship
Briefed status frequently

Source: From Muniz, E.J., Stout, R.J., Bowers, C.A., and Salas, E. A methodology for measuring team situational awareness: Situational Awareness Linked Indicators Adapted to Novel Tasks (SALIANT). *The First Annual Symposium/Business Meeting of the Human Factors and Medicine Panel on Collection Crew Performance in Complex Systems,* Edenburg, United Kingdom, 1998. With permission.

SOURCES

Muniz, E.J., Salas, E., Stout, R.J., and Bowers, C.A. The validation of a team Situational Awareness Measure. *Proceedings for the Third Annual Symposium and Exhibition on Situational Awareness in the Tactical Air Environment*. Patuxent River, MD: Naval Air Warfare Center Aircraft Division, 183–190, 1998.

Muniz, E.J., Stout, R.J., Bowers, C.A., and Salas, E. A methodology for measuring team situational awareness: Situational Awareness Linked Indicators Adapted to Novel Tasks (SALIANT). *The First Annual Symposium/Business Meeting of the Human Factors and Medicine Panel on Collection Crew Performance in Complex Systems*, Edenburg, United Kingdom, 1998.

4.1.3 Temporal Awareness

General description. Temporal awareness has been defined as "the ability of the operator to build a representation of the situation including the recent past and the near future"(Grosjean and Terrier, 1999, p. 1443). It has been hypothesized to be critical to process management tasks.

Strengths and limitations. Temporal awareness has been measured as the number of temporal and ordering errors in a production line task, number of periods in which temporal constraints were adhered to, and the temporal landmarks reported by the operator to perform his or her task. Temporal landmarks include relative ordering of production lines, and clock and mental representation of the position of production lines.

Data requirements. Correct time and order must be defined with tolerances for error data. Temporal landmarks must be identified during task debriefs.

Thresholds. Not stated.

SOURCE

Grosjean, V. and Terrier, P. Temporal awareness: Pivotal in performance? *Ergonomics*, 1999, 42(11), 1443–1456.

4.2 SUBJECTIVE MEASURES OF SA

Subjective measures of SA share many of the advantages and limitations of subjective measures of workload discussed in Section 3.2. Advantages include inexpensiveness, easy administration, and high face validity. Disadvantages include inability to measure what the subject cannot describe well in words and requirement for well-defined questions.

4.2.1 China Lake Situational Awareness

General description. The China Lake Situational Awareness (CLSA) is a five-point rating scale (see table 36) based on the Bedford Workload Scale. It was designed at the Naval Air Warfare Center at China Lake to measure SA in flight (Adams, 1998).

Strengths and limitations. Jennings, Craig, Cheung, Rupert, and Schultz (2004) reported a significant increase in CLSA ratings associated with a Tactile Situational Awareness System that provided tactile cues for maintaining aircraft position. The subjects were 11 pilots flying a Bell 205 Helicopter.

TABLE 36
China Lake SA Rating Scale

SA Scale Value	Content
VERY GOOD 1	Full knowledge of a/c energy state/tactical environment/mission Full ability to anticipate or accommodate trends
GOOD 2	Full knowledge of a/c energy state/tactical environment/mission Partial ability to anticipate or accommodate trends No task shedding
ADEQUATE 3	Full knowledge of a/c energy state/tactical environment/mission Saturated ability to anticipate or accommodate trends Some shedding of minor tasks
POOR 4	Fair knowledge of a/c energy state/tactical environment/mission Saturated ability to anticipate or accommodate trends Shedding of all minor tasks as well as many not essential to flight safety/mission effectiveness
VERY POOR 5	Minimal knowledge of a/c energy state/tactical environment/mission Oversaturated ability to anticipate or accommodate trends Shedding of all tasks not absolutely essential to flight safety/mission effectiveness

Bruce Hunn (personal communication, 2001) argued that the CLSA "fails to follow common practice in rating scale design, does not provide diagnostic results and in general, is unsuitable for assessing SA in a test environment." He argues that the scale cannot measure SA because it does not include the three components of SA: perception, comprehension, or projection. Further, Hunn argues that the terminology is not internally consistent, includes multiple dimensions and compound questions, and has not been validated.

Data requirements. Points in the flight during which the aircrew are asked to rate their SA using CLSA Rating Scale must not compromise safety.

Thresholds. Very good (1) to very poor (5).

SOURCES

Adams, S. Practical considerations for measuring Situational Awareness. *Proceedings for the Third Annual Symposium and Exhibition on Situational Awareness in the Tactical Air Environment*, 157–164, 1998.

Jennings, S., Craig, G., Cheung, B., Rupert, A., and Schultz, K. Flight-test of a tactile situational awareness system in a land-based deck landing task. *Proceedings of the Human Factors and Ergonomics Society 48th Annual Meeting*. 142–146, 2004.

4.2.2 Crew Awareness Rating Scale

General description. The Crew Awareness Rating Scale (CARS) has eight scales (see table 37) that are rated from 1 (the ideal case) to 4 (the worst case).

Strengths and limitations. McGuinness and Foy (2000) report that the CARS has been successfully used in studies of airline pilot eye movements, automation trust on flight decks of commercial aircraft, and military command and control.

TABLE 37
Definitions of CARS Rating Scales

Perception—The Assimilation of New Information

1. The content of perception—is it reliable and accurate?
2. The processing of perception—is it easy to maintain?

Comprehension—The Understanding of Information in Context

3. The content of comprehension—is it reliable and accurate?
4. The processing of comprehension—is it easy to maintain?

Projection—The Anticipation of Possible Future Developments

5. The content of projection—is it reliable and accurate?
6. The processing of projection—is it easy to maintain?

Integration—The Synthesis of the Preceding with One's Course of Action

7. The content of integration—is it reliable and accurate?
8. The processing of integration—is it easy to maintain?

McGuinness and Ebbage (2002) used CARS to evaluate the SA of seven commanding officer/operations officer teams in the Royal Military College of Science. All seven teams completed two 2-h land reconnaissance missions; one with a standard radio and one with a digital map and electronic text messaging. Content ratings gradually improved over the exercise. The processing ratings were not significantly different between the standard and digitized versions of the mission. Content, however, was higher in the digital version except for the commanding officers, who rated comprehension higher with the standard radio.

Data requirements. Use of the standard CARS rating scales.

SOURCES

McGuinness, B. and Ebbage, L. Assessing human factors in command and control: Workload and Situational Awareness metrics. http://www.dodccrp.org/Activities/Symposia/2002CCRTS/Proceedings/Tracks/pdf/060.PDF.

McGuinness, B. and Foy, L. A subjective measure of SA: The Crew Awareness Rating Scale (CARS). *Proceedings of the Human Performance, Situation Awareness, and Automation Conference.* 2000.

4.2.3 Crew Situational Awareness

General description. Mosier and Chidester (1991) developed a method for measuring situational awareness of air transport crews. Expert observers rate crew coordination performance, and identify and rate performance errors (type 1, minor errors; type 2, moderately severe errors; and type 3, major, operationally significant errors). The experts then develop information transfer matrices, identifying time and source of item requests (prompts) and verbalized responses. Information is then classified into decision or nondecision information.

Strengths and limitations. The method was sensitive to type of errors and decision prompts.

Data requirements. The method requires open and frequent communication among aircrew members. It also requires a team of expert observers to develop the information transfer matrices.

Thresholds. Not stated.

SOURCE

Mosier, K.L. and Chidester, T.R. Situation assessment and situation awareness in a team setting. In R.M. Taylor (Ed.) *Situation awareness in dynamic systems (IAM Report 708).* Farnborough, UK: Royal Air Force Institute of Aviation Medicine, 1991.

4.2.4 Human Interface Rating and Evaluation System

General description. The Human Interface Rating and Evaluation System (HiRes) is a generic judgment-scaling technique developed by Budescu, Zwick, and Rapoport (1986).

Strengths and limitations. HiRes has been used to evaluate SA (Fracker and Davis, 1990). These authors reported a significant effect of the number of enemy aircraft in a simulation and HiRes rating.

Data requirements. HiRes ratings are scaled to sum to 1.0 across all the conditions to be rated.

Thresholds. 0 to 1.0

SOURCES

Budescu, D.V., Zwick, R., and Rapoport, A. A comparison of the Eigen value and the geometric mean procedure for ratio scaling. *Applied Psychological Measurement.* 10, 69–78, 1986.

Fracker, M.L. and Davis, S.A. Measuring operator Situation Awareness and mental workload. *Proceedings of the fifth Mid-Central Ergonomics/Human Factors Conference,* Dayton, OH, May 23–25 1990.

4.2.5 Situational Awareness Rating Technique

General description. An example of a subjective measure of SA is the Situational Awareness Rating Technique (SART) (Taylor, 1990). SART is a questionnaire method that concentrates on measuring the operator's knowledge in three areas: (1) demands on attentional resources, (2) supply of attentional resources, and (3) understanding of the situation (see figure 25 and table 38). The reason that SART measures three different components (there is also a 10-dimensional version) is that the SART developers feel that, similar to workload, SA is a complex construct; therefore, to measure SA in all its aspects, separate measurement dimensions are required. Because information processing and decision making are inextricably bound with SA (because SA involves primarily cognitive rather than physical workload), SART has been tested in the context of Rasmussen's Model of skill-, rule-, and knowledge-based behavior.

		LOW						HIGH
		1	2	3	4	5	6	7
Demand	Instability of Situation							
	Variability of Situation							
	Complexity of Situation							
Supply	Arousal							
	Spare Mental Capacity							
	Concentration							
	Division of Attention							
Under	Information Quantity							
	Information Quality							
	Familiarity							

FIGURE 25 SART Scale.

Selcon and Taylor (1989) conducted separated studies looking at the relationship between SART and rule- and knowledge-based decisions, respectively. The results showed that SART ratings appear to provide diagnosticity in that they were significantly related to performance measures of the two types of decision making. Early indications are that SART is tapping the essential qualities of SA, but further validation studies are required before this technique is commonly used.

Strengths and limitations. SART is a subjective measure and, as such, suffers from the inherent reliability problems of all subjective measures. The strengths are

TABLE 38
Definitions of SART Rating Scales

Demand on Attentional Resources

Instability: Likelihood of situation changing suddenly
Complexity: Degree of complication of situation
Variability: Number of variables changing in situation

Supply of Attentional Resources

Arousal: Degree of readiness for activity
Concentration: Degree of readiness for activity
Division: Amount of attention in situation
Space capacity: Amount of attention left to spare for new variables

Understanding of the Situation

Information quantity: Amount of information received and understood
Information quality: Degree of goodness of information gained

Source: From Taylor, R.M. and Selcon, S.J. Subjective measurement of situational awareness. In R.M. Taylor (Ed.) *Situational awareness in dynamic systems (IAM Report 708).* Farnborough, UK: Royal Air Force Institute of Aviation Medicine, 1991, p. 10. With permission.

that SART is easily administered and is developed in three logical phases: (1) scenario generation, (2) construct elicitation, and (3) construct structure validation (Taylor, 1989). SART has been prescribed for comparative system design evaluation (Taylor and Selcon, 1991). SART is sensitive to differences in performance of aircraft attitude recovery tasks and learning comprehension tasks (Selcon and Taylor, 1991; Taylor and Selcon, 1990). SART is also sensitive to pilot experience (Selcon, Taylor, and Koritsas, 1991), timeliness of weather information in a simulated flight task (Bustamante, Fallon, Bliss, Bailey, and Anderson, 2005), and field of view and size of a flight display in a fixed-base flight simulator (Stark, Comstock, Prinzel, Burdette, and Scerbo, 2001).

Vidulich, McCoy, and Crabtree (1995) reported that increased difficulty of a PC-based flight simulation increased the Demand Scale on the SART but not the Supply or Understanding scales. Providing additional information increased the Understanding Scale but not the Demand or Supply scales.

However, Taylor and Selcon (1991) state that "there remains considerable scope for scales development, through description improvement, interval justification and the use of conjoint scaling techniques to condense multi-dimensional ratings into a single SA score" (p. 11). These authors further state that "The diagnostic utility of the Attentional Supply constructs has yet to be convincingly demonstrated" (p. 12).

Selcon, Taylor, and Shadrake (1992) used SART to evaluate the effectiveness of visual, auditory, or combined cockpit warnings. Demand was significantly greater for the visual than for the auditory or the combined cockpit warnings. Neither supply nor understanding ratings were significantly different, however, across these conditions. Similarly, Selcon, Hardiman, Croft, and Endsley (1996) reported significantly higher SART scores when a launch success zone display was available to pilots during a combat aircraft simulation. Understanding, information quantity, and information quality were also significantly higher with this display. There were no effects on Demand or Supply ratings.

Crabtree, Marcelo, McCoy, and Vidulich (1993) reported that the overall SART rating discriminated SA in a simulated air-to-ground attack, whereas a simple overall SA rating did not. However, the test-retest reliability of the overall SART was not significant.

Taylor, Selcon, and Swinden (1995) reported significant differences in both 3-D and 10-D SART ratings in an ATC task. Only three scales of the 10-D SART did not show significant effects as the number of aircraft being controlled changed (Information Quantity, Information Quality, and Familiarity).

See and Vidulich (1997) reported significant effects of target and display type on SART. The combined SART as well as the Supply and Understanding scales were significantly correlated to workload ($r = -0.73$, -0.75, and -0.82, respectively). Wilson, Hooey, Foyle, and Williams (2002) reported a significant difference in SART scores associated with alternate head-up display symbologies. Their subjects were 27 pilots performing taxi maneuvers in a fixed-based simulator.

Brown and Galster (2004) reported no significant effect of varying the reliability of automation in a simulated flight task on SART. Their subjects were eight male pilots. Huf and French (2004) reported that SART indicated low understanding of the situation in a virtual submarine environment, whereas SAGAT indicated high understanding, as evidenced by correct responses.

SART was modified to measure cognitive compatibility (CC-SART) to assess the effects of color coding in the Gripen fighter aircraft (Derefeldt, Skinnars, Alfredson, Eriksson, Andersson, Westlund, Berggrund, Holmberg, and Santesson, 1999). CC-SART has 3 primary and 10 subsidiary scales. Only the three primary scales were used: (1) level of processing, (2) ease of reasoning, and (3) activation of knowledge. Subjects were seven Swedish fighter pilots performing a simulated tracking of an adversarial aircraft using the head-up display and detecting a target on the head-down display. The CC-SART Index was lowest for the monochromatic displays and highest for the dichrome color displays. However, the shortest RT to detecting the head-down target occurred with polychromatic displays.

Data requirements. Data are on an ordinal scale; interval or ratio properties cannot be implied.

Thresholds. The data are on an ordinal scale and must be treated accordingly when statistical analysis is applied to the data. Nonparametric statistics may be the most appropriate analysis method.

SOURCES

Brown, R.D. and Galster, S.M. Effects of reliable and unreliable automation on subjective measures of mental workload, situation awareness, trust and confidence in a dynamics flight task. *Proceedings of the Human Factors and Ergonomics Society 48th Annual Meeting.* 147–151, 2004.

Bustamante, E.A., Fallon, C.K., Bliss, J.P., Bailey, W.R., and Anderson, B.L. Pilots' workload, situation awareness, and trust during weather events as a function of time pressure, role assignment, pilots' rank, weather display, and weather system. *International Journal of Applied Aviation Studies.* 5(2), 348–368, 2005.

Crabtree, M.S., Marcelo, R.A.Q., McCoy, A.L., and Vidulich, M.A. An examination of a subjective situational awareness measure during training on a tactical operations trainer. *Proceedings of the Seventh International Symposium on Aviation Psychology.* 2, 891–895, 1993.

Derefeldt, G., Skinnars, O., Alfredson, J., Eriksson, L., Andersson, P., Westlund, J., Berggrund, U., Holmberg, J., and Santesson, R. Improvement of tactical situation awareness with colour-coded horizontal-situation displays in combat aircraft. *Displays.* 20, 171–184, 1999.

Huf, S. and French, H.T. Situation awareness in a networked virtual submarine. *Proceedings of the Human Factors and Ergonomics 48th Annual Meeting.* 663–667, 2004.

See, J.E. and Vidulich, M.A. Assessment of computer modeling of operator mental workload during target acquisition. *Proceedings of the Human Factors and Ergonomics Society 41st Annual Meeting.* 1303–1307, 1997.

Selcon, S.J., Hardiman, T.D., Croft, D.G., and Endsley, M.R. A test-battery approach to cognitive engineering: To meta-measure or not to meta-measure, that is the question! *Proceedings of the Human Factors and Ergonomics Society 40th Annual Meeting.* 228–232, 1996.

Selcon, S.J. and Taylor, R.M. Evaluation of the situational awareness rating technique (SART) as a tool for aircrew systems design. *AGARD Conference Proceedings No. 478.* Neuilly-sur-Seine, France, 1989.

Selcon, S.J. and Taylor, R.M. Decision support and situational awareness. In R.M. Taylor (Ed.) *Situational awareness in dynamic systems (IAM Report 708).* Farnborough, UK: Royal Air Force Institute of Aviation Medicine, 1991.

Selcon, S.J., Taylor, R.M. and Koritsas, E. Workload or situational awareness?: TLX vs. SART for aerospace systems design evaluation. *Proceedings of the Human Factors Society 35th Annual Meeting*. 62–66, 1991.

Selcon, S.J., Taylor, R.M., and Shadrake, R.A. Multi-modal: Pictures, words, or both? *Proceedings of the Human Factors Society 36th Annual Meeting*. 57–61, 1992.

Stark, J.M., Comstock, J.R., Prinzel, L.J., Burdette, D.W., and Scerbo, M.W. A preliminary examination of situation awareness and pilot performance in a synthetic vision environment. *Proceedings of the Human Factors and Ergonomics Society 45th Annual Meeting*. 40–43, 2001.

Taylor, R.M. Situational awareness rating technique (SART): The development of a tool for aircrew systems design. *Proceedings of the NATO Advisory Group for Aerospace Research and Development (AGARD) Situational Awareness in Aerospace Operations Symposium (AGARD-CP-478)*, October 1989.

Taylor, R.M. *Situational awareness: Aircrew constructs for subject estimation (IAM-R-670)*, 1990.

Taylor, R.M. and Selcon, S.J. Understanding situational awareness. *Proceedings of the Ergonomics Society's 1990 Annual Conference*. Leeds, England, 1990.

Taylor, R.M. and Selcon, S.J. Subjective measurement of situational awareness. In R.M. Taylor (Ed.) *Situational awareness in dynamic systems (IAM Report 708)*. Farnborough, UK: Royal Air Force Institute of Aviation Medicine, 1991.

Taylor, R.M., Selcon, S.J., and Swinden, A.D. Measurement of situational awareness and performance: A unitary SART index predicts performance on a simulated ATC task. *Proceedings of the 21st Conference of the European Association for Aviation Psychology*. Chapter 41, 1995.

Vidulich, M.A., McCoy, A.L., and Crabtree, M.S. The effect of a situation display on memory probe and subjective situational awareness metrics. *Proceedings of the 8th International Symposium on Aviation Psychology*, 2, 765–768, 1995.

Wilson, J.R., Hooey, B.L., Foyle, D.C., and Williams, J.L. Comparing pilots' taxi performance, situation awareness and workload using command-guidance, situation-guidance and hybrid head-up display symbologies. *Proceedings of the Human Factors and Ergonomics Society 46 Annual Meeting*. 16–20, 2002.

4.2.6 Situational Awareness Subjective Workload Dominance

General descriptions. The Situation Awareness Subjective Workload Dominance Technique (SA SWORD) uses judgment matrices to assess SA.

Strengths and limitations. Fracker and Davis (1991) evaluated alternate measures of SA on three tasks: (1) flash detection, (2) color identification, and (3) location. Ratings were made of awareness of object location, color, flash, and mental workload. All ratings were collected using a paired comparisons technique. Color inconsistency decreased SA and increased workload. Flash probability had no significant effects on the ratings. However, Snow and Reising (2000) found no significant correlation between SAGAT and SA-SWORD and, further, only SA-SWORD showed statistically significant effects of visibility and synthetic terrain type in a simulated flight.

Data requirements. There are three required steps: (1) a rating scale listing all possible pairwise comparisons of the tasks performed must be completed, (2) a judgment matrix comparing each task to every other task must be filled in with each subject's evaluation of the tasks, and (3) ratings must be calculated using a geometric means approach.

Thresholds. Not stated.

SOURCES

Fracker, M.L. and Davis, S.A. *Explicit, implicit, and subjective rating measures of Situation Awareness in a monitoring task (AL-TR-1991-0091).* Wright-Patterson Air Force Base, OH, October 1991.

Snow, M.P. and Reising, J.M. Comparison of two situation awareness metrics: SAGAT and SA-SWORD. *Proceedings of the IEA 2000/HFES 2000 Congress.* 3, 49–52, 2000.

4.2.7 Situational Awareness Supervisory Rating Form

General descriptions. Carretta, Perry, and Ree (1966) developed the Situational Awareness Supervisory Rating Form to measure the SA capabilities of F-15 pilots. The form has 31 items that range from general traits to tactical employment (table 39).

Strengths and limitations. Carretta et al. (1996) reported that 92.5% of the variance in peer and supervisory ratings were due to one principal component. The best predictor of the form rating was flying experience (r = +0.704). Waag and Houck (1996) evaluated the Situational Awareness Supervisory Rating Form as one of a set of three SA rating scales. The other two were peer and self-ratings. The data were collected from 239 F-15C pilots. Reliabilities on the three scales ranged from +0.97 to +0.99. Interrater reliabilities was +0.84. Correlations between supervisory and peer ratings ranged from +0.85 to +0.87. Correlations with the self-report were smaller (+0.50 to +0.58).

Data requirements. Supervisors and peers must make the rating.

TABLE 39

Situational Awareness Supervisory Rating Form

Rater ID#: ______________________

Pilot ID#: ______________________

	Relative Ability Compared with Other F-15C Pilots					
	Acceptable		Good		Outstanding	
Item Ratings	**1**	**2**	**3**	**4**	**5**	**6**
General Traits						
1. Discipline						
2. Decisiveness						
3. Tactical knowledge						
4. Time-sharing ability						
5. Reasoning ability						
6. Spatial ability						
7. Flight management						
Tactical Game Plan						
8. Developing plan						
9. Executing plan						
10. Adjusting plan on the fly						
System Operation						
11. Radar						
12. TEWS						
13. Overall weapons system proficiency						
Communication						
14. Quality (brevity, accuracy, timeliness, and completeness)						
15. Ability to effectively use communication/information						
Information Interpretation						
16. Interpreting VSD						
17. Interpreting RWR						
18. Ability to effectively use AWACS/GCI						
19. Integrating overall information (cockpit displays, wingman communication, controller communication)						
20. Radar sorting						
21. Analyzing engagement geometry						
22. Threat prioritization						
Tactical Employment: BVR Weapons						
23. Targeting decision						
24. Fire-point selection						

Continued

TABLE 39 (Continued)
Situational Awareness Supervisory Rating Form

Rater ID#: ______________________

Pilot ID#: ______________________

	Relative Ability Compared with Other F-15C Pilots					
	Acceptable		Good		Outstanding	
Item Ratings	1	2	3	4	5	6
Tactical Employment: Visual Maneuvering						
25. Maintain track of bogeys/friendlies						
26. Threat evaluation						
27. Weapons employment						
Tactical Employement-General						
28. Assessing offensiveness/defensiveness						
29. Lookout (VSD interpretation, RWR monitoring, visual lookout)						
30. Defensive reaction (chaff, flares, maneuvering, etc.)						
31. Mutual support						
Overall situational awareness[a]						
Overall fighter ability						

[a] Items 1 through 31 are used for supervisory ratings. The overall fighter ability and situational awareness items are completed by both supervisors and peers. (From Carretta, T.R., Perry, D.C., and Ree, M.J. Prediction of situational awareness in F-15 pilots. *International Journal of Aviation Psychology*. 6(1), 40–41, 1996.With permission.)

SOURCE

Carretta, T.R., Perry, D.C., and Ree, M.J. Prediction of situational awareness in F-15 pilots. *International Journal of Aviation Psychology*. 6(1), 21–41, 1996.

Waag, W.L. and Houck, M.R. Development of criterion measures of Situation Awareness for use ion operational fighter squadrons. *Proceedings of the Advisory Group for Aerospace Research and Development Conference on "Situation Awareness"limitation and enhancement in the aviation environment (AGARD-CP-575)*. Neuilly-sur-Seine, France: AGARD, 8-1–8-8, January 1996.

4.3 SIMULATION

Shively, Brickner, and Silberger (1997) developed a computational model of SA. The model has three components: (1) situational elements, i.e., parts of the environment that define the situation, (2) context-sensitive nodes, i.e., semantically-related collections of situational elements, and (3) a regulatory mechanism that assesses the situational elements for all nodes.

See and Vidulich (1997) reported that a Micro Saint model of operator SA during a simulated air-to-ground mission matched SART predictions with the closest correlation with the understanding scale of the SART.

SOURCES

See, J.E. and Vidulich, M.A Assessment of computer modeling of operator mental workload during target acquisition. *Proceedings of the Human Factors and Ergonomics Society 41st Annual Meeting.* 2, 1303–1307, 1997.

Shively, R.J., Brickner, M., and Silbiger, J. A computational model of Situational Awareness instantiated in MIDAS, http://caffeine.arc.nasa.gov/midas/Tech_ Reports.html, 1997.

Acronym List

3D	Three Dimensional
a	number of alternatives per page
AET	Arbeitswissenschaftliches Erhebungsverfahren zur Tatigkeitsanalyze
AGARD	Advisory Group for Research and Development
AGL	Above Ground Level
AHP	Analytical Hierarchy Process
arcmin	arc minute
ATC	Air Traffic Control
ATWIT	Air Traffic Workload Input Technique
AWACS	Airborne Warning And Control System
BAL	Blood Alcohol Level
BVR	Beyond Visual Range
c	computer response time
C	Centigrade
CARS	Crew Awareness Rating Scale
CC-SART	Cognitive Compatibility Situational Awareness Rating Scale
cd	candela
CLSA	China Lake Situational Awareness
cm	centimeter
comm	communication
CTT	Critical Tracking Task
d	day
dBA	decibels (A scale)
dBC	decibels (C scale)
EAAP	European Association of Aviation Pscyhology
F	Fahrenheit
FOM	Figure of Merit
FOV	Field of View
ft	Feet
GCI	Ground Control Intercept
Gy	Gravity y axis
G_z	Gravity z axis
h	hour
HPT	Human Performance Theory
HSI	Horizontal Situation Indicator
HUD	Head Up Display
Hz	Hertz
i	task index

ILS	Instrument Landing System
IMC	Instrument Meteorological Conditions
in	inch
ISA	Instantaneous Self Assessment
ISI	Inter-stimulus interval
j	worker index
k	key press time
kg	kilogram
kmph	kilometers per hour
kn	knot
KSA	Knowledge, Skills, and Ability
LCD	Lquid Crystal Display
LED	Light Emitting Diode
LPS	Landing Performance Score
m	meter
m^2	meter squared
mg	milligram
mi	mile
min	minute
mm	millimeter
mph	miles per hour
msec	milliseconds
MTPB	Multiple Task Performance Battery
nm	nautical mile
NPRU	Neuropsychiatric Research Unit
OW	Overall Workload
PETER	Performance Evaluation Tests for Environmental Research
POMS	Profile of Mood States
POSWAT	Pilot Objective/Subjective Workload Assessment Technique
PPI	Pilot Performance Index
ppm	parts per million
PSE	Pilot Subjective Evaluation
rmse	root mean squared error
r	total number of index pages accessed in retrieving a given item
RT	reaction time
RWR	Radar Warning Receiver
s	second
SA	Situational Awareness
SAGAT	Situational Awareness Global Assessment Technique
SALIENT	Situational Awarements Linked Instances Adapted to Novel Tasks
SART	Situational Awareness Rating Technique
SA-SWORD	Situational Awareness Subjective Workload Dominance
SD	standard deviation
SPARTANS	Simple Portable Aviation Relevant Test Battery System

st	search time
STOL	Short Take-Off and Landing
STRES	Standardized Tests for Research with Environmental Stressors
SWAT	Subjective Workload Assessment Technique
SWORD	Subjective WORkload Dominance
t	time required to read one alternative
TEWS	Tactical Electronic Warfare System
TLC	Time to Line Crossing
TLX	Task Load Index
t_z	integration time
UAV	Uninhabited Aerial Vehicle
UTCPAB	Unified Tri-services Cognitive Performance Assessment Battery
VCE	Vector Combination of Errors
VDT	Video Display Terminal
VMC	Visual Meteorological Conditions
VSD	Vertical Situation Display
WB	bottleneck worker
WCI/TE	Workload/Compensation/Interference/Technical Effectiveness

Author Index

Subject Index

D

E

F

G

H

I

J

K

L

M

N

O

P

Q

R

S

T

U

V

W

Y

Z